D0087740

Major Basic Units

Quantity	Unit	SI Symbol
Length	meter	m
Mass	kilogram	kg
Time	second	s
Electric current	ampere	A
Temperature	degree kelvin	°K
Amount of substance	mole	mol

Prefixes for Fractions and Multiples

Fraction	Prefix	Sumbol	Multiple	Prefix	Symbol
10^{-1}	deci	d	10	deka	da
10^{-2}	centi	c	10^2	hecto	h
10^{-3}	milli	m	10^3	kilo	k
10^{-6}	micro	μ	10^6	mega	M
10^{-9}	nano	n	10^9	giga	G
10^{-12}	pico	p	10^{12}	tera	T

Derived SI Units

Quantity	Unit	SI Symbol	Formula
Area	square meter	—	m^2
Density	kilogram/cubic meter	—	kg/m^3
Energy	joule	J	$N \cdot m$
Force	newton	N	$kg \cdot m/s^2$
Power	watt	W	J/s
Stress	pascal	Pa	N/m^2

Conversion of Customary Units to SI Units

$$1 \text{ dyne} = 10^{-5} \text{ N}$$
$$1 \text{ erg} = 10^{-7} \text{ J}$$
$$1 \text{ calorie} = 4.184 \text{ J}$$
$$1 \text{ psi} = 6.8948 \text{ kN/m}^2 \text{ (kPa)}$$
$$1 \text{ ksi} = 6.8948 \text{ MN/m}^2 \text{ (MPa)}$$
$$1 \text{ kg/(mm)}^2 = 9.8066 \text{ MN/m}^2 \text{ (MPa)}$$
$$1 \text{ hbar} = 10 \text{ MN/m}^2 \text{ (MPa)}$$

Fundamentals of Physical Metallurgy

JOHN D. VERHOEVEN
Department of Metallurgy
Iowa State University
Ames, Iowa

JOHN WILEY & SONS
New York • Chichester • Brisbane • Toronto • Singapore

Library of Congress Cataloging in Publication Data

Verhoeven, John D 1934–
    Fundamentals of physical metallurgy.

    Includes bibliographical references and index.
    1. Physical metallurgy. I. Title.

TN690.V46   669'.9   75–4600
ISBN 0–471–90616–6

Printed in the United States of America

20 19 18 17 16 15 14 13 12

To Liz

PREFACE

This book has evolved from a junior level course on physical metallurgy that I have taught for several years. The purpose of the book is to present a sound introduction to the fundamentals of physical metallurgy to students who have had an introductory course in metallurgy or materials science using such books as Van Vlack's or the series by Wulff et al. (see reference list for Chapter 1). It has been our experience at Iowa State, and I believe it is quite common in the United States, that a large number of students entering graduate study in metallurgy come from the fields of physics, chemistry, mechanical engineering, or chemical engineering. It is hoped that this book will prove useful to these students in an introductory course or for self-study, as well as being useful in undergraduate metallurgy curricula, to present students with a sound understanding of the fundamentals of physical metallurgy at the junior or senior level.

The major goal of the book is to provide a theoretical base for understanding how structure is controlled; also, the way structure in turn controls the mechanical properties of metals is explained. Consequently, the book will serve as a basis for a more theoretical than applied approach to physical metallurgy. However, the theoretical treatments have been aimed at presenting simple first-order treatments as clearly and simply as possible; applications of the ideas developed to the control of mechanical properties have been included in the later chapters.

The book attempts to present the subject of physical metallurgy in a more structured manner than previous physical metallurgical textbooks. It is my experience that such an approach serves as a very useful pedagogical aid. The first half of the book (Chapters 1–8) is devoted to the basic ideas of physics and chemistry upon which physical metallurgy builds. The first chapter presents some atom packing concepts that are used in later chapters. The stereographic projection is also included in order to be able to use this important tool later and to give the student some experience in its use. Most students do not learn this subject until they study x-ray diffraction, which is often at a later time. The chapter on structure determination emphasizes the Laue technique by illustrating ideas concerning crystal orientation, a subject that is often important in physical metallurgy applications. An introduction to transmission and scanning electron microscopes is also included because of their growing importance in evaluating the structure of metals. The chapter on dislocations has been preceded by a chapter on single-crystal plasticity because I have found that a study of dislocations is more meaningful to the student if he can first see a macroscopic phenomenon that these defects control. For similar reasons the chapter on diffusion proceeds from the phenomenological to the atomistic picture. A chapter on interfaces has been included in order to emphasize the importance of this topic. The ideas of this chapter are usually dispersed throughout most physical metallurgy textbooks. I feel that treating them as a distinct topic helps the student see their wide applicability to all phase transformations. The last half of the book (the last six chapters) is devoted to phase transformations, with emphasis on how they may be used to control the mechanical properties of metals.

The very important subjects of crystal structure description and phase diagrams have not been covered in the early chapters because these subjects are generally a major topic of the introductory books upon which this book builds. However, I have found that students generally have not mastered these subjects after an introductory course, but are quickly able to do so by working problems. A number of problems involving use of direction numbers, Miller indices, planes, and directions of a form are included in the first four chapters, and several problems involving phase diagram use are included in Chapters 6, 9, 11, and 12. I strongly recommend that students without a previous introduction to these two subjects study them in introductory books, such as Refs. 1 and 2 of Chapter 1.

Several problems are included at the end of each chapter because I have found problem working by the student to be essential to a sound mastery of this subject. In several chapters (e.g., Chapter 6) I have used the problems to present practical applications of the ideas presented in the chapter. To reduce publication costs the number of problems presented in the book was halved, but the additional problems will be supplied upon request. Publication costs also required the number of photographs used in the book to be minimized. Therefore, schematic representations are used where they are adequate, and the student is referred to one or more references for actual photographs.

A chapter on the physics of metals has not been included as is often the case in physical metallurgy textbooks because I feel that it is impossible to do justice to this complex subject in a single chapter. Rather than discourage the probing student by presenting this material too briefly for him to achieve a sound understanding, it is omitted in favor of encouraging him to take an introductory solid-state physics course. In connection with this point it will be noted that I have emphasized the application of physical metallurgy to the control of mechanical properties with very little emphasis on control of electromagnetic properties. This limitation is pointed out in the Introduction (p. 2) and partially justified there. It is also the result of some personal bias on my part in believing that the major contributions to the development of the control of mechanical properties by those educated in the United States has been made by men trained in traditional metallurgy curricula, whereas the major developments in the control of the electromagnetic properties of metals have been made by men trained in physics.

It now appears evident that the scientific and technological communities will achieve the adoption of a standardized system of units. These are the International System of Units (Systeme International d'Unites) which are commonly referred to as SI units. A brief discussion of the SI units commonly encountered in physical metallurgy is presented in Appendix A.

I am particularly indebted to Iowa State University and its Department of Metallurgy for having provided me the opportunity to teach physical metallurgy to undergraduate students for several years, an opportunity that led to the development of this book. I would also like to acknowledge several of my colleagues who have reviewed different chapters for me, Dr. Tom Scott, Dr. Rohit Trivedi, Dr. Ken Kinsman, Dr. Frank Kayser, Dr. Monroe Wechsler, and Dr. John Patterson. In addition, I would like to acknowledge three of my former professors at the University of Michigan who kindled my interest in physical metallurgy, Dr. Edward E. Hucke, Dr. David V. Ragone, and Dr. Maurice J. Sinnott. Finally, I am most grateful to my excellent secretary, Miss Verna Thompson for preparation and editing of the typed manuscript, and to Mr. Harlan Baker for preparation of several of the photomicrographs that appear in the book.

Ames, Iowa   JOHN D. VERHOEVEN
October 1974

CONTENTS

FUNDAMENTALS OF PHYSICAL METALLURGY

INTRODUCTION

Metals have always played a very dominant role in the development of mankind, as is well illustrated by the fact that archeologists refer to a bronze age and an iron age in their classification of the stages in man's development. This dominant role is particularly true for the industrialization of the past 120 years that has so drastically changed our daily lives from those of our great grandfathers and grandmothers. To illustrate this fact consider our vehicles of transportation. For example, how would you manufacture any of the following without the use of metals: cars, trains, airplanes, tractors, ships, bicycles, skate boards, and so on. It is clear that metals play a critical role in these vehicles, particularly in the engines that power them. This critical dependence upon metals is also quite obvious when one walks through any industrial plant, be it chemical, mechanical, electronic, nuclear, or whatever. However, unless one reflects a bit it is not nearly as obvious that metals have played a key role in the changes that have occurred right in our homes. Consider the following items that we take for granted, which were unknown in the homes of our great grandparents of the nineteenth century:

Plumbing systems
Wiring and electric lights
Electric stoves
Refrigerators
Washers and driers
Dishwashers and garbage disposals
Furnace and ducts (or pipes)
Radio and TV
Aluminum window frames
Air conditioners

At some point in all of these items, metals fulfill a critical role so that it would be very difficult and frequently impossible to produce them without the use of metals. Of course, other materials besides metals play a key role in the manufacture of many items in our modern technology. Consider in automobiles, for instance, the role of rubber in the tires and glass for transparent windows.

The design engineer selects those materials that have the desired *physical properties* and are the most economical. There are essentially three main physical properties possessed by metals that account for their key role in man's development. These are:

1. Formability.   Notice that practically all of the metal objects that you use have been mechanically shaped. Only about 20% are used directly in their as-cast form.

2. Strength without brittleness.   Many materials, such as glass, are very strong but also extremely brittle. Hence, one would never dream of designing an airplane wing out of glass even though its yield strength is over 7 times higher than our best aluminum alloys.

3. Electrical and magnetic properties.   The relatively low electrical resistivity of metals and the magnetization of the ferrous metals are the key properties here.

Physical metallurgy is concerned primarily with the effect of the atomic arrangement and the microstructure of the metal upon its physical properties. One may broadly summarize these relationships as in the following table.

Physical Property		Influence of Atomic Arrangement and Defects	Influence of Microstructure
Mechanical	Strength Ductility	Very strong	Very strong
Electrical and thermal	Resistivity Magnetization Thermal con- ductivity	Slight	Slight → strong

Chemical	Corrosive
	  resistance
	Catalytic
	  potential
		Slight	Slight → moderate

It can be seen from the table that by control of the atomic arrangement and the microstructure one may exercise some control over all of the physical properties of a metal, but particularly its mechanical properties. Consequently, we will emphasize the control of the mechanical properties of metals in the later chapters of this book. The metallurgist is able to exercise control over atomic arrangement and microstructure through such processes as casting, working, and heat treatment. A major goal of this book is to provide the theoretical base that allows one to understand how and why these processes control the atomic arrangement and microstructure in the way they do.

CHAPTER 1

DESCRIPTION OF CRYSTALS

If you examine most all of the nonmetallic solid materials with which you come into daily contact, you find that they have no characteristic difference in outward shape from most metal objects. Hence, it is quite surprising for most people to learn that metals have a crystalline structure while such materials as wood, plastics, paper, glass, and so on, do not. Since this crystalline structure of metals exerts a very strong influence upon their properties, it is essential to have a good grasp of some of the elementary concepts of crystals.

To say that metals are crystals means that their atoms are arranged in a periodically repeating geometric array. As is shown in introductory materials textbooks,[1,2] crystal structures are described by reference to their Bravais lattice. Metals have particularly simple crystal structures. Of the 14 Bravais lattices, all except four metals form in one of the following very simple structures: (1) face-centered cubic—fcc, (2) body-centered cubic—bcc, (3) hexagonal close packed, hcp, and (4) tetragonal. Many metals exist in more than one crystal structure depending on temperature, but, in most cases, the transitions are between these four simple crystal structures.

It is assumed here that you are already familiar with the geometry of these crystal structures. It is also assumed that you are familiar with the rotational symmetry of these structures, the use of direction numbers $[xyz]$ and Miller indices (hkl) to describe directions and planes in the crystals, and directions of the form $\langle xyz \rangle$ and planes of the form $\{hkl\}$ (see, for example, Ref. 1, pp. 74–89, or Ref. 2, pp. 45–50 and 208–214). The first seven problems are presented to provide you with a review of these concepts.

1.1 ATOM PACKING IN FCC AND HCP CRYSTALS

In the above approach the atomic structure of metal crystals is described by referencing the atom array to a space lattice. One may also describe the atomic structure of crystals from a different approach that is quite useful for simple crystal structures such as those possessed by metals. We will consider the atoms of the crystal to be spheres of the same size. Considering the atoms to be spheres is, of course, an approximation, but it does provide us with a good first-order approximation that in some cases is quite useful.

The metallic bonds that hold metal atoms together have very little preferred directionality. Therefore, you might expect this attractive force that pulls the atoms together to cause them to pack together equally in all directions so as to leave the minimum amount of void space between them. Considering the atoms as spheres we then ask: In how many ways can we pack equal-sized spheres together and minimize the void space between them? The resultant structure is termed a "closest-packed" structure for obvious reasons. To answer the question we proceed as follows. First we determine how spheres can be packed together in two dimensions to give a closest-packed planar array. Then we determine how these closest-packed planes may be stacked together in the closest way to build up a closest packed three-dimensional array. Consider the two rows of atoms in Fig. 1.1(a). It is apparent that if the two rows are held together and the upper row is pushed into the position of Fig. 1.1(b), the two rows will be as close together as is possible. If one redraws Fig. 1.1(b) as in Fig. 1.2, it is seen that the atoms in a two-dimensional close packing have their centers arranged on a hexagonal grid. In order to consider the stacking of these close-packed planes on top of each other, consider the hexagon region shown in Fig. 1.2(b). Notice that this hexagonal grid unit can be thought of as six equilateral triangles and that the centers of the six triangles coincide with the centers of six voids in the packing. Figure 1.3(a) shows that the six voids may be divided into two groups, B and C, which each form equilateral triangles; and the distance between the voids in each group exactly matches the distance between atoms on the grids. Hence, by stacking the second grid such that its atoms fall at the B voids one obtains a closest-packed three-dimensional stacking as shown in Fig. 1.3(b). Alternatively, one can stack the second grid at positions C to obtain the

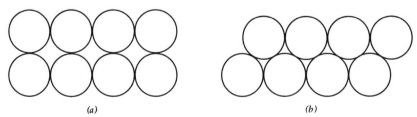

(a)   (b)

Figure 1.1 Two-dimensional packing.

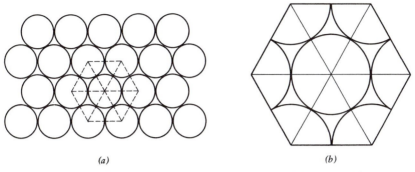

Figure 1.2 Hexagonal form in two-dimensional closest packing.

three-dimensional closest packing shown in Fig. 1.3(c). If the positions of the atoms on the lower grid are called A positions, then Fig. 1.3(b) is A-B stacking and that of Fig. 1.3(c) is A-C stacking. To obtain a crystal structure model the grids of atoms must be stacked together so as to obtain long-range order. Four possibilities are obvious, (1) -A-B-A-B-A-B-, (2) -A-C-A-C-A-C-, (3) -A-B-C-A-B-C-, (4) -A-C-B-A-C-B. The difference between stacking sequence (1) and (2) in two different crystals would not be distinguishable to us. The same is true of (3) and (4). Hence, there are two simple types of close packing. In the first the grids repeat every second plane, -A-B-A-B-A-B-, and in the second the grids repeat every third plane, -A-B-C-A-B-C-.

The hcp crystal structure and the fcc crystal structure are both close packed; they correspond to the types of stacking sequences discussed here. In the hcp structure the basal planes are the close-packed planes. It may be seen from Fig. 1.9 that the atoms on these planes are stacked directly over one another with only one close-packed plane between them, the (0002) plane. Hence, the stacking sequence is -A-B-A-B- in the hcp structure. In the fcc structure the (111) planes are the close-packed planes. Figure

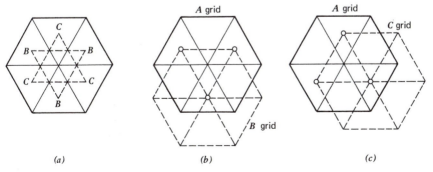

Figure 1.3 Void location and stacking of two-dimensional grids.

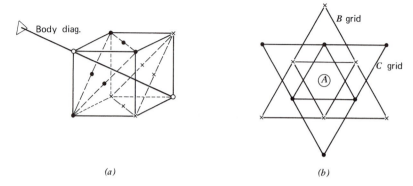

Figure 1.4   Close-packed planes in fcc crystals.

1.4(*a*) shows the trace of two of these planes on the unit cell. If one views these two planes down the body diagonal shown, the result is as shown in Fig. 1.4(*b*). It is apparent that these two planes are neighboring close-packed planes, and we will call them *B* and *C* grids. The two corner atoms on the body diagonal lie in parallel (111) planes and it should be clear that these planes would be in *A* locations since the atoms lie over the remaining void in *B* and *C*. Hence, the stacking sequence of the fcc structure is -*A*-*B*-*C*-*A*-*B*-*C*-.

The bcc structure is not a close-packed structure. In this structure the atoms touch along the body diagonal, and using this fact you may show that this structure contains more void space than the close-packed structures (see problem 1.7).

A.   VOID LOCATIONS IN THE STACKING

The effective radius of a carbon atom in steel is 0.7 Å compared to 1.24 Å for the radius of the iron atoms. Carbon may dissolve in iron by replacing the iron atoms on the lattice sites (substitutional solute) or by squeezing into the interstitial void space between the iron atoms (interstitial solute). Because of its smaller size, carbon dissolves interstitially as do nitrogen, hydrogen, and oxygen. These atoms produce a very strong effect upon the mechanical properties of metals, particularly bcc metals, and so it is quite useful to have a good understanding of the size and locations of the interstitial void spaces between the spheres of these sphere-model crystals.

The voids will be characterized by their coordination number. The coordination number is defined here as the number of nearest-neighbor atoms. By considering the body-center atom of a bcc unit cell it is apparent that each atom of the crystal has a coordination number of 8. The coordination number of a void is determined in a similar manner by first locating the center of the void and then determining the number of

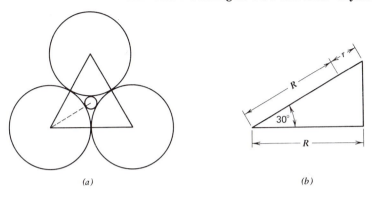

(a)   (b)

***Figure* 1.5** Maximum-size sphere in a two-dimensional void.

nearest-neighbor atoms about this point. As a simple example consider the voids in the close-packed planes of Fig. 1.2. Each void has three nearest neighbor atoms and, hence has threefold coordination. It might therefore be called a triangular void. A question of obvious interest to us is: What is the largest sized sphere that could be placed in the void, as shown in Fig. 1.5? Letting the radius of the packed spheres be R and of the interstitial sphere be r, one may show by simple geometry, as in Fig. 1.5(b), that the maximum value of r is $0.155\,R$.

In the two close-packed crystal structures there exist just two interstitial voids, as shown in Table 1.1. The centers of the four atoms surrounding the tetragonal void connect together to form a regular tetrahedron (all sides equal). Similarly, if one connects the centers of the six atoms surrounding an octahedral void a regular octahedron is formed. These polyhedrons are sketched in Fig. 1.6 to help illustrate the geometry. The void itself is at the center of these figures, and it is an interesting problem in geometry to determine the maximum size sphere that fits the voids, as given in Table 1.1. Be sure you realize that we are using the polyhedrons only to locate and characterize the void. The voids are not polyhedrons but the hole at the center of the polyhedrons.

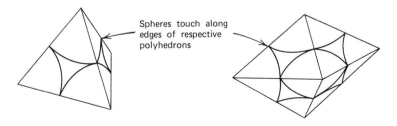

Spheres touch along
edges of respective
polyhedrons

***Figure* 1.6** The tetrahedral and octahedral voids in closest packing.

Table 1.1 The Voids in fcc, hcp, and bcc Crystals

Crystal	Void	Coordination No.	Maximum Value of r
Close-packed	Tetragonal	4	$0.225R$
Close-packed	Octahedral	6	$0.414R$
bcc	Tetragonal	4	$0.291R$
bcc	Octahedral	6	$0.154R$

If one simply removes an atom from a close-packed structure, a void having 12-fold coordination is obtained. This is not an interstitial void since it lies directly on a lattice site. This type of void is a defect in the crystal structure and it is called a *vacancy*.

The location of the tetrahedrons surrounding two tetrahedral voids in the fcc unit cell are shown in Fig. 1.7. By examining the figure it can be seen that relative to atom A the center of the upper void is located at $-\frac{1}{4}$, $-\frac{1}{4}$, $-\frac{1}{4}$, and, hence, is on the body diagonal. We now ask, how many of these tetragonal voids are contained within the unit cell? This question is easily answered using the fourfold rotation symmetry of the fcc crystal. By applying the fourfold rotation operation about the vertical axis, both of the voids shown will generate three more voids within the unit cell for a total of eight.

Crystals that are fcc have an octahedral void located directly at the center of the unit cell as shown in Fig. 1.8. The other octahedral void shown in this figure is located at the center of an edge of the unit cell. Again using the fourfold symmetry of fcc crystals, it is easy to show that an

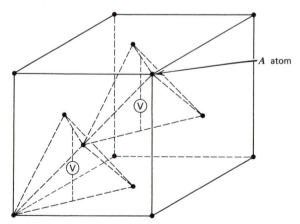

Figure 1.7 Location of two of the tetrahedral voids within an fcc unit cell.

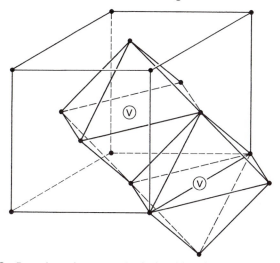

***Figure* 1.8**   Location of two octahedral voids relative to the fcc unit cell.

octahedral void must be centered at each of the 12 edges of the unit cell. We now ask how many octahedral voids are contained within the fcc unit cell? Show yourself that the answer is four.

It may be seen in Fig. 1.8 that a tetrahedral void is fitted between the two octahedral voids in the lower right-hand front of the unit cell. If you sketch the octahedrons for the other octahedral voids on Fig. 1.8 you will see that the tetrahedrons of the tetrahedral voids fit snugly between the octahedrons so that no free space is left. This shows that space may be completely filled by stacking regular tetrahedrons and octahedrons of the same side length together. When the octahedrons touch edge to edge the stacking corresponds to the void arrangement in fcc crystals.

The tetrahedrons surrounding three tetrahedral voids within an hcp unit cell are shown in Fig. 1.9. By applying the sixfold rotation inversion of hcp crystals you may quickly show that eight tetrahedral voids are contained within the unit cell, two of which touch along one face of their surrounding tetrahedrons.

By a figure similar to Fig. 1.9 one may locate the octahedral voids in the hcp crystal and again one finds that the tetrahedrons and octahedrons fit together to fill space. However, in the hcp case the octahedrons touch face to face in the direction of the c axis and edge to edge in other directions. It is apparent from this discussion that space can be completely filled by stacking regular tetrahedrons and octahedrons together in two different ways. One way corresponds to the void arrangement in hcp crystals and the other to the void arrangement in fcc crystals.

Figure 1.10 locates the octahedral and tetrahedral void in bcc crystals.

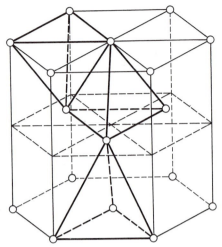

Figure 1.9 Location of three tetrahedral voids in the hcp unit cell.

Notice that the polyhedrons formed by connecting the nearest-neighbor atoms are irregular polyhedrons, some of the sides having length a and others length $0.866a$. The octahedral voids are located at the center of each unit cell face and the center of each unit cell edge and four tetrahedral sites are located in each unit cell face. Notice that, in contrast with close-packed crystals, the tetrahedrons are actually contained within the octahedrons. One then might wonder why the tetrahedral void is not simply considered as a portion of the octahedral void. The reason is that if one places a maximum sized sphere in a tetrahedral site it becomes trapped there. It cannot move to the nearby octahedral site unless it pushes its near-neighbor atoms apart. It can be seen from Table 1.1 that the tetrahedral void will hold a larger maximum sized sphere than the octahedral void.

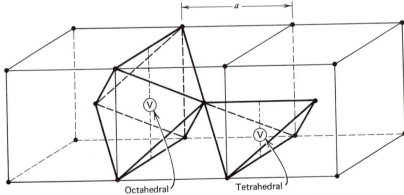

Figure 1.10 Location of the octahedral and tetrahedral voids in bcc crystals.

From a practical point of view, the voids in bcc metals are probably more important than in fcc metals since interstitial atoms play a more important role in the mechanical properties of bcc metals. This is partly because the near-neighbor atoms form irregular polyhedra and hence when an interstitial atom moves between voids it causes a nonsymmetric straining of the crystal.

B. FAULTS IN THE STACKING

In close-packed crystals a disruption of the long-range stacking sequence can produce two common types of crystal defects: (1) a twin region and (2) a stacking fault. A change in the stacking sequence over many atom spacings produces a twin whereas a change over a few atom spacings produces a stacking fault. We will illustrate this for fcc crystals.

1. Twins. Consider a stacking sequence that goes

$$-A\text{-}B\text{-}C\text{-}A\text{-}B\text{-}C\text{-}A\text{-}C\text{-}B\text{-}A\text{-}C\text{-}B\text{-}A\text{-}C\text{-}B\text{-}$$

←Point where stacking
sequence changes

In order to picture the effect of this change of stacking sequence upon the geometry of the crystal structure it is useful to draw an edge view of the close-packed (111) planes. Figure 1.11(a) shows the *trace* of the (111) and $(\frac{1}{2}\frac{1}{2}\frac{1}{2})$ planes on the fcc unit cell. If one views this unit cell down the $[\bar{1}10]$ direction, an edge view of the (111) planes is seen and one views the $(\bar{1}10)$ plane at right angles. This view of the $(\bar{1}10)$ plane is shown in Fig. 1.11(b). The (111) plane will be called a B plane and the $(\frac{1}{2}\frac{1}{2}\frac{1}{2})$ plane will be called a C plane. Then the parallel planes through atoms 1 and 3 of Fig. 1.11(b) will be A planes. We now take the rectangular section of the $(\bar{1}10)$ plane of Fig. 1.11(b) and rotate it so that the (111) planes are horizontal as in Fig. 1.12.

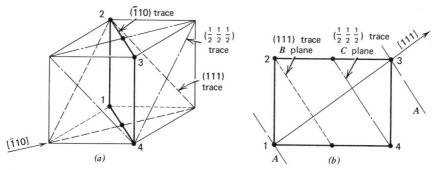

Figure 1.11 (a) An fcc unit cell locating the $(\bar{1}10)$ trace. (b) The $(\bar{1}10)$ trace giving an edge view of the close-packed planes.

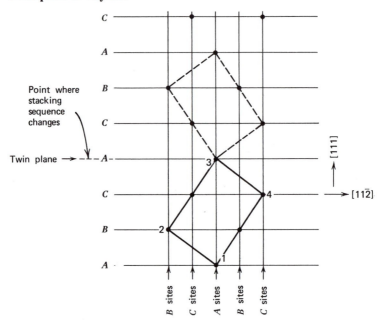

Figure **1.12**   A change in stacking sequence.

Consider first only the four lower (111) planes of Fig. 1.12. These are the four (111) planes that touch the unit cell of Fig. 1.11(*b*), and the atoms on these planes define the location of the *A*, *B*, and *C* sites. Our problem now is to produce a change in the stacking sequence as shown on the left of Fig. 1.12. To make any plane a *B* plane we simply put atoms on this plane directly above the *B* sites. Hence, placing the upper atoms as shown produces the desired change in stacking. Notice that the atoms above the plane labeled twin plane form a mirror image of the atoms below this plane. Hence, the crystal structure above this plane is a twin of the one below and the change in the stacking sequence has produced a twin region above the twin plane.

2.   Stacking Faults.   Consider a stacking sequence that goes

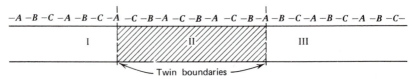

It should be clear from above that region II will have a twin orientation relative to regions I and III, which have identical orientations. Region II is separated from regions I and III by two boundaries called twin boundaries.

If these boundaries are separated by a few millimeters region II will be called a twin. However, if these boundaries are separated by only a few atom spacings, as in the model shown above, we call region II a stacking fault. In other words, if the boundary separation is so small as to not be physically resolvable the region is a stacking fault. In a sense the stacking fault is two twin boundaries. The extreme case of a stacking fault is where one simply deletes one plane in an otherwise perfect stacking: -A-B-C-A-B ↓ A-B-C-A-B-C-.

1.2 THE STEREOGRAPHIC PROJECTION

In order to be able to discuss ideas that involve specific directions and planes within a crystal, it is very useful to have some sort of a map on which we can show these directions and planes of crystals. The stereographic projection provides us with such a map and it is widely used in the metallurgical literature. The overall scheme of the stereographic projection is as follows:

1. We represent the planes in a crystal by their normals.
2. The crystal is placed at the center of a sphere and these normals are projected out to intersect the sphere.
3. The intersections of the normals on the sphere are then mapped onto a plane to obtain the desired projection.

The resulting stereographic projection is a very useful tool that, among other things, provides us with a means to:

1. visualize and discuss the relationship between planes and directions in a crystal,
2. aid in the analysis of x-ray and electron-beam-produced diffraction patterns,
3. schematically represent the symmetry of a lattice, and
4. determine angles between planes and directions in the crystal.

In order to introduce the stereographic projection it is very useful to use the unit cell as a "reference cube." The procedure is as follows:

1. Place the origin of a coordinate system at the center of a unit cell.
2. Construct lines from this origin perpendicular to each of the major planes of the crystal, thus forming their normals.
3. Mark the intersection of these plane normals on the unit cell with the symmetry symbol of the plane represented.

Figure 1.13 presents this scheme with the intersection of each normal labeled with the plane represented and its symmetry symbol. The advantage of using the reference cube is that the cube has an extremely simple

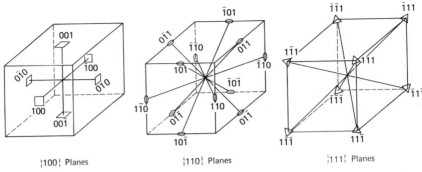

|100| Planes |110| Planes |111| Planes

Figure 1.13   Reference cubes showing the three major sets of planes in cubic crystals.

geometry, and by displaying the plane normals on this simple geometry it allows one to easily visualize the spatial relationship between the planes represented.

The reference cube is now placed at the center of a sphere and the plane normals are extended out to intersect the surface of the sphere, as shown for an octant of the sphere in Fig. 1.14. The intersections of the normals on the sphere are called *poles*. From the geometry of the cube it should be obvious that the three {101} poles shown will lie on lines at 45° from the coordinate axes.

Our problem now is to map this reference sphere onto a plane. Three different methods of mapping the upper half of the reference sphere onto a

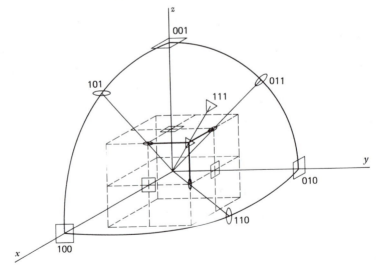

Figure 1.14   Projection of plane normal onto sphere.

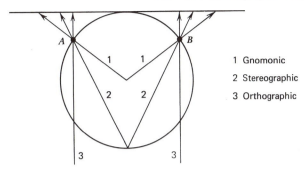

1 Gnomonic

2 Stereographic

3 Orthographic

Figure **1.15**   Different methods of projection.

plane are shown in Fig. 1.15. In this figure the two points *A* and *B* are seen to project to different points on the plane depending on the location of the projection point. When the projection point is located at the sphere center the projection is called *gnomonic*, at the sphere base it is called *stereographic*, and at infinity it is called *orthographic*. Most engineering students are very familiar with orthographic projections from courses in graphics. The stereographic projection is quite similar to the orthographic projection in appearance. A significant property of the stereographic projection that we will find useful is: Angles between the poles on the projection are always the true angles between the normals of the planes represented by the poles. The angle between plane normals is also the angle between the planes represented by these normals. Hence, by determining angles between poles we obtain the angles between planes. To be able to easily project the sphere stereographically and also measure angles on the projection, the following technique is used. The latitude and longitude lines are ruled onto a sphere (usually at 2° intervals). These latitude and longitude lines are then projected stereographically onto the plane to obtain the grid shown on page 19. This grid is frequently called a Wulff net. Because of our familiarity with the globe, the top of the net will be called the north pole and the horizontal axis (west-east axis) will be called the equator. Then the vertically running lines are longitudes and the horizontal lines are latitudes. With the use of this grid we can locate any point on the projection if we know its longitude and latitude coordinates. Also, we can use the grid lines to measure angles between poles.

Angles can be measured by counting degrees along a latitude or along a longitude, and it is very important to realize the difference that is involved here. Consider the three points *A*, *B*, *C* shown in Fig. 1.16(*a*) located on the 0° long., 40°N lat., and 60°E long., 40°N lat., and 60°E long., 10°N lat. Figure 1.16(*b*) shows a three-dimensional view of the upper right-hand quadrant of Fig. 1.16(*a*) with the same three points located. With the Wulff

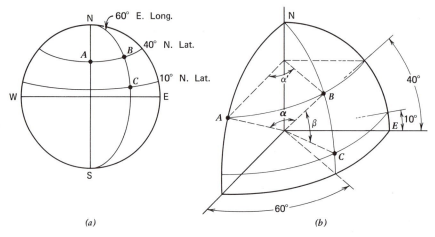

Figure 1.16   Relation of Wulff net (a) to reference sphere (b).

net in the position of Fig. 1.16(a) one may measure an angle between A and B by moving along the 40°N latitude and counting the degree marks on the net. One obtains an angle of 60°, and it should be clear from Fig. 1.16(b) that this is the angle α'. Notice that if A and B are poles then this angle α' is *not* the angle between the planes represented by the poles. That angle is shown as α on Fig. 1.16(b), the angle between the plane normals, which must originate at the sphere center. Now consider the points B and C. With the Wulff net in the position of Fig. 1.16(a) we may easily measure an angle between B and C by moving along the 60°E long. and counting the degree marks on the net. We obtain an angle of 30°, which corresponds to the angle β in Fig. 1.16(b). If points B and C are poles then this angle $\beta = 30°$ does correspond to the true angle between the planes of these poles because it is the angle between their normals. Notice that in the first case we measured the angle between A and B by moving along a *latitude*, whereas in the second case the angle between B and C was measured by moving along a *longitude*. We now ask, how could we measure the angle α between A and B with a Wulff net? Suppose we were to pass a plane through the sphere of Fig. 1.16(b), which contained the origin and points A and B. The trace of this plane on the sphere would be a great circle. It would be a great circle because the center of this trace circle would also be the center of the sphere. If we could measure degrees along this great circle then we could measure α. It should be clear from Fig. 1.16(b) that all of the longitudes are great circles but none of the latitudes are great circles except the equator. Hence, we have an important point concerning the use of Wulff nets: *True angles between planes represented by poles may be measured along great circles (longitudes and equator) but not along small*

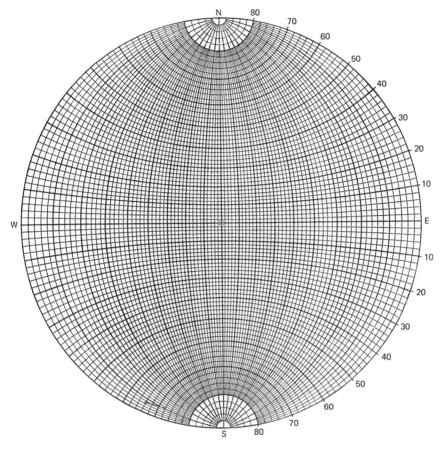

The Wulff Net

circles (*latitudes*). To measure the angle between poles *A* and *B* we first place a piece of tracing paper over the Wulff net and locate *A* and *B* on this paper. We then place a pin at the origin of the Wulff net and rotate the tracing paper over the Wulff net until both points fall on some longitudinal line. The true angle, α, may then be measured along that longitude.

In order to illustrate the use of the Wulff net three examples will be presented.

1. Construct a Standard (001) Stereographic Projection. An *hkl* stereographic projection is a projection of the reference sphere with the *hkl* pole at the center of the projection and usually with at least the {100}, {111}, and {011} poles displayed. To proceed we place a tracing paper over a Wulff net and put the (001) pole at the center. From Fig. 1.17(*a*) it can be

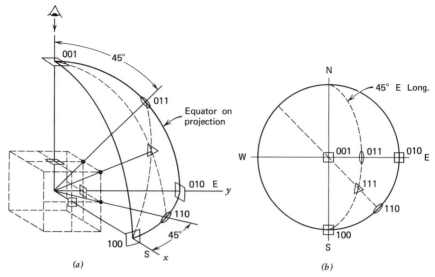

Figure 1.17   Construction for example Problem 1.

seen that the (100) pole would be at the south pole of the projection and the (010) pole at the east end of the equator as shown in Fig. 1.17(*b*). Next, consider the plane in Fig. 1.17(*a*) that contains the origin, the (100) pole, and the (011) pole. It should be clear from the geometry of Fig. 1.17(*a*) that this plane makes a trace on the sphere that will show on the projection as the 45°E longitude as shown in Fig. 1.17(*b*). It should also be clear that

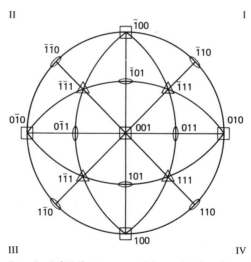

Figure 1.18   Standard (001) stereographic projection for cubic crystals.

the (011) pole will be located at the intersection of this longitude with the equator as shown in Fig. 1.17(*b*). The (111) pole lies on this 45°E longitude as may be seen from Fig. 1.17(*a*). The (111) pole also must lie on the great circle through the (001) and (110) poles. This circle will be a straight line on the projection and the (111) pole is located at the intersection of this line and the 45°E longitude as shown in Fig. 1.17(*b*). From the fourfold symmetry of the cubic crystals the rest of the (001) projection is easily constructed; the result is shown in Fig. 1.18. It is quite useful to realize that any pole in quadrant I must have indices of sign {$\bar{h}kl$}; and for the other quadrants we have, II—{$\bar{h}\bar{k}l$}, III—{$h\bar{k}l$}, and IV—{hkl}.

2. Locate the ($\bar{1}\bar{3}2$) Pole on a Standard (001) Projection. To solve this problem we will first locate the ($\bar{1}\bar{3}0$) pole and then the ($\bar{1}\bar{3}2$) pole, for reasons that will become obvious as we proceed. Figure 1.19 shows a sketch of both of these planes on a unit cell. Notice that the two planes intersect along the [$\bar{3}10$] direction. The ($\bar{1}\bar{3}0$) plane is parallel to the [001] direction and consequently its pole must lie on the outside circle of the stereographic projection. Notice in Fig. 1.18 that all poles on the outside circle have an *l* Miller indice of zero. To locate the pole we will make use of Table 1.2, which lists the angles between planes in the cubic system. It may be seen that each {310} plane makes angles of 18.4°, 71.6°, and 90° with the {100} planes. From above it should be clear that the ($\bar{1}\bar{3}0$) pole will lie on the outside edge in the II quadrant of the stereographic projection as shown by the dark line in Fig. 1.20(*a*). Any pole on this line will be 90° from the (001) pole. Hence, the ($\bar{1}\bar{3}0$) must be located on this line 18.4° from either the (0$\bar{1}$0) or the ($\bar{1}$00) pole. It should be clear from a top view of Fig. 1.19 that the ($\bar{1}\bar{3}0$) pole must be closer to the (0$\bar{1}$0) than ($\bar{1}$00) pole. Therefore, we locate the pole 18.4° from the (0$\bar{1}$0) by counting off degree marks on the outside circle of the Wulff net.

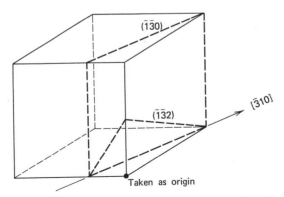

Figure 1.19 Construction for example Problem 2.

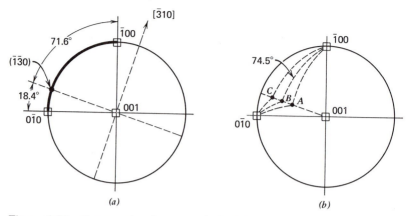

Figure 1.20 Construction for example Problem 2.

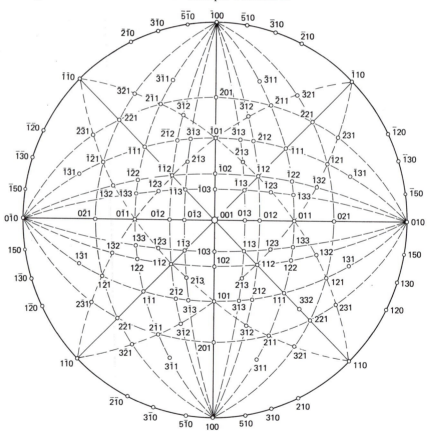

Figure 1.21 A (001) stereographic projection of planes in the cubic system.

22

To locate the $(\bar{1}\bar{3}2)$ pole notice on Fig. 1.19 that if you rotate the $(\bar{1}30)$ plane 90° about the $[\bar{3}10]$ direction, its pole will move up the reference sphere along the dashed line of Fig. 1.20(a) until it falls on the (001) pole. After a rotation of something less than 90° the plane will coincide with the $(\bar{1}\bar{3}2)$ plane. Therefore, the $(\bar{1}\bar{3}2)$ pole must lie along this dashed line somewhere in quadrant II. From Table 1.2 we see that each {321} plane makes angles of 36.7°, 57.7°, and 74.5° with the {100} planes. Therefore the pole must lie on the dashed line at either 36.7°, 57.7°, or 74.5° from the (001) pole as shown by points A, B, and C in Fig. 1.20(a). We now locate these three points using the Wulff net. Next we measure the angles between these three points and the $(\bar{1}00)$ pole by measuring along a longitude and we find that only point B makes one of the remaining angles, 74.5°. Further measurement shows that point B makes the 36.7° angle with the $(0\bar{1}0)$ pole and, hence, point B is the location of the $(\bar{1}\bar{3}2)$ pole.

In this manner one may locate all of the poles on the projection. However, it is extremely useful to have a complete standard (001) projection at your disposal when using the stereographic projection, and such a projection is provided in Fig. 1.21. Note the location of the $(\bar{1}30)$ and $(\bar{1}\bar{3}2)$ poles on this projection.

Table 1.2 Angles Between Crystallographic Planes in the Cubic System (in degrees)

HKL	hkl				
100	100	0.00	90.00		
	110	45.00	90.00		
	111	54.74			
	210	26.56	63.43	90.00	
	211	35.26	65.90		
	221	48.19	70.53		
	310	18.43	71.56	90.00	
	311	25.24	72.45		
	320	33.69	56.31	90.00	
	321	36.70	57.69	74.50	
110	110	0.00	60.00	90.00	
	111	35.26	90.00		
	210	18.43	50.77	71.56	
	211	30.00	54.74	73.22	90.00
	221	19.47	45.00	76.37	90.00
	310	26.56	47.87	63.43	77.08
	311	31.48	64.76	90.00	
	320	11.31	53.96	66.91	78.69
	321	19.11	40.89	55.46	67.79	79.11

Table 1.2 *Continued*

111	111	0.00	70.53					
	210	39.23	75.04					
	211	19.47	61.87	90.00				
	221	15.79	54.74	78.90				
	310	43.09	68.58					
	311	29.50	58.52	79.98				
	320	36.81	80.78					
	321	22.21	51.89	72.02	90.00			
210	210	0.00	36.87	53.13	66.42	78.46	90.00	
	211	24.09	43.09	56.79	79.48	90.00		
	221	26.56	41.81	53.40	63.43	72.65	90.00	
	310	8.13	31.95	45.00	64.90	73.57	81.87	
	311	19.29	47.61	66.14	82.25			
	320	7.12	29.74	41.91	60.25	68.15	75.64	82.87
	321	17.02	33.21	53.30	61.44	68.99	83.14	90.00
211	211	0.00	33.56	48.19	60.00	70.53	80.40	
	221	17.72	35.26	47.12	65.90	74.21	82.18	
	310	25.35	40.21	58.91	75.04	82.58		
	311	10.02	42.39	60.50	75.75	90.00		
	320	25.06	37.57	55.52	63.07	83.50		
	321	10.89	29.20	40.20	49.11	56.94	70.89	77.40	83.74
		90.00						
221	221	0.00	27.27	38.94	63.51	83.62	90.00	
	310	32.51	42.45	58.19	65.06	83.95		
	311	25.24	45.29	59.83	72.45	84.23		
	320	22.41	42.30	49.67	68.30	79.34	84.70	
	321	11.49	27.02	36.70	57.69	63.55	74.50	79.74	84.89
310	310	0.00	25.84	36.87	53.13	72.54	84.26	
	311	17.55	40.29	55.10	67.58	79.01	90.00	
	320	15.26	37.87	52.12	58.25	74.74	79.90	
	321	21.62	32.31	40.48	47.46	53.73	59.53	65.00	75.31
		85.15	90.00					
311	311	0.00	35.10	50.48	62.96	84.78		
	320	23.09	41.18	54.17	65.28	75.47	85.20	
	321	14.76	36.31	49.86	61.09	71.20	80.72	
320	320	0.00	22.62	46.19	62.51	67.38	72.08	
	321	15.50	27.19	35.38	48.15	53.63	58.74	68.24	72.75
		77.15	85.75	90.00				
321	321	0.00	21.79	31.00	38.21	44.41	49.99	64.62	69.07
		73.40	85.90					

3.   Locate the Position of the ($\overline{1}30$) Pole after a 90° Clockwise Rotation about the ($\overline{1}\overline{3}2$) Pole.   In order to rotate one pole about another it is very useful to bring one of the poles to the (001) position of an (001) projection. In this problem we will rotate the ($\overline{1}\overline{3}2$) up to the (001) pole position and then rotate the ($\overline{1}30$) pole about it. It should be clear that with the ($\overline{1}\overline{3}2$) pole at the (001) position the rotation will cause the ($\overline{1}30$) pole to move along a circle on the projection whose center is at the center position of the projection. This problem may be solved by the following procedure:

   1.   Locate the ($\overline{1}30$) and ($\overline{1}\overline{3}2$) poles on the projection as described above and then rotate the *projection* about its center to bring the ($\overline{1}\overline{3}2$) pole over the equator of the Wulff net as shown in Fig. 1.22.
   2.   The ($\overline{1}\overline{3}2$) pole is 57.7° from the (001) pole. Moving the ($\overline{1}\overline{3}2$) pole up the Wulff net equator to the (001) position will bring the ($\overline{1}30$) pole up the equator by 57.7° so that it now is placed along the Wulff net equator at 32.3° from the (001) pole as shown in step (2) on Fig. 1.22.
   3.   The ($\overline{1}30$) pole is now rotated 90° clockwise about the (001) pole position with the use of a compass. Since the angle is exactly 90° the new position of the ($\overline{1}30$) pole will be on the N-S axis of the underlying Wulff net.
   4.   We now rotate the ($\overline{1}\overline{3}2$) pole back down the Wulff net equator 57.7° to its original position. In making this rotation it is useful to consider what is happening on the reference sphere. This rotation causes the reference

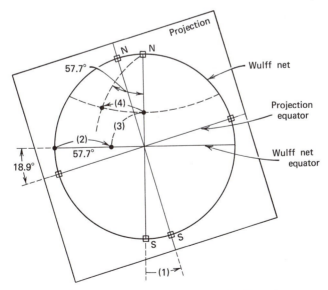

Figure 1.22   Construction for example Problem 3.

sphere to rotate about its corresponding N-S axis (see Fig. 1.17 to locate N-S axis on sphere and projection). Such a rotation will cause the projected points to move from east to west along the *latitudes* of the Wulff net. Consider, for example, how the points on a globe move as you rotate it. Hence, this rotation causes the position of the ($\bar{1}30$) pole to move to the west 57.7° along the latitude passing through it as shown in Fig. 1.22.

5. The projection is now removed from the Wulff net and the ($\bar{1}30$) pole is in the proper rotated position.

A common error in rotating poles is to move the poles along a longitude. Notice that rotation of a globe about its north-south axis causes all points to move along latitudes. It might be helpful for rotations to imagine that the projection is a sphere that you are observing from a distance. Rotation of a point is achieved by rotation of the sphere about some one of its axes. If you rotate the sphere about its east-west axis how do the points on the sphere move in the projection plane? They cannot move along the longitude of the underlying Wulff net because this would cause all points on the sphere to collapse into the north pole. Rotate the north-south axis of the Wulff net to lie under the east-west axis of your projection. Rotation about the east-west projection axis now causes the points of the projection to move along the latitudes of the underlying Wulff net.

Up to now we have restricted our attention to describing crystal planes with the stereographic projection. The stereographic projection may be conveniently used to show directions in crystals. A line in the desired direction is extended out to the reference sphere and its intersection is projected onto the plane. Therefore, any direction in a crystal will be represented by a point on the stereographic projection.

Frequently just a small section of the stereographic projection is used.

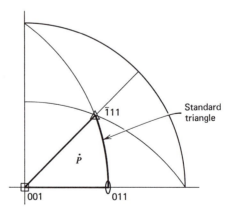

Figure 1.23 A standard triangle in the cubic system.

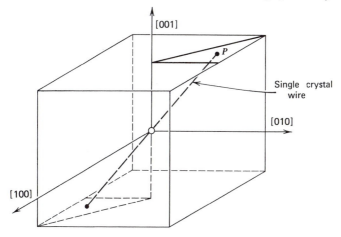

Figure 1.24 Projection of the standard triangle on the reference cube.

The small section usually displayed is the triangular region shown in Fig. 1.23 and it is called the *standard triangle*. The standard triangle is often used to indicate the orientation of crystals or grains. Suppose you were told that the axis of a single crystal wire was indicated by the pole P on the standard triangle in Fig. 1.23. How could you visualize the orientation of this crystal? One convenient way to do so is with the use of the reference cube. Figure 1.24 shows the position of the standard triangle upon the reference cube with the direction P indicated upon it. If one now imagines that the reference cube of Fig. 1.24 is a large single crystal with the orientation shown, then the single crystal wire would be a thin cylinder cut from this block along the line from P through the center, as is shown on Fig. 1.24.

It is frequently convenient to talk about the trace of a plane on the stereographic projection. If any plane, (hkl), is extended out to the reference sphere it will intersect the sphere along a great circle. This great circle is the trace of the plane and it may simply be defined as the locus of points $90°$ away from the pole of the plane. Figure 1.25 shows the poles and the traces of the $(\bar{1}\bar{1}1)$ and the $(0\bar{1}1)$ planes. It should be clear that the pole is $90°$ away from each point along the trace.

Planes that intersect along a common line are said to belong to the same zone and are called *planes of a zone*. The common line is called a zone axis. Suppose you were asked to determine the zone axis direction for the two planes $(\bar{1}\bar{1}1)$ and $(0\bar{1}1)$. By drawing the trace of these two planes on a stereographic projection as in Fig. 1.25 the intersection point becomes immediately apparent and the zone axis is seen to be the [011] direction. Notice that the 45°W longitude is $90°$ away from the [011] direction.

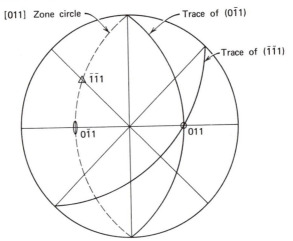

[011] Zone circle

Trace of (0$\bar{1}$1)

Trace of ($\bar{1}\bar{1}$1)

$\bar{1}\bar{1}$1

0$\bar{1}$1

011

Figure 1.25 Trace of planes upon stereographic projection.

Hence, every pole on this 45°W longitude represents a plane of the [011] zone, and so this longitude may be called the [011] zone circle. The lines drawn on the standard projection in Fig. 1.18 are often referred to as the major zone circles.

A property of the stereographic projection that is sometimes useful is the following. Any circle on the reference sphere will appear on the stereographic projection as a true circle. A discussion of this point as well as a good discussion of the gnomonic and stereographic projections may be found in Refs. 3 and 4.

REFERENCES

1. L. H. Van Vlack, *Materials Science for Engineers*, Addison-Wesley, Reading, Mass., 1970, Chapter 5.

2. W. G. Moffatt, G. W. Pearsall, and J. Wulff, *The Structure and Properties of Materials*, Vol. I, Wiley, New York, 1964, Chapter 3 and Appendix 3.

3. B. D. Cullity, *Elements of X-ray Diffraction*, Addison-Wesley, Reading, Mass., 1956, Chapter 2.

4. C. S. Barrett, *Structure of Metals*, McGraw-Hill, New York, 1943, Chapter 2.

PROBLEMS

1.1 The atoms of a crystal lie on the points of a tetragonal lattice with $a=b$ and $c=\frac{1}{2}a$. A plane intercepts the fourth atom from the origin on the Z axis, the second atom on the Y axis, and the sixth atom on the X axis. What are the Miller indices of this plane?

1.2   Show that the [111] direction is perpendicular to the (111) planes in the cubic system. Would this relation hold in the tetragonal system?

1.3   The {111} planes of the cubic system form an octahedron and are therefore sometimes called the octahedral planes. List all of the Miller indices of these {111} planes. By means of a sketch show the octahedron that these {111} planes form within the unit cell of the cubic lattice. Label the planes on your sketch.

1.4   (a) Make a sectioned drawing through the (111) plane of the fcc unit cell showing the atom positions.

   (b) On the section show the $[10\bar{1}]$, $[01\bar{1}]$, $[11\bar{2}]$, and $[\bar{2}11]$ directions.

1.5   Show a trace of the $(10\bar{1}2)$ plane on a full-size unit cell of the hexagonal system. Also show the $[11\bar{2}0]$ and $[10\bar{1}\bar{1}]$ directions. List the Miller indices of all of the $\{10\bar{1}2\}$ planes.

1.6   The following information is given for a description of the crystal structure of β-tin on page 235 of volume 8 of the ASM Metals Handbook: Body-centered tetragonal, four atoms per cell at 0, 0, 0; $\frac{1}{2}, \frac{1}{2}, \frac{1}{2}$; 0, $\frac{1}{2}, \frac{1}{4}$; and $\frac{1}{2}$, 0, $\frac{3}{4}$. $a = 5.83$ Å and $c = 3.18$ Å.

   (a) From this information draw a unit cell for β-Sn.

   (b) List all of the directions in the Sn lattice that have twofold rotation symmetry.

1.7   Calculate the packing efficiency (percent volume of a crystal occupied by atoms) for the bcc and fcc crystals assuming the hard-sphere model.

1.8   Show that the c/a ratio in a hcp crystal of perfect spheres is 1.633.

1.9

Packing	Coordination	Void Radius/Sphere Radius
Closest packing	3-fold	0.155
	4-fold (tetrahedral)	0.225
	6-fold (octahedral)	0.414
	12-fold	1.0
bcc packing	Tetrahedral void	0.291
	Octahedral void	0.154
	8-fold	1.0

The above table gives the ratio of the void radius to sphere radius for two different kinds of packing. Derive the given ratios for the tetrahedral voids in both closest packing and bcc packing.

1.10   (a) Steel is a solution of carbon in iron. Since carbon atoms are considerably smaller than iron atoms, one expects the carbon to lie in the *interstitial* voids (interstitial solid solution) of the iron lattice. Based on your answer to Problem 1.7, would you expect α iron (bcc) or γ iron (fcc) to hold more carbon?

   (b) From x-ray data there is good evidence that the radius of iron and carbon are given as

$$r_{Fe} \ (\gamma \text{ phase}) = 1.25 \text{ Å}$$
$$r_{Fe} \ (\alpha \text{ phase}) = 1.24 \text{ Å}$$
$$r_c = 0.7 \text{ Å}$$

Based on the hard-sphere model,
1. In which *interstitial* voids would you expect to find the C atoms in α and γ iron?
2. Would you expect the α or the γ phase to dissolve more carbon?

1.11 In a number of problems with metal crystals it is useful to know the density of atoms whose centers lie on the planes (hkl). This density may be characterized by an area packing efficiency, which is calculated as the percent area of the plane (hkl) occupied by atoms whose centers lie on that plane. Using the hard sphere model one then obtains the following result:

	Area Packing Efficiency		
Lattice	(100)	(110)	(111)
bcc	58.9	83.4	33.5
fcc	78.4	55.5	90.6

Calculate the area packing efficiency for the' (100) planes of the bcc lattice.

1.12 The following data are given in Table 1.2.
1. The angle between the (210) plane and the (100) plane is 26.56°.
2. The angle between the (211) and the (100) is 35.26° and that between (211) and (110) is 30.00°.

Construct a standard (001) stereographic projection by placing a transparent paper over a Wulff net. Let the south pole be the (100) and the east pole be the (010). Using the above information and your knowledge of the crystal structure, locate the poles of the (210) and (211) planes on your projection. What is the angle between these two planes? Identify and locate the pole of the zone axis to which these two planes belong.

1.13 Pole A, whose coordinates are 20°N lat., 50°E long., is to be rotated about the axes shown below. In each case, find the coordinates of the final position of pole A and show the path traced out during rotation.
  (a) 60° rotation about an axis normal to the plane of projection, clockwise to observer.
  (b) 60° rotation about an inclined axis B whose coordinates are 10°S lat., 30°W long. clockwise to the observer.
  Answer: (a) 27°S, 48°E; (b) 39°S, 61°E.

1.14 White tin has the tetragonal structure with $c/a = 0.545$. List the planes of the form {001}, {100}, {110}, {011}, {111}. Draw the standard 001 projection for white tin and indicate the poles of the above planes with their proper symmetry symbols.

Show the important zone circles. (The major zone circles for the cubic case are the great circles shown on Fig. 1.18.)

1.15   Pole A is located at 20°S lat. and 40°W long., and pole B is located at 10°N lat. and 10°E long. Locate the position of pole B after it is rotated 90° clockwise to the observer about pole A. Work the problem in three ways by rotating pole A to the origin of the projection in three different ways.

(a) First rotate pole A to the equator of your *projection* and then along equator to the origin. Do not use a compass for the first rotation, but rotate along either a latitude or longitude (whichever is correct) underneath your projection.

(b) Repeat part a using a compass for the first rotation.

(c) Rotate the Wulff net about the origin (leave projection stationary) until the Wulff net equator lies directly under A. Now rotate A on your projection directly to the origin along the underlying Wulff net equator.

Note that the position of pole B after A is rotated to the origin is different in each of the three cases, but the angle between the two poles is the same. This illustrates the angle true character of the stereographic projection.

CHAPTER 2
STRUCTURE DETERMINATION

It is illustrated in the table on page 2 of the Introduction that the physical properties of metals are influenced by both the microstructure of the metal and the atom arrangement in the metal.

Microstructure. When a metal is examined on the microscopic scale with an optical microscope after suitably polishing and etching the surface, widely varying structures are observed depending on the alloy composition and preparation. An example of a cast structure is shown in Fig. 2.1. This picture is a lead–tin solder composed of two phases; the light-colored phase is bct tin solid solution and the dark-colored phase is fcc lead solid solution. The microstructure of the alloy is described by identification of the types of phases present and description of their shape and size distributions. The most useful metallurgical tool for characterizing microstructure is the optical microscope. Two other very useful microscopes have been developed in recent years, the transmission electron microscope (TEM) and the scanning electron microscope (SEM). These electron microscopes provide the metallurgist with a means of characterizing microstructures of much finer details than is possible with the optical microscope. A brief introduction to the principles of the TEM and SEM will be presented in this chapter. However, the optical microscope will not be discussed on the assumption that the student is familiar with the use of optical microscopes.

Atom Arrangement. Each of the individual white regions on Fig. 2.1 is a nearly perfect single crystal of tin, and the larger white regions (dendrites) all have a $\langle 110 \rangle$ axis nearly normal to the plane of polish. The crystallographic orientation of such microconstituents often has a strong influence upon the physical properties of the alloy. In addition, the presence and the extent of defects in the individual metal crystals play a strong role in controlling physical properties. The most useful metallurgical tool for

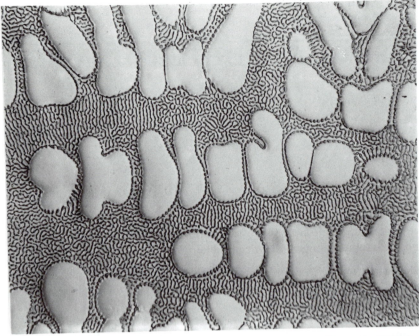

Figure 2.1 Lead–tin alloy showing tin dendrites in a eutectic matrix (125×).

determining crystallographic orientations and the nature and extent of crystalline imperfection has been x-ray diffraction. However, with the advent of the TEM and the SEM, electron diffraction directly within the microscope has become a very powerful tool for these purposes. Both x-ray and electron diffraction will be discussed in this chapter.

2.1 X-RAY DIFFRACTION

When an x-ray beam is impinged on a metal a number of different and interesting interactions occur. One of these interactions, known as Thomson scattering, causes the electrons of the crystal atoms to emit electromagnetic radiation of the same wavelength, λ, as the impinged beam, and this produces a diffraction effect.

Figure 2.2 shows an edge view of a crystal having atom planes at a spacing of d_{hkl}. The planes are perpendicular to the page. A beam of radiation of wavelength λ is impinged upon the crystal at some angle θ. The diffraction due to Thomson scattering will produce a strong reflection from the crystal at an angle α only if the following conditions are all met:

    1. Only one angle of α is possible, $\alpha = \theta$.

2.   The Bragg equation must be satisfied, namely

$$\frac{n\lambda}{2d_{hkl}} = \sin\theta \qquad n = \text{integer} = 1, 2, 3, \ldots \qquad (2.1)$$

At the higher values of n the intensity of the reflection quickly drops off. Hence, we will take $n = 1$ unless otherwise noted.

3.   It is possible that the atoms between the diffracting planes may be positioned in such a way as to destroy the diffracted beam. Hence, for any crystal structure certain planes will not diffract even though conditions 1 and 2 are met. The diffracting planes for cubic crystals are listed in Table 2.1.

A derivation of the Bragg equation may be found in a number of textbooks.[1-3] Reference 1 provides an excellent presentation. For our purposes we are interested in the use of the Bragg equation to analyze structures and we will not consider the theory of diffraction here. X-ray diffraction has proven extremely useful in metallurgy. Consider the following examples.

1.   The Bragg law may be written $d_{hkl} = n\lambda/(2 \cdot \sin\theta)$. When λ is known and the value of θ producing reflection is determined, the interplanar spacing, d_{hkl}, may be found.

2.   The Bragg law may be written $\theta = \sin^{-1}(n\lambda/2d_{hkl})$. When d_{hkl} and λ are both known, the measured values of θ allow one to determine the orientation of the crystal planes relative to the beam.

3.   Analysis of the shape of the diffracted beam provides information on the perfection of the crystal.

4.   There are numerous other examples; see Ref. 1.

There are three main techniques of x-ray analysis commonly used by metallurgists.

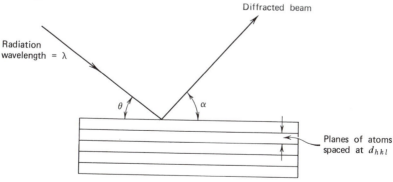

Figure 2.2   Diffraction.

Name of Technique	Wavelength, λ	θ
Laue	Variable	Fixed
Powder	Fixed	Variable
Diffractometer	Fixed	Variable

If the radiation has only one wavelength it is termed *monochromatic*; if it is variable it is called *white radiation*.

A.   LAUE METHOD

In this technique white radiation is used and the diffracted beams are measured as spots on a photographic plate. When the plate is on the same side of the sample as the source of the beam, the radiation first passes through the plate and the diffracted beams travel back to the plate. This is called a back-reflection picture, and when the plate is on the opposite side from the source it is called a transmission picture. To illustrate this technique consider the transmission arrangement shown in Fig. 2.3. The white radiation is directed parallel to the [010] direction of the fcc cubic crystal as shown. We now ask: What is the geometric pattern that will appear on the plate due to diffraction from {111} planes? First we must determine whether or not diffraction occurs from the four {111} planes. Since λ is a variable here, it is apparent that the Bragg equation will be satisfied for all values of θ. Reference to Table 2.1 shows that the {111} planes produce reflections in fcc crystals, and so we expect a spot from each of the four sets of {111} planes. It should be apparent from examination of Fig. 2.2 that the incident and diffracted beams lie in a plane perpendicular

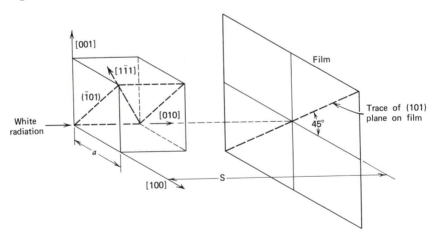

Figure 2.3   Transmission Laue arrangement.

Table 2.1 Indices of the Reflecting Planes for Cubic Structures

Simple Cubic	Body-Centered Cubic	Face-Centered Cubic
{100}	—	—
{110}	{110}	—
{111}	—	{111}
{200}	{200}	{200}
{210}	—	—
{211}	{211}	—
{220}	{220}	{220}
{221}	—	—
{300}	—	—
{310}	{310}	—
{311}	—	{311}
{222}	{222}	{222}
{320}	—	—
{321}	{321}	—
{400}	{400}	{400}
{322}	—	—
{410}	—	—
{330}	{330}	—
{411}	{411}	—
{331}	—	{331}
{420}	{420}	{420}
{421}	—	—
{332}	{332}	—

to the diffracting planes. This fact is very useful in visualizing diffraction: *The incident and diffracted beams must be coplanar with the normal of the diffracting planes.*

We now ask: What plane contains the incident beam and the normal to the $(1\bar{1}1)$ plane? This plane would be defined by the [010] and $[1\bar{1}1]$ directions, and it can be seen from Fig. 2.3 that these two lines define the $(\bar{1}01)$ plane. Hence, the diffracted beam must lie somewhere in the $(\bar{1}01)$ plane. If we extend this $(\bar{1}01)$ plane out so that it intersects the photographic film, it makes a straight-line trace on the plate at an angle 45° up from the horizontal as shown in Fig. 2.3. Now in order to locate the position on this trace where the diffracted beam appears, we must determine the angle at which it leaves the crystal. Figure 2.4(*a*) locates the $(1\bar{1}1)$ and $(\bar{1}01)$ planes on the unit cell; their intersection is marked by the dashed line, 1-2. Figure 2.4(*b*) is a view normal to the $(\bar{1}01)$ plane. In this view the diffracting $(1\bar{1}1)$ plane is perpendicular to the page and runs from

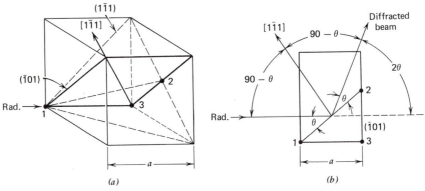

Figure 2.4 Geometric relation between x-ray beam and diffracting planes.

point 1 to 2. Hence, the incident and diffracted angle, θ, must appear directly on this view as is shown. Notice that the diffracted beam is projected up at an angle of 2θ from the line of the incident radiation. From the geometry of the unit cell, one sees that the tangent of θ is half the face diagonal divided by the cube edge so that $2\theta = 70.6°$.

In Laue pictures using x-radiation most of the incoming radiation passes straight through the sample, producing a strong spot at the center of the film, which we may use as a reference point. We will let R be the distance away from this center spot at which the diffracted beam strikes the film. Letting the distance from the sample to the film be S, this distance R is calculated as shown in Fig. 2.5(a). We now know that the diffraction spot from the $(1\bar{1}1)$ plane must lie on the 45° trace at a distance R from the center spot, so we locate it on the film as shown in Fig. 2.5(b). By an analysis similar to that above, you may show yourself that the other three {111} spots are located as shown in Fig. 2.5(b).

We will now show that this problem may be solved very simply using the stereographic projection. From Fig. 2.4(b) we note the following general

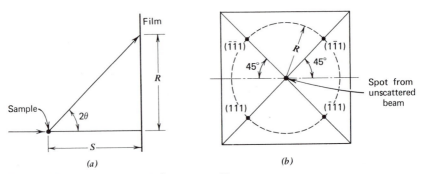

Figure 2.5 Location of {111} spots on film.

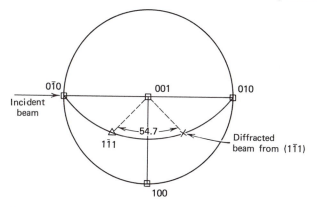

Figure 2.6 Use of stereographic projection to locate diffraction spots.

condition: *The incident and diffracted beams both make an angle of* $90 - \theta$ *with the normal of the diffracting planes.*

To begin, imagine that the crystal sits at the center of the reference sphere and the radiation enters the sphere along the [010] direction at the $(0\bar{1}0)$ pole as shown in Fig. 2.6. We will consider that the crystal acts as a point source and the diffracted beams will appear as points on the sphere. The diffracted beam from the $(1\bar{1}1)$ planes will first be located. We know that the plane containing this diffracted beam will be defined by the W-E axis of the projection (beam direction) and the $(1\bar{1}1)$ pole (normal of diffracting planes). This plane is the great circle shown on Fig. 2.6 passing through the W-E axis and the $(1\bar{1}1)$ pole. Since the incident beam enters the sphere at the $[0\bar{1}0]$ pole, we know that the angle between the $(1\bar{1}1)$ pole and the $[0\bar{1}0]$ pole is equal to $90 - \theta$. From Table 2.1 we determine that this angle is 54.7°, so that 2θ has been determined to be 70.6° simply by inspection of the stereographic projection. We also note that the diffracted beam must lie along the longitude of Fig. 2.6 at an angle of $90 - \theta = 54.7°$ from the $(1\bar{1}1)$ pole as shown in Fig. 2.6.

A back-reflection Laue photograph of a single crystal of vanadium (bcc, $a = 3.039$ Å) is shown in Fig. 2.7. The standard film-to-sample distance of 3 cm was used and the crystal was oriented with a ⟨011⟩ axis parallel to the beam. The planes producing some of the spots are identified and you may identify some of the remaining ones. (Note: Photo reduced by 0.66.)

B. POWDER METHOD

The powder method may be explained with the use of the stereographic projection. In this case a powdered sample is imagined at the center of the reference sphere and monochromatic radiation is directed at it along the W-E axis of the reference sphere. Each small particle in the powder will act

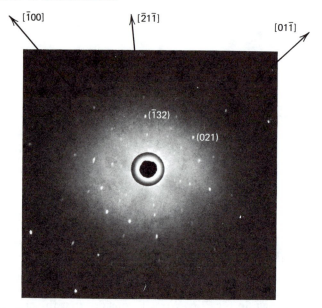

Figure 2.7   Laue pattern of a vanadium crystal (bcc) oriented with an [011] direction nearly parallel to the x-ray beam (courtesy Don Bailey).

as a single crystal and make its own contribution to the diffraction pattern. Ideally there will be a sufficient number of randomly oriented particles in the sample so that all possible orientations of these small crystals relative to the beam are represented. We now ask: What is the locus of points on the reference sphere made by the beams diffracted from the {111} planes? Notice that for this problem both λ and d_{hkl} are fixed. Therefore, we write the Bragg equation as

$$\theta = \sin^{-1}\left(\frac{n\lambda\sqrt{3}}{2a}\right) \tag{2.2}$$

where a is the lattice parameter and we have determined the distance between (111) planes from the relation

$$d_{hkl} = a\left(\frac{1}{h^2+k^2+l^2}\right)^{1/2} \tag{2.3}$$

which holds in cubic crystals. Equation 2.2 tells us that diffraction will only occur for some fixed θ, say 30°. Hence all those particles whose {111} planes make a 30° angle with the incident beam will produce a diffracted beam from {111} planes and the remaining particles will not. The normals of these diffracting {111} planes will be at $90-\theta$ or 60° from the incident beam. Hence the normals of the {111} planes in those crystals giving

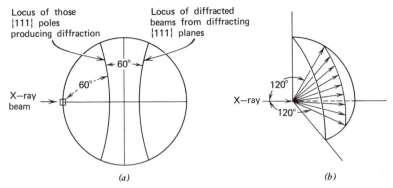

Locus of those {111} poles producing diffraction

Locus of diffracted beams from diffracting {111} planes

X—ray beam

60°

60°

X—ray

120°

120°

(a)

(b)

Figure 2.8 Use of stereographic projection to locate diffracted beams from a powder sample.

diffraction must all lie along the latitude line 60° from the point where the beam enters the reference sphere, as shown in Fig. 2.8(a). Notice on Fig. 2.4(b) that the diffracted beam is located at $(180° - 2\theta)$ away from the incident beam, which for our case is 120° since $\theta = 30°$. Hence the spot from any given {111} plane producing diffraction will lie on the sphere 120° away from the incident beam spot, in that plane defined by the incident beam and the {111} normal. It should be clear that the locus of such spots will be the latitude at 120° from the incident beam spot as shown on Fig. 2.8. Since the diffracted beams all originate from the sphere center, the locus of the diffracted beams is a cone, whose cone axis is colinear with the x-ray beam. Figure 2.8(b) is an attempt to illustrate this. If one had the specimen surrounded by a spherical film, the diffracted beams from each {hkl} set would produce a circle on this film. Spherical film is impractical and the usual powder technique places the powdered sample at the center of a cylinder with a film strip running along the inside surface of the cylinder. The film strip intercepts a small segment of the diffraction cone and produces two lines for each {hkl} set as shown in Fig. 2.9. Knowing

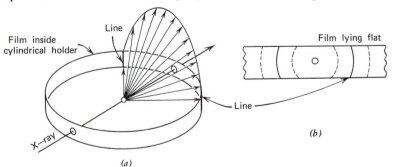

Line

Film inside cylindrical holder

X—ray

Film lying flat

Line

(a)

(b)

Figure 2.9 The powder technique.

the film radius and measuring the distance between the pairs of lines on the film one may calculate the apex angle of the cone and thus determine θ. From the Bragg equation then one has a measure of d_{hkl}.

C.   DIFFRACTOMETER METHOD

In this method one may use a powder or a single crystal. The 2θ value that the diffracted beam makes with the incident beam is simply determined by direct measurement with a Geiger tube moving on an arc. The method is very straightforward and the student is referred to Refs. 1–3 for a description.

2.2   TRANSMISSION ELECTRON MICROSCOPE (TEM)

In the TEM the image of the specimen is formed by a focused beam of electrons that is transmitted directly through the specimen. The electron beam is generated by accelerating electrons through a potential, V; they emerge as an essentially monochromatic beam with a wavelength given approximately as $\lambda \approx \sqrt{150/V}$ (where V is in volts and λ is in angstroms). Since the electrons are charged they may be focused by a symmetric electrical or magnetic field. Electron microscopes utilize symmetric magnetic fields for lenses, and their focal lengths can be changed by simply changing the current in the coils producing the magnetic field. A schematic view of a typical TEM is shown in Fig. 2.10. After the electron beam emerges from the gun it passes through two condenser lenses. These lenses are adjusted to focus the electron beam upon the specimen and to control the size of the electron beam on the specimen. The focal length of the objective lens is adjusted to form a magnified image of the specimen in the plane of the intermediate aperture. The function of the two remaining projector lenses is simply to magnify further the image from the objective lens and to focus it on the photographic plate. The use of two projector lenses rather than one, and the use of two condenser lenses rather than one, both provide the TEM with greater flexibility of operation. (For further explanation see Sections 7.5 and 7.9 of Ref. 5.) In essence, then, the TEM consists of condenser lenses that serve to condense the electron beam down to a small diameter upon the specimen, an objective lens that forms an enlarged image of the specimen, and finally, projector lenses that further magnify and project the image of the specimen onto the photographic plate or the viewing screen.

The ultimate useful magnification of the TEM and of the optical microscope is inversely proportional to the wavelength of the radiation (see p. 69 of Ref. 6 and p. 8 of Ref. 7). Using green light [$\lambda = 5200$ Å (520 nm)], the optical microscope is capable of resolving a minimum distance of

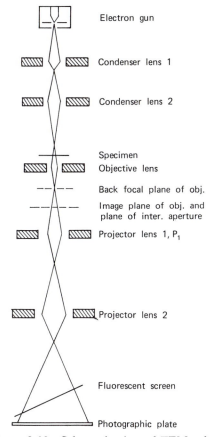

Figure **2.10**   Schematic view of TEM column.

0.2 μm.[6,7] Most TEMs currently used in metallurgical laboratories operate at either 100 or 200 kV. At these voltages the wavelength is in the range of 0.025–0.037 Å (2.5–3.7 pm) and it is primarily because of these much smaller wavelengths that higher magnifications can be achieved than with optical microscopes. A well-adjusted TEM in the research laboratory is generally capable of resolving distances on metallurgical specimens down to 1 nm (10 Å).

A.   MICROSTRUCTURE

Since it is necessary for the electron beam to go through the specimen, it is not possible to examine directly the etched surface of a bulk metal sample. However, one may examine such surfaces by an indirect technique referred to as the replica technique. In this technique one first forms a thin film of either plastic, carbon, or oxide upon the surface. This film is then carefully stripped from the surface, thereby forming a replica of the surface topography. The variation in thickness and/or shape of the replica due to the surface topography produces contrast variations in the TEM that reveal the surface topography of the original specimen (see pp. 191 and 277 of Ref. 6). The replica technique will probably be largely replaced by the more recently developed SEM because pictures from the latter are easier to take and interpret, although resolution is not quite as good.

It is possible to examine thin foils of metals in the TEM by producing foils thin enough to allow penetration by the electron beam. In 100-kV machines the foils must be thinned down to thicknesses on the order of 0.05–0.5 μm, depending on the atomic number of the metal (for comparison, commercial aluminum foil has a thickness on the order of 50 μm). This thin-film technique, which was developed subsequent to the replica technique in the mid-1950s, has proven to be an extremely powerful research tool. Not only does it reveal microstructural detail, but information is also obtained on the defect structure and crystallographic orientation of quite small regions of the specimen. Since the metal foil is crystalline, the electron beam may be diffracted as it passes through the foil. Just as with x-ray diffraction, the beam is diffracted if any set of (hkl) planes satisfy the Bragg equation. Figure 2.11 illustrates how such diffraction produces contrast in the image. The rays that are diffracted by the foil are blocked out of the transmitted image by the objective aperture, which is located at the back focal plane of the objective lens. Consequently, if one were viewing a number of small grains, those grains oriented such that an (hkl) set satisfied the Bragg relation would appear darker in the image due

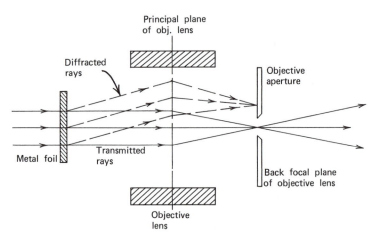

***Figure* 2.11** Ray diagram through objective lens illustrating origin of diffraction contrast.

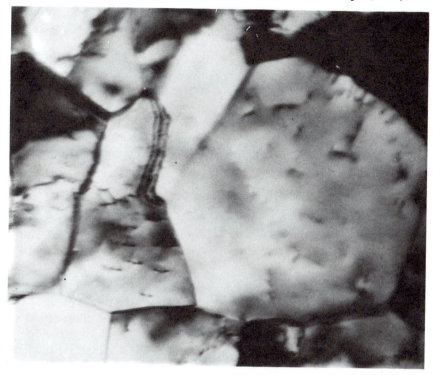

Figure 2.12 A TEM picture of a thin film of an Fe–Si alloy showing subgrain structure containing several dislocations (50,000×).

to the loss of the diffracted rays from the image. This effect is illustrated by a few of the subgrains shown in the TEM picture of Fig. 2.12. The short, dark lines within some of the subgrains are dislocations. These dislocation images are also produced by diffraction contrast and will be discussed further in Section 4.7.

B. DIFFRACTION

Consider a single-crystal region in the specimen. Suppose one set of (hkl) planes in this region diffracts the beam off at an angle as shown in Fig. 2.13. These diffracted rays will all be focused in the back focal plane of the objective lens as shown in Fig. 2.13 for rays diffracted from one (hkl) set. Hence, the back focal plane contains an array of diffraction spots, each spot being produced by one set of (hkl) planes that satisfies the Bragg equation. To observe this diffraction pattern on the view screen it is only necessary to change the intermediate projector lens, P_1. Normally P_1 is set to project the image plane of the objective lens onto the view screen. The diffraction

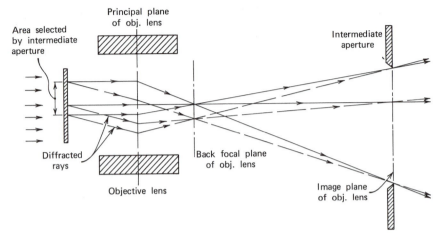

Figure 2.13 Ray diagram through objective lens illustrating area selection for diffraction by intermediate aperture.

pattern is observed by increasing the focal length of P_1 until it focuses on the back focal plane of the objective rather than on the image plane of the objective. A small aperture is inserted into the image plane of the objective lens in order to restrict the area on the specimen from which diffraction is observed, as illustrated by the ray diagram of Fig. 2.13. With this technique it is possible to obtain diffraction patterns from selected areas of the specimen having dimensions as small as 0.5–1 μm. Figure 2.14(a) presents

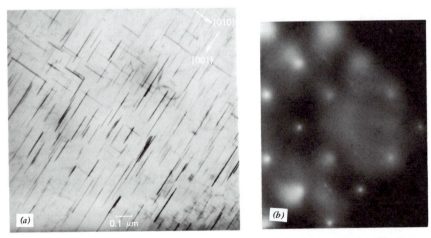

Figure 2.14 (a) The θ' precipitate (Cu–Al) in the aluminum matrix. (b) Corresponding selected area diffraction pattern from (a) (courtesy David Pearson).

a micrograph of an aluminum matrix containing the θ' precipitate (Cu–Al), and Fig. 2.14(b) presents a selected area diffraction pattern from this region. Just as the Laue patterns discussed in the above x-ray section determine crystal orientation, an analysis of diffraction patterns such as that of Fig. 2.14(b) permits one to determine the crystallographic orientation of the selected area producing the pattern. (See p. 222 of Ref. 6 and p. 148 of Ref. 7.) Hence, one may determine the crystallographic orientation of interesting features that appear in the TEM micrograph, such as interface boundaries and dislocations. For example, from the diffraction pattern of Fig. 2.14(b) the crystallographic directions and planes in the foil were determined as shown on Fig. 2.14(a). It is thus found that the θ' precipitate lies along {001} planes of the Al matrix. The very small spots in the diffraction pattern arise from a double diffraction between the Al matrix and the θ' precipitates.

2.3 SCANNING ELECTRON MICROSCOPE (SEM)

The SEM operates in a fashion very similar to commercial television; see Fig. 2.15. A very small diameter electron beam is made to trace out (scan) a square region (a raster) on the specimen surface. At the same time a square raster is traced out in synchronism on a cathode-ray tube (CRT). Low-energy electrons are ejected from the specimen surface by the electron beam, the intensity being a function of the angle of the beam with the local specimen surface. The intensity of the beam on the CRT is varied electronically in proportion to the number of low-energy electrons ejected from the specimen. This causes an image of the surface topography to be built up on the CRT as the raster is traced out, and this image may be

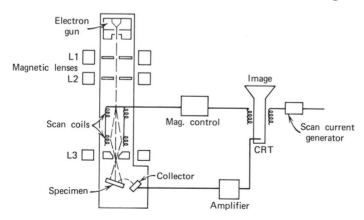

Figure **2.15**   Schematic description of an SEM.

viewed and photographed. Hence, whereas the TEM produces a direct image of the specimen, in an SEM one views an indirect image of the specimen.

The resolution of the SEM is determined by the size of the electron beam upon the specimen surface. The minimum resolvable distance is roughly equal to the beam size at the surface, so that one desires the smallest possible beam size. The electron gun acts as an electrostatic lens forming an image of the filament roughly 60 μm in diameter. This image acts as the object of the first condenser lens and it is the function of all three lenses in the SEM to demagnify the beam. (Some SEMs use only two lenses.) It is also the function of the third lens, L3, to focus the beam on the specimen surface, and at that point the beam diameter has been demagnified by the three lenses to sizes as small as 10 nm (100 Å). The beam is shown in Fig. 2.15 as a line and, consequently, the focusing action of the lenses does not appear. However, Fig. 2.15 does illustrate how the two sets of scan coils deflect the beam to generate the raster on the specimen surface. The magnification is equal to the width of the raster on the CRT divided by the width of the raster on the specimen surface. The width is constant on the CRT, while the width of the raster on the specimen surface is controlled by the current to the scan coils.

A. MICROSTRUCTURE

The etched surfaces of metal specimens can easily be examined in the SEM at magnifications from 10 to 40,000×. However, since most etched metallurgical samples are relatively flat, the SEM does not offer an advantage over the optical microscope on such samples except at magnifications of around 400× and above. At the present time the better research SEMs operating in the laboratory are generally capable of resolving minimum distances of 100 Å (10 nm). Table 2.2 illustrates the resolving power of the three microscopes of most use to metallurgists.

The SEM offers two major advantages over the optical microscope: (a) magnifications up to 20 times greater, and (b) a depth of focus about 300 times greater. This latter feature is illustrated in Fig. 2.16. Two 200-mesh

Table 2.2   Resolution Characteristics of Microscopes

Microscope	Minimum Resolvable Distance in Practice	Maximum Magnification Above Which No Additional Detail Is Revealed
Optical	0.2 μm (200 nm)	1,000×
SEM	100 Å (10 nm)	20,000×
TEM	10 Å (1 nm)	200,000×

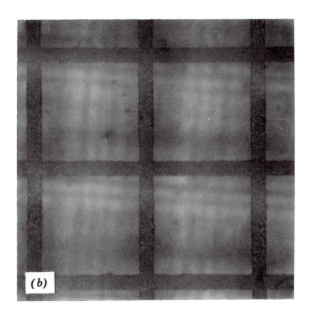

Figure 2.16 Micrographs of two 200-mesh screens lying on top of a 1000-mesh screen. (*a*) SEM focused on central screen (240×). (*b*) Optical microscope focused on central screen (240×).

screens are lying on top of a 1000-mesh screen. In Fig. 2.16(*a*) the SEM has been focused on the center screen and the remaining two screens remain well in focus. Figure 2.16(*b*) was taken on a standard metallurgical microscope focused on the central screen and the remaining two screens are nowhere near focus. The much greater depth of focus in the SEM is extremely valuable in examining surfaces with large relief, in particular, fracture surfaces. The larger depth of focus results from the mode of image formation with a beam of small divergence. For further discussion of this point and other features of the SEM see Refs. 8 and 9.

B.   CRYSTALLOGRAPHIC ORIENTATION

It has recently been found that a diffraction-type effect occurs in the SEM that permits one to determine crystallographic orientations of small surface features. Notice in Fig. 2.15 that the double-deflection scan coil arrangement causes the beam to cross the column axis at a fixed point in lens L3. By suitable adjustments of the scan coil current and lens L3, one can move this crossover point down to the surface of the specimen so that the beam rocks about a fixed point on the specimen as shown in Fig. 2.17. For (*hkl*) planes nearly colinear with the column axis, the beam will pass through the

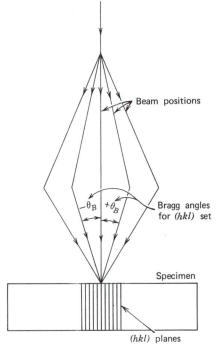

Figure 2.17   Scanning conditions for selected area channelling.

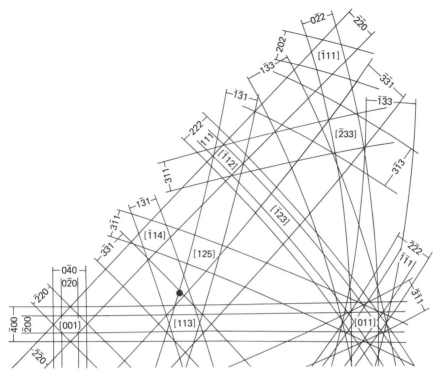

Figure **2.18** Channelling map showing expected channelling patterns for all orientations within the standard triangle. Copper, 20 kV.

Bragg angle on each side of the column axis as shown on Fig. 2.17. When this occurs the intensity of the ejected electrons changes due to a diffraction-related effect, and this change at the two Bragg angles produces a band on the CRT. Each (*hkl*) set causing the diffraction effect produces a band whose midpoint corresponds to the trace of the (*hkl*) plane on the image. The bandwidth is a function of the Bragg angle. Figure 2.18 shows the band patterns that would appear from the major diffracting planes of a fcc crystal whose crystallographic axis (which is colinear to the electron column axis) has any orientation within the standard triangle. For example, a crystal oriented with a ⟨110⟩ direction colinear with the column axis would produce a pattern similar to that shown around the [011] pole of Fig. 2.18. The change in intensity of ejected electrons that occurs at the Bragg angle is similar to certain channeling effects, and for this reason these band patterns have been termed channeling patterns (see p. 613, Ref. 8). Selected-area channeling patterns (SACP) taken by the scheme shown in Fig. 2.17 may be obtained from areas down to 2 μm in size. As an example,

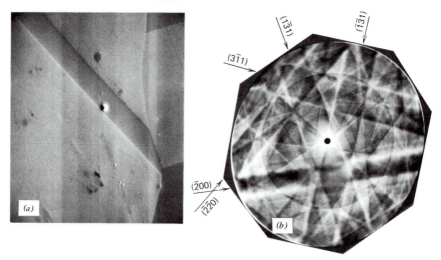

Figure 2.19 (a) Twin band in Cu. Picture taken with SEM in normal mode. (b) Corresponding selected area channelling pattern at 20 kV.

Fig. 2.19(a) shows a narrow twin band in Cu running up to the reader's left. The SACP technique may be used to measure the crystallographic orientation of the twin band relative to the neighboring grains. An SACP from the region is shown in Fig. 2.19(b). The crystallographic axis of this region is located near the [013] pole as shown by the dot on Fig. 2.18. The various bands in Fig. 2.19(b) may be identified by comparison to Fig. 2.18 and all other crystallographic directions of the region identified.

C. CHEMICAL COMPOSITION

In the mid-1950s electron beam instruments were developed that allowed measurements of chemical compositions from microscopic regions. These instruments, called electron microprobes, are now largely being replaced by SEMs because the same function can be carried out in an SEM. When the electron beam strikes the surface of the metal it can cause the inner electrons of the atoms (i.e., the K or L electrons) to be removed. As electrons drop back into these emptied orbits, x-radiation is emitted of a wavelength that is characteristic of the atomic number of the atom involved. An x-ray spectrometer is added to the specimen chamber of the SEM. By scanning the wavelengths of the x-rays produced by the electron beam in a given particle, one may determine the elements present in the particle. By comparing the intensity of characteristic radiation of an element within a particle to standards, one may quantitatively determine the amount of the element present in a particle. With this technique it is possible to determine compositions from regions as small as a few microns.

This technique is particularly useful in determining chemical composition variations in a given microstructure and in identifying the different phases in a microstructure.

REFERENCES

1. B. D. Cullity, *Elements of X-ray Diffraction*, Addison-Wesley, Reading, Mass., 1956, Chapter 3.

2. R. E. Reed-Hill, *Physical Metallurgy Principles*, D. Van Nostrand, New York, 1964, Chapter 2.

3. A. G. Guy, *Elements of Physical Metallurgy*, Addison-Wesley, Reading, Mass., 1960, Chapter 4.

4. E. A. Wood, *Crystal Orientation Manual*, Columbia University Press, New York, 1963.

5. C. E. Hall, *Introduction to Electron Microscopy*, McGraw-Hill, New York, 1966, Chapter 7.

6. H. Modin and S. Modin, *Metallurgical Microscopy*, Butterworths, London, 1973, Chapters 2 and 7.

7. R. E. Smallman and K. H. G. Ashbee, *Modern Metallography*, Pergamon, New York, 1966, Chapters 1 and 7–9.

8. S. Amelinckx et al. (editors), *Modern Diffraction and Imaging Techniques in Material Science*, pp. 553–682; G. R. Booker, *Scanning Electron Microscopy*, North-Holland, Amsterdam, 1970.

9. C. W. Oatley, *The Scanning Electron Microscope*, Cambridge Univ. Press, Cambridge, 1972.

PROBLEMS

2.1   A transmission Laue pattern is made of a cubic crystal having a lattice parameter of 4.0 Å. The x-ray beam is horizontal. The $[0\bar{1}0]$ axis of the crystal points along the beam toward the x-ray tube, the $[\bar{1}00]$ axis points vertically upward, and the [001] axis is horizontal and parallel to the photographic film. The film is 5 cm from the crystal.

(a) What is the wavelength of the radiation undergoing first-order diffraction from the $(3\bar{1}0)$ planes?

(b) Where will the $(3\bar{1}0)$ reflection strike the film?

2.2   A back-reflection Laue pattern is made of a cubic crystal in the orientation of Problem 2.1. Draw a standard (001) stereographic projection on tracing paper locating the (100) pole at the south pole. On this projection locate the poles of the $(\bar{1}20)$, $(\bar{1}23)$, and (121) planes using the tables of angles in Table 1.2. What is the zone axis of these three planes? On the stereographic projection locate the points where the diffracted beams from each of these planes intersect your reference sphere. One of the diffraction spots will lie on the bottom half of your reference sphere. Show this spot by simply projecting it orthographically up to the top half of your reference sphere. The diffracted beams from the planes of a zone fall on a cone whose axis is the zone axis. Draw the zone axis of the above three planes on your stereographic projection and show that the diffracted beams lie on a cone whose axis is the zone axis. What is the apex angle of the cone formed by the diffracted beams of this zone? Answer: 127°.

2.3   A simple cubic crystal ($a = 2.86$ Å) is arranged with its [001] direction vertical. A narrow beam of monochromatic x-radiation ($\lambda = 5.36$ Å) is impinged upon the crystal at right angles to the [001] direction, and a large photographic plate is placed perpendicular to the incoming beam at a distance of 2 in. from the crystal. The plate is placed on the same side of the crystal as the incoming beam so that we have a back-reflection type of arrangement. The crystal is now rotated about its [001] direction. This will cause spots or streaks to appear on the plate from all of the different parallel sets of planes. Considering only the {001} planes:

(a) Will spots or streaks appear on the plate from these planes? Explain your reasoning.

(b) Locate all of the markings on the plate (relative to the central beam) produced by the rays diffracted from these planes. Consider only first-order diffraction.

2.4   A beam of white radiation enters a simple cubic crystal along its [011] axis.

(a) On a standard (001) stereographic projection show the point of emergence of the diffracted beam from (1) the (111) planes, and (2) the (101) planes. Use tracing paper and a Wulff net.

(b) If a circular photographic plate having a 20-cm radius were placed 10 cm from the crystal, would the diffracted beams from either of these planes appear on the plate? Assume a transmission Laue arrangement.

2.5   A Laue pattern is made of a cubic single crystal. The [0$\bar{1}$0] direction of the crystal points along the x-ray beam toward the x-ray tube. On a standard (001) stereographic projection locate the position of the incoming x-ray beam. Also locate the positions of the ($\bar{1}\bar{1}$0), ($\bar{1}\bar{1}$2), and the (221) poles.

(a) The above three planes all belong to the same zone. Locate the zone axis on your stereogram and determine its direction.

(b) On your stereogram locate the position of the diffracted beams from each of the above three planes and show that they lie on the surface of a cone whose axis is the zone axis of the three planes.

(c) Which of the three diffracted beams might possibly be picked up with the back-reflection technique?

CHAPTER 3
THE PLASTIC DEFORMATION OF METAL CRYSTALS

It was mentioned earlier that of all the physical properties of metals, the strength and ductility are the properties most strongly controlled by the atomic structure and the microstructure of the metal. One of our main purposes is to develop an understanding of the mechanism of this control. Our approach to this problem will be first to consider the deformation of single crystals from the macroscopic, or phenomenological, point of view and then from the atomistic point of view in the next chapter.

To start, we will briefly review some elementary concepts. The mechanical properties of a metal may be represented by a stress–strain diagram, which, for polycrystalline metals, often appears similar to Fig. 3.1(a). If the applied stress is less than the elastic limit (which is closely measured by the yield stress), the deformation is said to be elastic. At stress levels equal to or greater than the yield stress, the metal deforms plastically, and any stress level above the yield stress is referred to as a flow stress. The distinction between elastic and plastic deformation may be formally defined as follows. (1) *Elastic deformation* (strain): Deformation (strain) completely recoverable upon release of stress. (2) *Plastic deformation:* Deformation not recoverable upon release of stress.

The stress level of the *elastic limit* is difficult to measure experimentally and so one defines a *yield stress* (sometimes called *proof stress*), which is more easily measurable. The difference between the elastic limit and the yield stress is small and will be neglected here (see p. 9, Ref. 1). At low stresses the stress–strain function is linear and is said to follow Hooke's law, $\sigma = E\varepsilon$, where the proportionality constant E is Young's modulus. For the case of shear stress, τ, rather than a tensile or compressive stress, σ, this relation is written, $\tau = G\gamma$, where G is the shear modulus and γ the shear strain. If one stresses a metal to point 1 of Fig. 3.1(a) and then releases the stress, the metal will return to essentially its original shape. However, if one were to release the stress at a level shown by point 2, the metal would

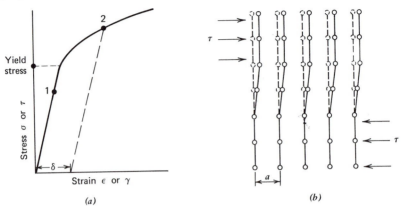

Figure 3.1 (a) Stress–strain diagram. (b) Atom displacement produced by a shear stress τ.

retain a permanent elongation shown as δ on Fig. 3.1(a). It is important to understand the difference between these two modes of deformation on an atomic scale. If one applies a shear stress to a crystal, the atoms will be displaced from their equilibrium positions (their normal lattice sites) as shown in Fig. 3.1(b). If the shear stress is small, the atoms are not displaced very far from their equilibrium positions, and upon releasing the stress the bonding between atoms causes them to return to their original lattice sites. Thus, we have elastic deformation with no permanent atom displacement. Suppose, however, that the shear stress were large enough to cause the upper region of atoms to be displaced a distance a. These atoms would now lie in new equilibrium positions (lattice sites), so that upon release of the stress they would simply retain their permanent displacement. This permanent displacement of metal atoms occurs during plastic deformation by four primary mechanisms:

1. Slip
2. Twinning
3. Grain boundary sliding
4. Diffusional creep

Mechanism 1 is by far the most important deformation mode and in this chapter we will restrict ourselves to this mode. Mechanism 2 generally becomes operative at low temperatures and is significant at low temperatures in hcp metals. Mechanisms 3 and 4 may become significant at high temperatures.

3.1 SLIP SYSTEMS

The slip mechanism may be formally defined as follows: The parallel movement of two adjacent crystal regions relative to each other across some plane (or planes).

If one electropolishes a smooth piece of aluminum sheet its surface will have the appearance of a good quality mirror. Bending the aluminum strip through an angle of 10°–30° causes the actual grain structure of the aluminum strip to appear before your eyes, and if the grain size is fairly large a structure such as Fig. 3.2(a) will appear. Examination under a microscope at about 100× shows many parallel lines within each grain; see Fig. 3.2(b). These parallel sets of lines in each grain are actually surface relief due to slip caused by the plastic deformation produced by the bending. It can be seen in Fig. 3.2(c) that the deformation of each grain is accomplished by small blocks of the crystal sliding past each other along parallel sets of planes. The slip does not occur on just one plane but over small regions of parallel planes called either slip bands or slip lines depending on their thickness; see Fig. 3.2(c). Since the slip lines all occur in parallel sets within each single crystal (each grain), they must correspond to the same set of occupied (hkl) planes of the particular grain. From measurements on single-crystal specimens of known orientations we can determine (1) the planes on which slip occurs, and (2) the direction of slip within these planes. Such experiments have revealed the very fascinating fact that slip in fcc crystals always occurs on {111} planes but only in the ⟨110⟩ directions. This means that if slip does occur on a (111) plane it will be in one of the three directions $\pm[10\bar{1}]$, $\pm[\bar{1}10]$, $\pm[0\bar{1}1]$; see Fig. 3.3(a). The following definition has become accepted.

Slip system: The combination of a plane and a direction lying in the plane along which slip occurs.

Figure 3.3(a) shows three slip systems for fcc crystals, $(111)[\bar{1}10]$, (111)-$[10\bar{1}]$, and $(111)[0\bar{1}1]$. Slip systems may be determined very easily from a stereographic projection. Figure 3.3(b) shows the trace of the (111) plane and it is seen that three ⟨110⟩ directions are on this trace and therefore represent directions lying in this plane. Because fcc crystals have four different {111} sets with three ⟨110⟩ directions in each, these crystals have 12 slip systems. The 12 slip systems may be quickly identified from Fig. 3.3(b). Examination of the right-hand grain of Fig. 3.2(b) reveals that slip has occurred on three of the four possible {111} sets, whereas slip in the left-hand grain has been predominately on only one set of {111} planes. It is interesting to note that slip has occurred across the boundary between the left-hand and upper grains, thus indicating that the angle of misorientation across that boundary is quite small.

In other metal crystal structures, the slip systems have more variability, as shown in Table 3.1. In the bcc metals the slip lines have a wavy appearance. In these crystals slip occurs predominantly on the $\langle 111 \rangle \{110\}$ systems, and the wavy appearance is apparently due to simultaneous slip on the other two possible systems of planes $\{211\}$ and $\{321\}$. In real hcp crystals the c/a ratio does not equal the ideal value of 1.633 for the hard-sphere model (see Problem 1.8). For metals having $c/a > 1.63$ there is some preference for basal plane slip, $\langle 11\bar{2}0 \rangle (0001)$, whereas for those metals having $c/a < 1.63$ the other two slip systems are usually preferred. A more detailed discussion of these points may be found in Ref. 1. Three general observations are of value:

1. The slip directions are always in the direction of closest packing. Apparently there are some exceptions, for example, solid mercury.

2. Slip *usually* occurs on the most close packed planes. This observation appears to be related to the fact that the most densely packed planes are also the set of occupied (hkl) planes having the widest spacing.

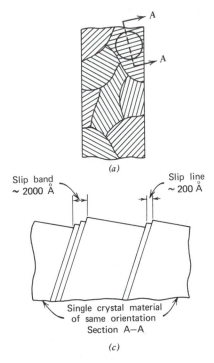

(a)

Slip band ~ 2000 Å   Slip line ~ 200 Å

Single crystal material of same orientation
Section A–A

(c)

Figure 3.2 (*a*) Macroview of grain structure revealed by slip lines. (*c*) Slip line intersections at surface of A1 strip.

Figure 3.2   (*b*) Micrograph at 100× showing slip lines and grain boundaries in Al.

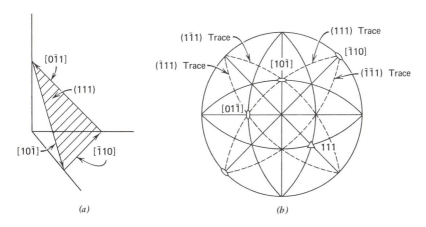

Figure 3.3   Slip systems in fcc crystals.

Table 3.1 Slip Systems Observed in the Common Metal Crystal Structures

Structure	Slip Direction	Slip Planes	Examples
FCC	$\langle 110 \rangle$	$\{111\}$	Cu, Al, Ni, Pb, Au, Ag, Fe
BCC	$\langle 111 \rangle$	$\{110\}$	Fe, W, Mo, brass, Nb, Ta
BCC	$\langle 111 \rangle$	$\{211\}$	Fe, Mo, W, Na
BCC	$\langle 111 \rangle$	$\{321\}$	Fe, K
HCP	$\langle 11\bar{2}0 \rangle$	(0001)	Cd, Zn, Mg, Ti, Be, Co
HCP	$\langle 11\bar{2}0 \rangle$	$\{10\bar{1}0\}$	Ti, Mg, Zr, Be
HCP	$\langle 11\bar{2}0 \rangle$	$\{10\bar{1}1\}$	Ti, Mg

Consequently, one might expect these planes to offer the least resistance to shear.

3. Slip occurs first on that slip system having the highest shear stress along its slip direction.

3.2 RESOLVED SHEAR STRESS (SCHMID FACTOR)

Suppose we take a single crystal of an fcc metal and pull it in tension as shown in Fig. 3.4. We know that slip will occur upon the (111) planes of this crystal, which are shown on the figure. The force that causes this slip to occur is not a tensile force but is a shear force in the (111) planes along one of the three directions of slip. Hence, we must resolve the tensile force into the (111) plane along the three $\langle 110 \rangle$ directions in that plane. Knowing the

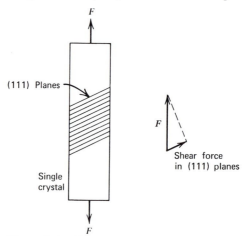

Figure **3.4** Single crystal pulled in tension.

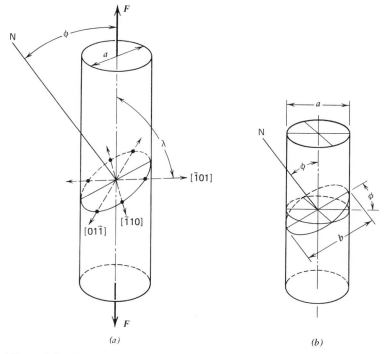

Figure 3.5 Geometry for determining resolved shear stress.

area of the (111) plane we may then determine the shear stress that might cause slip to occur along one of these three directions. This procedure is best illustrated by a three-dimensional view of a cylindrical single crystal as shown in Fig. 3.5. In this figure the orientation of the (111) planes is characterized by its normal, which makes an angle ϕ with the tensile axis. The three possible slip directions in the (111) plane are shown and the acute angle between any one of these slip directions and the tensile axis is called λ. The diameter of the cylinder is taken as a. We now ask: What is the shear stress in the [$\bar{1}01$] direction? Let τ_r be the shear stress (force/area) resolved into this direction. The force in the [$\bar{1}01$] direction is simply $F \cdot \cos \lambda$. Since the (111) planes intercept the cylinder at an angle other than 90°, their intersections form an ellipse of major axis b and minor axis a as shown on Fig. 3.5(b). Hence the slip-plane area is $(\pi/4)a \cdot b$, and we have $\tau_r = F \cdot \cos \lambda / (\pi/4)a \cdot b$. From the geometry illustrated in Fig. 3.5(b) it is apparent that $b = a/\cos \phi$. If we let the cross-sectional area of the cylinder, $(\pi/4)a^2$, be called A, the usual form of the resolved shear stress is obtained:

$$\tau_r = \frac{F}{A} (\cos \lambda \cos \phi) \tag{3.1}$$

This equation gives the shear stress along a direction at an angle λ from the tensile axis in the plane whose normal makes an angle ϕ with the tensile axis. We will refer to the term in parenthesis as the Schmid factor, after one of the authors who first formulated this equation.[3] The resolved shear stress is simply the applied tensile stress (or compressive stress), F/A, times the Schmid factor. Hence, the resolved shear stress of any slip system is proportional to its Schmid factor and, consequently, it is useful to recognize the allowable range of this factor. For any given ϕ the maximum Schmid factor occurs for $\lambda = 90 - \phi$. Hence, the maximum Schmid factor occurs at the maximum of the function $\cos(90 - \phi) \cdot \cos \phi$, which is obtained at $\phi = 45°$. This shows that the maximum resolved shear stress occurs directly up a plane at 45° from the tensile axis, and that the maximum possible value of the Schmid factor is $\frac{1}{2}$.

It has been found that slip will occur on a slip system when the resolved shear stress on that system reaches a certain critical value. The resolved shear stress required to initiate slip on a given slip system is often called the *critical resolved shear stress*, CRSS, and its value is strongly dependent on the purity of the metal. The values of the CRSS are well defined in

Table 3.2. The Critical Resolved Shear Stress for Various Metals

Metal	Lattice	CRSS (psi)	CRSS (MN/m^2)	Slip System
Ag	FCC	54	0.37	$\{111\}\langle110\rangle$
Al		114	0.79	"
Cu		71	0.49	"
Ni		470–1040	3.24–7.17	"
Mg	HCP	57–72	0.39–0.50	$\{0001\}\langle11\bar{2}0\rangle$
Mg		5900	40.7	$\{10\bar{1}0\}\langle11\bar{2}0\rangle$
Be		200	1.38	$\{0001\}\langle11\bar{2}0\rangle$
Be		7600	52.4	$\{10\bar{1}0\}\langle11\bar{2}0\rangle$
Co		93–100	0.64–0.69	$\{0001\}\langle11\bar{2}0\rangle$
Ti		1850	12.8	$\{10\bar{1}0\}\langle11\bar{2}0\rangle$
Zr		93–100	0.64–0.69	$\{10\bar{1}0\}\langle11\bar{2}0\rangle$
Fe	bcc	4,000	27.6	$\{110\}, \{112\}\langle111\rangle$
Mo		14,000	96.5	$\{110\}, \{112\}\langle111\rangle$
Nb		4,900	33.8	$\{110\}$ $\langle111\rangle$
Ta		6,000	41.4	$\{110\}$ $\langle111\rangle$

high-purity hcp crystals, but there is some ambiguity in the point of initial slip in cubic metals.[1] The existence of the CRSS shows that in a pure metal there is some inherent resistance of the lattice to slip, which is overcome at a reproducible shear stress. Various measured values of the CRSS are shown in Table 3.2.

If a single crystal of an fcc metal is pulled in tension, slip will be initiated on the first of the 12 slip systems that attains a resolved shear stress equal to the CRSS. Suppose the tensile axis is aligned along the [001] direction of the crystal and we want to determine the slip system on which slip will initiate. Our problem then is to determine which of the 12 slip systems has the maximum Schmid factor. All 12 slip systems are geometrically represented on the upper half of the octahedron shown in Fig. 3.6(a). The faces of the octahedron are the {111} planes and the edges are the ⟨110⟩ directions. Examination of the geometry shows that

1. ϕ is the same for all {111} planes, 54.7°.
2. λ is the same for [$\bar{1}$01], [101], [011], [0$\bar{1}$1], 45°.
3. λ is 90° and Schmid factor is 0 for other two ⟨110⟩ directions.

Note that these relationships are displayed quite obviously on an (001) stereographic projection. Hence, it is concluded that there are eight slip systems with the same Schmid factor and four with a zero Schmid factor.

Now consider a single crystal tensile specimen having the orientation A shown in Fig. 3.6(b). From the geometry shown in this figure one expects the slip direction to be either [011], [$\bar{1}$01], or [$\bar{1}$10] since the other three directions are nearly at right angles to the tensile axis. By calculating the Schmid factor for all 12 slip systems one can determine the particular slip

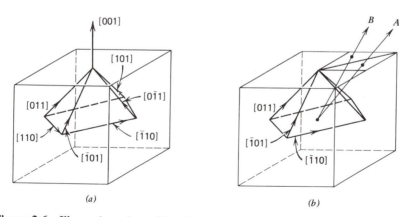

(a)   (b)

Figure 3.6 Illustration of possible slip systems for different orientations of the tensile axis.

system having the highest resolved shear stress. The results for the two directions A and B of Fig. 3.6(b) are

A: Highest Schmid factor system $= (111)[\bar{1}01]$
B: Highest Schmid factor system $= (\bar{1}\bar{1}1)[011]$

Notice that direction A may be represented by point A inside the lower triangle on the stereographic projection shown in Fig. 3.7(a). If one moves point A to any other spot inside of the lower triangle, the same slip system, $(111)[\bar{1}01]$, always has the highest Schmid factor, as indicated on Fig. 3.7(a). A single crystal whose tensile axis lies along the line from the (001) pole to the ($\bar{1}11$) pole will have two slip systems with a maximum Schmid factor, and these are the slip systems of the two surrounding triangles, that is, $(111)[\bar{1}01]$ and $(\bar{1}\bar{1}1)[011]$. One may show the active slip systems (slip system with highest Schmid factor) for all orientations on a stereographic projection as in Fig. 3.7(b).[2] The letters A to D specify the poles of the active slip plane and the numerals I to VI the direction of slip. At orientations along the zone circles of this stereogram, the two slip systems in the adjacent areas have the highest Schmid factor. At orientations where the zone circles intersect, the slip systems having the highest Schmid factor are identified from the areas adjacent to the intersection point. For example, we have shown above that a specimen in the [001] orientation has eight slip systems of equal and maximum Schmid factor. These eight slip systems are given by the eight areas surrounding the (001) pole on Fig. 3.7(b).

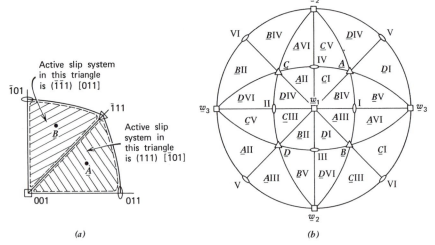

(a)   (b)

Figure 3.7 Active slip systems in fcc crystals. [Part (b) reprinted with permission of Macmillan Publishing Co., Inc., from Ref. 1, Copyright 1966 by William John McGregor Teegart.]

3.3 SINGLE-CRYSTAL TENSILE TESTS (FCC)

Suppose that four different single crystals of an fcc metal crystal are subjected to a tensile or a compressive test with the tensile axis being oriented along different crystallographic directions in each of the four crystals. The orientation of the tensile axis direction is shown by the four points A, B, C, and D on the standard triangle of Fig. 3.8(a) where point D is any location in the shaded area. The resultant stress strain diagram is shown for the four crystals in Fig. 3.8(b). Notice that for a specimen of D orientation a considerable amount of plastic flow has occurred before the metal begins to work harden, whereas the other three orientations show considerable work hardening as soon as the yield stress is exceeded and plastic flow begins. This difference is related to the fact that in orientations D the crystals have only one active slip system, whereas the other three orientations all have more than one active slip system, that is, more than one slip system with the same maximum Schmid factor. Figure 3.7(b) quickly shows that orientation A has eight active slip systems, B has six, and C has four active slip systems.

To analyze what is happening in specimen D, it is first necessary to understand what is physically happening to the single-crystal specimen in the tensile machine. Figure 3.9(a) locates the slip planes and the slip direction on the crystal before applying the tensile force. If the grips of the tensile machine could somehow move frictionlessly in the lateral direction, the specimen would deform as shown in Fig. 3.9(b). However, the tensile

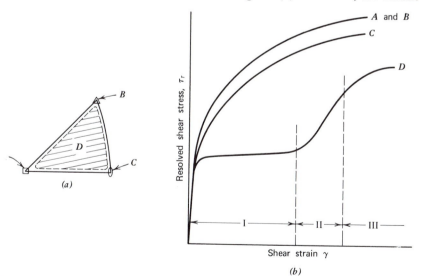

Figure 3.8 Stress–strain relationships for fcc single crystals.

grips allow no lateral motion of the ends of the specimen. Imagine that the specimen of Fig. 3.9(b) is quite long and then move the ends back in line with the original tensile axis. This motion may be accomplished in the central region by a simple rotation of the lattice, whereas some bending and consequent distortion of the lattice will occur near the grips as indicated in Fig. 3.9(c). We concern ourselves here only with the rotation of the lattice in the central section.

Specimen D has the active slip system $(111)[\bar{1}01]$. Figure 3.10(a) shows the location of the slip direction, $[\bar{1}01]$, the tensile axis, D, and the slip plane normal, (111), on a stereographic projection. Notice that the angles λ and ϕ of Fig. 3.5(a) are displayed directly on the projection. This shows that the stereographic projection provides a convenient means of determining the Schmid factor. The tensile force causes the slip direction of the crystal to rotate toward the tensile axis, and we would like to show this rotation on the projection. Rotation of the slip direction toward the tensile axis causes the angle λ to decrease. We may represent this rotation by (1) moving D toward the $[\bar{1}01]$ direction along the great circle through D and $[\bar{1}01]$, or by (2) leaving D stationary and moving the $[\bar{1}01]$ toward D along this great circle. Note that this second method moves all the crystal poles and directions. Method (1) will be used since this method requires that we only move one point on the projection, the tensile axis D. As the strain increases, the lattice rotation causes λ to decrease and Fig. 3.10(b) shows that after a rotation from D to point 2 the specimen has become oriented

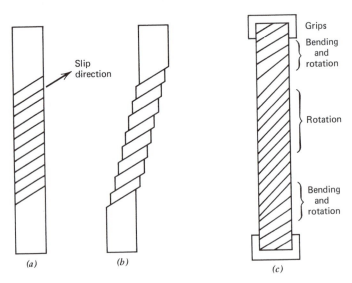

Figure 3.9   Physical changes in a single crystal deformed in a tensile machine.

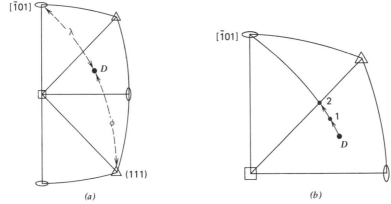

Figure 3.10 Representation on a stereographic projection of deformation-induced crystal rotation.

so that *two* slip systems, (111)[$\bar{1}$01] and ($\bar{1}\bar{1}$1)[011], have the same maximum Schmid factor. At this point, then, one expects slip to occur on two slip systems simultaneously. It is customary to divide the stress–strain diagram for a specimen having a D orientation into three sections called stages I, II, and III, as shown on Fig. 3.8(*b*). Experiments have shown that stage II begins at the point where slip commences on the second slip system. Stage I is usually called the "easy glide" stage because very little stress produces considerable plastic flow, and stages II and III are called the linear and parabolic work hardening stages, respectively. It should be clear that work hardening requires slip on more than one slip system. This seems reasonable because one intuitively suspects that slip on intersecting slip planes might interfere with and perhaps block additional slip.

Two conclusions based on the above may now be drawn:

1. Slip occurs relatively easily (low τ_r) on a single slip system.

2. Slip is much more difficult when it occurs on many different slip systems simultaneously.

The above discussion has been limited to fcc crystals for ease of presentation. Similar behavior has been found for bcc and hcp crystals. The student is referred to Ref. 1 for a more detailed discussion and also for consideration of the effects of temperature and purity.

3.4 RELATIONSHIP TO POLYCRYSTALLINE DEFORMATION

In the vast majority of applications, metals are used in the polycrystalline form. As a matter of fact, single crystals of metals are obtained only with some difficulty. The grain size in polycrystalline metals usually runs from around 0.025 to 0.150 mm. When such a polycrystalline metal undergoes plastic deformation the path along which slip occurs through the metal is extremely complex by comparison to single-crystal deformation. In the absence of grain-boundary sliding, which generally occurs only at high temperatures, the grains remain coherent across their grain boundaries. This means that an individual grain must deform to accommodate the deformation of each of its neighboring grains. It has been shown that in order for a crystal to undergo such a homogeneous strain by slip, at least five independent slip systems are required.[1,4] We have shown that there are 12 slip systems in fcc crystals, and it is left as an exercise to show that there are also 12 slip systems of the form $\{110\}\langle111\rangle$ in bcc crystals and three slip systems of the form $(0001)\langle11\bar{2}0\rangle$ in hcp crystals. Not all of these slip systems are independent[1]; nevertheless, it turns out, as might be expected, that at least five independent slip systems are present in fcc and bcc crystals but not in hcp crystals. It is because of this lack of slip systems in hcp crystals that twinning is an important mode of deformations in these crystals. It is in large part due to their hcp crystal structure that iitanium alloys have proven to be so difficult to form.

Experiments have shown that even at very low strains each grain is clearly deforming on several slip systems.[4] Slip probably initiates on that slip system in a given grain that has the highest Schmid factor (SF) (see Eq. 3.1), but then slip is soon required on other slip systems within that grain in order to accommodate deformation in neighboring grains. If one assumes that slip occurs in a given grain at the CRSS as measured on single crystals, the yield stress of the polycrystal is given from Eq. 3.1 as

$$\sigma = \frac{\mathrm{CRSS}}{\overline{\mathrm{SF}}} \tag{3.2}$$

where $\overline{\mathrm{SF}}$ is a Schmid factor averaged over the orientations of all of the grains. Theoretical treatments[1,4] give $\overline{\mathrm{SF}} = 1/3.1$ for fcc metals and $\frac{1}{2}$ for bcc metals. You will recognize the value of $\frac{1}{2}$ in bcc metals as the maximum value of the SF. This indicates that in these metals there is a high probability that each grain will have one of its slip systems oriented very close to the maximum SF. This result is due to the high number of slip systems and the ease of cross slip to the other slip systems in bcc metals. These theoretical treatments[1,4] assume that the total strains are exclusively plastic, which is not a very good assumption at low strains.[5] Nevertheless, to a fairly good approximation the yield stress of polycrystalline fcc metals should be on the order of three times the values of the CRSS. For example, the yield stresses of commercial aluminum alloys range from values of around 6000 psi (41 MN/m^2) for certain casting alloys up to around

80,000 psi (552 MN/m^2) for the 7000 series heat-treatable alloys. The corresponding values of the CRSS as calculated from Eq. 3.2 for these alloys would range from 2000 to 27,000 psi (14–186 MN/m^2). The value of the CRSS for pure aluminum is listed in Table 3.2 as only 114 psi (0.79 MN/m^2). The much higher strength of the commercial alloys is achieved through a variety of strengthening mechanisms that inhibit slip. These include such things as the blockage of slip at grain boundaries and at precipitate particles and other mechanisms that will be studied later.

A. TEXTURE

As discussed above, when a single crystal is plastically deformed the crystal rotates such that the slip direction approaches the tensile axis. Similarly, when a polycrystalline metal is plastically deformed all of the individual grains will rotate. However, the rotation is restricted because of the requirement of coherency at the grain boundaries. As deformation occurs the grains elongate and it has been found that a preferred orientation of the grains is evident in fcc and bcc metals at strains exceeding 40%, and in hcp metals at strains exceeding 10%.[1] These preferred orientations are usually referred to as textures. In wires the grains tend to rotate so as to align a specific crystallographic direction parallel to the wire axis, which is often called the *fiber axis*. The bcc metals have a [110] fiber texture, and the fcc metals often have a double fiber texture[6] with some grains having the [111] directions and others the [100] directions parallel to the axis. Deformation by rolling causes the grains to develop a preferred crystallographic direction parallel to the rolling direction and a preferred crystallographic plane parallel to the rolling plane. In some fcc metals the principle rolling texture is (110)[$\bar{1}$12],[6] which means the {110} planes lie parallel to the rolling plane and the ⟨112⟩ directions lie parallel to the rolling direction. Table 3.3, taken from Ref. 4, illustrates some common textures produced by cold work.

A texture is often described with a pole figure on a stereographic projection. Suppose that all of the grains were randomly oriented. If we were to plot the {100} poles of these grains on a stereographic projection we would obtain a "shotgun" pattern as shown on Fig. 3.11(a) where RD and TD refer to rolling direction and transverse direction, respectively. Suppose that rolling caused a (100)[001] texture. The (100) pole figure would then appear as shown in Fig. 3.11(b). This texture is called the cube texture.

Figure 3.12 gives a {111} pole figure for α brass.[4] The ideal {110}⟨112⟩ texture of α brass is shown by the triangles on this figure. Not all of the grains have their {111} normals in these ideal locations. The contours on the pole figure represent the locus of points along which the intensity of

Table 3.3 Common Rolling and Wire Textures

Crystal Structure	Mode of Working		Texture
fcc	Wire drawing and extrusion	$\langle 111 \rangle$ $\langle 100 \rangle$	Parallel to wire axis
bcc	Wire drawing and extrusion	$\langle 110 \rangle$	Parallel to wire axis
hcp	Wire drawing and extrusion	$\langle 10\bar{1}0 \rangle$	Parallel to wire axis
fcc	Rolling	$\{110\}$	Parallel to rolling plane
		$\langle 112 \rangle$	Parallel to rolling direction
bcc	Rolling	$\{001\}$	Parallel to rolling plane
		$\langle 110 \rangle$	Parallel to rolling direction
hcp	Rolling	$\{0001\}$	Parallel to rolling plane
		$\langle 11\bar{2}0 \rangle$	Parallel to rolling direction

Note. These textures are idealized. In practice, considerable scatter occurs, and very frequently other textures are present.
(From Ref. 4, used with permission of St. Martin's Press, Inc., Edward Arnold Ltd.)

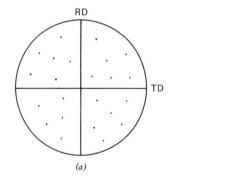

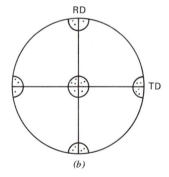

Figure **3.11** Schematic (100) pole figures.

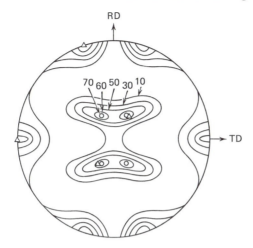

Figure 3.12 A (111) pole figure measured on α brass (From Ref. 4, used with permission of St. Martin's Press Inc., and Edward Arnold Ltd.).

diffracted x-rays from {111} planes was equal. The numbers labeling the contours are relative x-ray intensities from {111} diffraction; they provide a qualitative idea of the percentage of the grains having {111} normals in the {110}⟨112⟩ orientation. These contours show that the {111} normals of the grains exhibit a strong preference for the {110}⟨112⟩ texture. Reference 7 provides a good explanation of the x-ray technique used to generate pole figures such as Fig. 3.12.

The formation of texture is an extremely important phenomenon from a commercial point of view because it results in highly anisotropic physical properties in wire and sheet materials. This anisotropy may be detrimental or beneficial and will be discussed in Chapter 10, where the effect of recrystallization upon texture is considered.

3.5 THEORETICAL STRENGTH OF METALS

It is possible to estimate the theoretical strength of a metal from a knowledge of its shear modulus by means of a very simple model. Figure 3.13 shows schematically how slip is generated by a shear stress. The applied shear stress, τ, in Fig. 3.13(a) produces the offset shown in Fig. 3.13(b). It was pointed out earlier in this chapter that this slip causes all the atoms above (or below) the slip plane to be permanently displaced from one set of lattice sites to a new set of lattice sites. The question we now ask is: What is the theoretical shear stress necessary to cause a permanent displacement of the atoms?

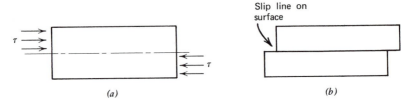

Figure 3.13   Offset produced by a shear stress.

When an atom is located directly on its lattice site it is in a minimum-energy position. If the atom is moved off its lattice site its energy increases, so that one may represent the variation of potential energy of an atom in a lattice as shown in Fig. 3.14(a). The force required to move the atoms is given by the derivative of the energy curve, $F = dE/dx$, so that one obtains the force curve shown at the bottom of Fig. 3.14(a). Consider now the shear stress applied across the four atoms shown in Fig. 3.14(b). The variation in shear stress required to displace the upper atoms relative to the lower atoms must follow a curve similar to the force curve of Fig. 3.14(a).

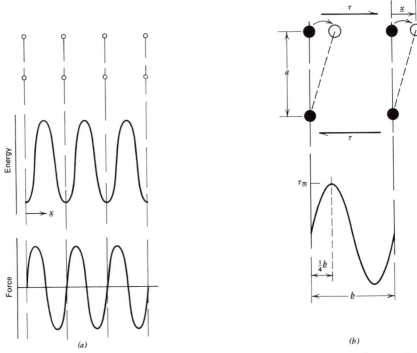

Figure 3.14   The periodic force function for relative displacement of atoms in a crystal.

Assuming such a variation as shown at the bottom of Fig. 3.14(b), it is seen that if the shear stress attains a value equal to or greater than τ_m the atoms above the slip plane will be able to move to the new lattice site to their right. Our problem then is to determine the stress function of Fig. 3.14(b). We will use the approximation made by Frenkel in 1926.[8] He assumed that the shear stress is a sine function of position, x, and that the shear stress attains its maximum value at a displacement of $\frac{1}{4}$ of b:

$$\tau = \tau_m \sin\left(2\pi \frac{x}{b}\right) \tag{3.3}$$

At sufficiently small values of displacement we may write

$$\tau = \tau_m \cdot 2\pi \frac{x}{b} \tag{3.4}$$

since the sine of an argument equals the argument at small values. According to the model of Fig. 3.14(b) the shear strain is $\gamma = x/a$. At sufficiently small values of displacement, Hooke's law will hold so that we may write

$$\tau = G\gamma = G\frac{x}{a} \tag{3.5}$$

Comparing Eqs. 3.4 and 3.5 we have

$$\tau_m = \frac{G}{2\pi}\frac{b}{a} \tag{3.6}$$

Table 3.4 lists the shear modulus in the close-packed directions (slip directions) for some metals and the value of τ_m calculated from Eq. 3.6 assuming $b = a$. The CRSS is the measured value of τ_m, and it is apparent that these measured values are lower than the theoretical values by three to four orders of magnitude. One may question the validity of assuming a

Table 3.4 Comparison of Theoretical Shear Stress for Slip to Measured Values of CRSS

Metal	Shear Modulus in Slip Direction ($\times 10^6$ psi)	τ_m (psi)	CRSS (psi)
Al	3.54	556,000	114
Ag	3.63	577,000	54
Cu	5.91	940,000	71
α-Fe	~10	~1,590,000	4000
Mg	2.39	381,000	57

simple sine function for the shear stress variation with position. Detailed models have been considered for this stress function and the results give a value of τ_m lower than Eq. 3.6 by a factor of 10 at most.[9] Hence the theoretical yield stress in shear for metals is larger than the measured values by a factor of 100 to 1000. This question of why metals are so much weaker than their theoretical strengths has been solved with the prediction and observation of dislocations. However, the problem of how one could design a metal to achieve its theoretical strength by control of dislocations remains a challenging and intriguing problem for metallurgists.

REFERENCES

1. W. J. M. Tegart, *Elements of Mechanical Metallurgy*, Macmillan, New York, 1966, Chapters 5 and 6.

2. L. M. Clarebrough and M. E. Hargreaves, *Progress in Metal Physics*, Vol. 8, B. Chalmers and R. King, Eds., Pergamon, New York, 1959.

3. E. Schmid and G. Siebel, *Z. Electrochem.* **37**, 447 (1931).

4. R. W. K. Honeycombe, *The Plastic Deformation of Metals*, St. Martin's Press, New York, 1968, Chapters 9 and 12.

5. J. E. Dorn and J. D. Mote, *Materials Science Research*, Vol. 1, H. H. Stadelmaier and W. W. Austin, Eds., Plenum, New York, 1963.

6. C. S. Barrett, *Structure of Metals*, McGraw-Hill, New York, 1952, Chapter 9.

7. B. D. Cullity, *Elements of X-Ray Diffraction*, Addison-Wesley, Reading, Mass., 1956, Sec. 9-9.

8. J. Frenkel, *Z. Phys.* **37**, 572 (1926).

9. A. H. Cottrell, *Dislocations and Plastic Flow in Crystals*, Oxford Univ. Pr., Oxford, England, 1952, Section 2.4.

ADDITIONAL READING

1. H. W. Hayden, W. G. Moffatt and J. Wulff, *The Structure and Properties of Materials*, Vol. III, *Mechanical Behavior*, Wiley, New York, 1965, Chapter 5.

2. R. E. Smallman, *Modern Physical Metallurgy*, Butterworths, London, 1962, Chapter 5.

PROBLEMS

3.1 (a) An fcc single crystal is pulled in tension along its [131] axis. By use of a stereographic projection, determine the resolved shear stress on the following slip systems, $(111)[0\bar{1}1]$, $(111)[10\bar{1}]$, and $(111)[1\bar{1}0]$, for a tensile stress of 100 psi.

(b) With the aid of Fig. 3.7(b) specify the slip system (or systems) upon which this crystal would undergo initial slip. Calculate the resolved shear stress for this system.

3.2   We know that in the bcc crystals, slip occurs in the $\langle 111 \rangle$ directions and that one set of active slip planes under certain conditions is the $\{112\}$ planes. By examination of the geometry of the bcc lattice, determine the total number of different slip systems for slip in the $\langle 111 \rangle$ direction on $\{112\}$ planes. This problem is most easily solved using the stereographic projection.

3.3   Suppose you were doing a compression test on a single crystal of Be and you wanted to be sure that the crystal slipped on the prism planes, $\{10\bar{1}0\}$, and not the basal planes, $\{0001\}$. You decide for the initial test to orient the specimen axis so that it lies in the plane formed by the (0001) and $(10\bar{1}0)$ normals. What angles could the specimen axis make with the basal plane to ensure slip on the $(1\bar{1}00)[11\bar{2}0]$ slip system? Use data of Table 3.2.

3.4   (a) Using a Wulff net make a stereographic projection for a hcp crystal as follows:

1. Locate the (0001) pole at the center of the projection.
2. Locate the a_1, a_2, and a_3 directions on the projection.
3. Locate the $(1\bar{1}00)$, $(10\bar{1}0)$, and $(01\bar{1}0)$ poles on the projection.

(b) A single crystal of a hcp metal is oriented with its compression axis in the plane formed by the normals of the (0001) and $(1\bar{1}00)$ planes. The compression axis makes an angle α with the basal plane. For $\alpha = 30°$ locate the position of the compression axis on the above projection, and using the projection with a Wulff net determine the Schmid factor for slip on the $(01\bar{1}0)[2\bar{1}\bar{1}0]$ slip system.

(c) Suppose that the critical resolved shear stress for slip on this $(01\bar{1}0)[2\bar{1}\bar{1}0]$ system is 400 psi and the CRSS for slip on the $(0001)[2\bar{1}\bar{1}0]$ system is only 100 psi. For what angles of α would slip occur on the $(01\bar{1}0)[2\bar{1}\bar{1}0]$ system?

3.5   (a) Locate the following poles on a standard (001) projection of a cubic crystal, (111), $(\bar{1}11)$, $(1\bar{1}1)$, $(\bar{1}\bar{1}1)$, $(\bar{1}12)$, $(\bar{1}\bar{1}0)$.

(b) By suitable rotations locate the positions of the four $\{111\}$ poles when the $(\bar{1}\bar{1}0)$ pole is at the center of your projection and the $(\bar{1}12)$ pole at the north pole (top) of your projection.

(c) Alpha brass has a $\{110\}\langle 112 \rangle$ rolling texture. A pole figure is made with the rolling direction at the north pole and the transverse direction at the east pole (this means the plane of the sheet lies in the plane of the projection). On your stereographic projection show clearly all of the positions where $\{111\}$ poles would appear on a 111 pole figure for a rolled sheet of α brass having a perfect $\{110\}\langle 112 \rangle$ texture. [Hint: With the $[\bar{1}12]$ direction along the rolling direction, the (110) pole may also lie at the center of your projection.]

3.6   Suppose a surface of a single crystal of Cu (fcc) is exactly parallel to the (001) planes of the crystal. The crystal is made to slip on all possible slip planes and the corresponding slip lines appear on the above surface. Show with a sketch the pattern of slip lines you would expect to see on the surface and indicate the angles between the lines. Repeat the problem for the case of $\{111\}$ planes parallel to the surface.

CHAPTER 4
DISLOCATIONS

In Chapters 1 and 2 we discussed the structure of metal crystals assuming a perfectly repeating three-dimensional array. Real metal crystals contain a number of different defects from this perfect array of atoms, and these defects account for many of the interesting properties of metals. The dislocation is probably the most important of these defects. The concept of the dislocation was introduced in 1934, not entirely independently, in three different papers.[1-3] The history of the development of the ideas leading up to these works is presented in Ref. 4. Many metallurgical engineers in the United States were quite reluctant to admit the dislocation to be anything more than an improbable theoretical concept as late as the early 1950s. However, with the development of the transmission electron microscope technique in the late 1950s, experimental evidence has conclusively shown that the strength and ductility of metals are controlled by these defects called dislocations.

4.1 THE EDGE DISLOCATION

The edge dislocation may be quite easily visualized as an extra half-plane of atoms in a lattice. This is shown in two dimensions in Fig. 4.1(a) and in three dimensions in Fig. 4.1(b). The dislocation is called a *line defect* because the locus of defective points produced in the lattice by the dislocation lie along a line. This line runs along the bottom of the extra half-plane shown in Fig. 4.1(b).

If a shear stress τ is applied to a crystal containing a dislocation as shown in Fig. 4.1(a) the extra half-plane of atoms is pushed to the right until it eventually pops out on the surface forming the beginning of a slip line as shown in Figs. 4.8(b) and 4.12(a) below. We now consider what the motion of the dislocation really entails. The dark circles of Fig. 4.2 show the location of atoms when the extra half-plane is located at position 1. In this

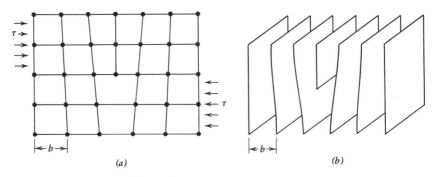

Figure 4.1 The edge dislocation.

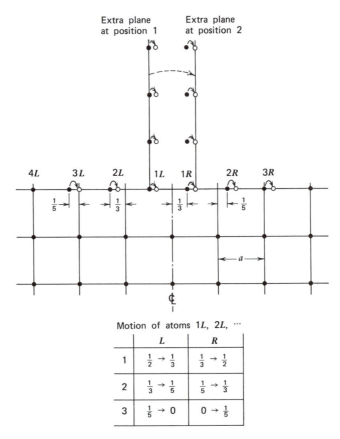

Motion of atoms 1L, 2L, ···

	L	R
1	$\frac{1}{2} \to \frac{1}{3}$	$\frac{1}{3} \to \frac{1}{2}$
2	$\frac{1}{3} \to \frac{1}{5}$	$\frac{1}{5} \to \frac{1}{3}$
3	$\frac{1}{5} \to 0$	$0 \to \frac{1}{5}$

Figure 4.2 A simplified picture of the atom motion associated with movement of an edge dislocation.

simplified picture the extra half-plane of atoms lies on a plane symmetrically located between the two lower atoms and it has pushed its first nearest neighbors, 2L and 1R, off their lattice sites by $\frac{1}{3}a$, the second nearest neighbors, 3L and 2R, off by $\frac{1}{5}a$, and the third neighbors, 4L and 3R, are unaffected. If a shear stress causes all of the atoms to be shifted just a small amount into the open circle positions shown on the figure, the extra half-plane will now be located one lattice parameter length, a, to the right at position 2. Notice that after the extra half-plane of atoms has shifted to position 2 it is composed of all different atoms. The atoms that made up the extra half-plane at position 1 have shifted only a small amount, $\frac{1}{6}a$, and by so doing they have moved off the symmetry plane, while the atoms at 1R shifted $\frac{1}{6}a$ onto the symmetry plane and they now make up the extra half-plane at position 2. Consider atom 3R. As the extra half-plane moves to the right, this atom will be shifted progressively to its right. Atom 3R will be shifted to the right by $\frac{1}{2}a$ when the extra plane lies on top of it, and will be shifted to the right by exactly a after the extra plane has moved well past it. Consequently, the net effect of the passage of the dislocation is to shift the upper atoms to the right by a. Notice that this shift is accomplished by a local disturbance, restricted to the region immediately around the dislocation, similar to the motion of a wave running along the surface of a pond. Consider by analogy a person trying to pull a heavy rug across the floor. By exerting a steady pull the rug may be moved with great effort. If, however, one flicks the rug as he pulls, a running wave will pass down the rug and it may be moved in short jerks with considerably less effort. Millions of people make daily use of running waves to straighten the sheets and blankets of beds. Two other interesting analogies to dislocations have been pointed out by Orowon.[4] A worm propels itself by stretching out a section of its body and thus moving its head forward as shown in Fig. 4.3(a). By moving the tension section along its body to the tail the entire body is propelled forward. The snake pulls its tail forward by generating a compressive region, which it then runs up to its head thus moving its entire body forward as shown in Fig. 4.3(b). Nature seems to have many examples of motion analogous to dislocation motion.

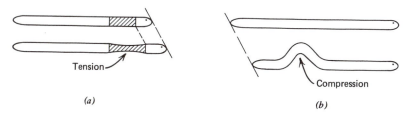

Tension

Compression

(a)

(b)

Figure 4.3 Dislocation analogies.

We will now consider the amount of energy required to move a dislocation. Return again to Fig. 4.2 and notice that the motion of the atoms is symmetrical about the centerline as shown in the box at the bottom of the figure. For example, the motion of atom 1L relative to the surrounding lattice is equal and opposite to the motion of atom 1R. Hence, if a chemical bond of atom 1L is stretched by a certain amount the corresponding bond to atom 1R is contracted by the same amount. Therefore, the energy required to stretch bonds during this motion will be compensated by the energy released in contracting bonds. Hence, this simple model shows that to a first approximation the energy to move a dislocation will be zero. This result explains why metals are so much weaker than their theoretical strengths. If dislocations exist on the slip systems, then their motion will occur at a very low shear stress and their motion will produce slip.

That the motion of dislocations does require a finite shear stress may be illustrated with Fig. 4.4. The upper part of this figure shows a lattice with a dislocation moving to the right. The upper three configurations shown for the extra plane of atoms correspond to positions 1 and 2 of Fig. 4.2 and also to a position intermediate between these two positions. Hence the

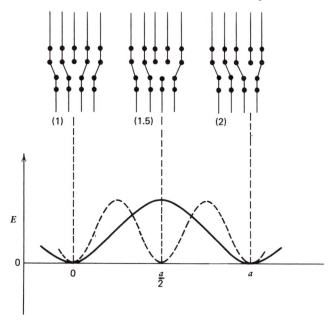

Figure 4.4 Variation of lattice energy with position of an edge dislocation. (Reprinted with permission of Macmillan Publishing Co., Inc., from Ref. 6., Copyright Macmillan Publishing Co., Inc., 1964.)

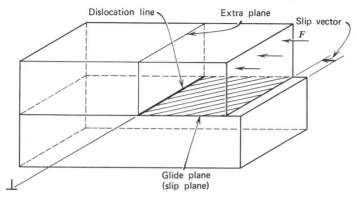

Figure 4.5 The edge dislocation.

three configurations represent a motion of the extra plane to the right by one lattice vector, a. The energy associated with the dislocation will be the same at positions 1 and 2. The energy will only be the same at an intermediate position if the motion of the atoms has been completely symmetrical as the dislocation moved to the intermediate position. The symmetrical atom motion described above (Fig. 4.2) is only a first-order approximation. Consequently, the dislocation energy at intermediate locations will be increased over their energy at positions 1 and 2. The curves of Fig. 4.4 show two possible energy curves giving the energy of the dislocation at all intermediate positions from 1 to 2. If the energy function is a minimum with the dislocation located at positions 1 and 2, as is shown for both curves, then a force will be required to move it out of its symmetrical locations at positions 1 or 2. This force that is required to move a dislocation is called either the Peierls force or the Peierls–Nabarro force. Calculated values of the Peierls force agree reasonably well with measured values of the CRSS in ionic crystals, but are found to be somewhat higher than the measured CRSS in metals.

We now consider the edge dislocation from a slightly different approach. Suppose that a force is applied to the upper half of the right-hand side of the single crystal shown in Fig. 4.5. The force causes the atoms above the right-hand surface to slip to the left by one atomic spacing and the motion is shown by the slip vector on the figure. Since the atoms above the left-hand surface have not moved to the left, this motion must necessarily generate an extra half-plane of atoms as shown in the figure. Consider the following definitions:

Dislocation line: The line running along that edge of the extra plane of atoms which terminates within the crystal.

Glide plane: The plane defined by the dislocation line and the slip vector.

If the dislocation moves in the direction of the slip vector it is said to move by glide, and the dislocation line moves along the glide plane.

Symbol: Edge dislocations are conveniently symbolized with a perpendicular sign, ⊥. When the sign points up, ⊥, the extra plane of atoms is above the glide plane and the dislocation is called positive. When the sign points down, ⊤, the extra plane is below the glide plane and the dislocation is negative.

Slip vector: The slip vector is usually called the Burgers vector and it is formally defined as follows.

A. Define a positive direction along the dislocation line. This direction is just arbitrarily chosen.

B. Construct a plane perpendicular to the dislocation line, as for example in Fig. 4.6.

C. Trace out a path around the dislocation line in this plane moving *n* lattice vectors in each of the four mutually perpendicular directions. Go in a clockwise direction when sighting down the positive sense of the dislocation line.* This is frequently called a Burgers circuit.

D. Failure of this path to close indicates a dislocation. The Burgers vector, *b*, is the necessary vector to give closure; see Fig. 4.6. The Burgers vector equals the slip vector.

As mentioned above, dislocation motion in the direction of the Burgers vector is called *glide motion.* When the dislocation line moves at right angles to the Burgers vector the motion is called *dislocation climb.* It should

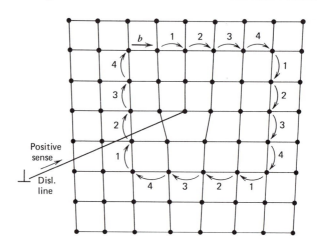

Figure 4.6   The Burgers circuit for an edge dislocation.

* There is disagreement on the clockwise direction. Reference 6 uses this definition while Ref. 7 uses a counterclockwise rotation.

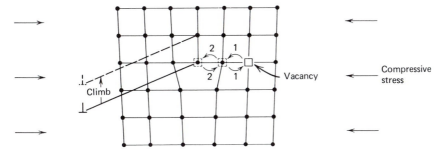

Figure 4.7 Climb motion of edge dislocations.

be clear from considering Fig. 4.5 that movement of the line normal to the slip vector will cause the extra half-plane to become either shorter or longer. Figure 4.7 shows the extra half-plane becoming shorter. The lattice vacancy migrates to the bottom of the extra plane of atoms and by so doing causes it to move up one lattice vector. Notice that to move the entire extra plane of atoms upward a whole row of vacancies must cooperatively migrate to the dislocation line.

1. Dislocation motion that requires the motion of atoms and vacancies is called *nonconservative* motion. Hence, climb motion of edge dislocations is nonconservative and glide motion is conservative.

2. Climb motion requires more energy than glide motion because of the vacancy migration required.

3. Climb motion reducing the size of the half-plane is called *positive climb*, while motion increasing its size is called *negative climb*.

4. Positive climb causes vacancies to be annihilated, while negative climb causes vacancies to be generated.

5. Climb motion may be aided by stress. Consider for example the application of the compressive stress to either side of Fig. 4.7. This stress tends to squeeze the dislocation upward. It actually accomplishes this by biasing the vacancy motion toward the dislocation line. Hence, a compressive stress causes positive climb and a tensile force causes negative climb.

There is one final point to emphasize before discussing screw dislocations.

Dislocation glide causes the atoms on one side of the glide plane to be displaced one Burgers vector relative to the atoms on the opposite side of the glide plane.

Figure 4.8(a) shows the location of an edge dislocation and its glide plane in a crystal. The letter A is inscribed on the surface so that it cuts across the glide plane as shown and a shear stress is applied causing the dislocation to

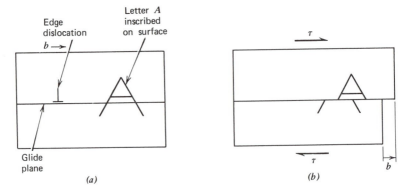

Figure 4.8   Relative lattice displacement produced by glide motion of an edge dislocation.

move to the right by glide. Figure 4.8(b) illustrates that the net effect of the passage of this dislocation is simply a shift of the affected upper half of the crystal by one b vector relative to the bottom half of the crystal.

We now summarize some of the more significant properties of the edge dislocation.

1. The edge dislocation may be visualized as an extra half-plane of atoms.

2. The Burgers vector is perpendicular to the dislocation line.

3. The glide plane is defined by the dislocation line and the Burgers vector.

4. Glide motion causes the atoms above the glide plane to be displaced one Burgers vector relative to atoms below the glide plane.

5. Climb motion occurs when the size of the extra half-plane changes, and it is accompanied by vacancy generation or annihilation.

4.2   THE SCREW DISLOCATION

There is a second fundamental type of dislocation, the *screw dislocation*, which is more difficult to visualize geometrically. The origin of the screw dislocation is frequently attributed to Burgers in 1939.[8] Suppose one were to take a plate and apply a shear stress across the end as shown in Fig. 4.9(a). This shear stress could cause the plate to rip, much as a piece of paper is ripped in half as is shown in Fig. 4.9(b). The displacement shown in Fig. 4.9(b) is one representation of the geometry of the screw dislocation. Notice that the top half of the block is shifted (dislocated) with respect to the bottom half by a fixed *slip vector* across some *glide plane*. The screw

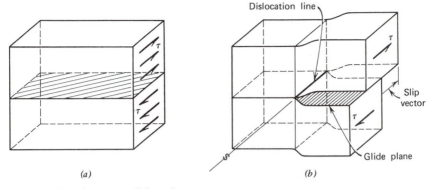

Figure 4.9 The screw dislocation.

dislocation line is located at the leading edge of the shift and it is parallel to the slip vector. The screw dislocation is sometimes symbolized with an S through the line as shown in Fig. 4.9(b). It is not obvious from Fig. 4.9 why the dislocation is termed a screw dislocation. Figure 4.10(a) shows a cylinder with a screw dislocation running down its center. The planes of the cylinder perpendicular to its axis and spaced at a distance b will have been connected together to form a helix as a result of the dislocation displacement, that is, a screw thread.

The Burgers vector (slip vector) is defined as described for the edge dislocations. This may be illustrated nicely by using the front face of the crystal in Fig. 4.9 as the plane perpendicular to the dislocation line. Figure 4.10(b) shows the Burgers circuit on this plane in a clockwise direction looking down the positive sense of the dislocation line. The Burgers vector

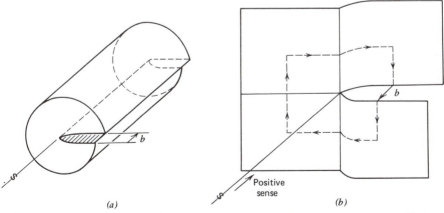

Figure 4.10 (a) The helical displacement produced by screw dislocations. (b) A Burgers circuit for a screw dislocation.

is parallel to the line pointing toward the observer. Notice that looking down the positive sense of the dislocation line, a clockwise motion on the planes perpendicular to the line causes advancement away from the observer. Hence the dislocation of Fig. 4.10 is sometimes called a right-hand screw dislocation. For our clockwise convention of the Burgers circuit we have:

Right-hand screw dislocation: Burgers vector points in negative sense of line
Left-hand screw dislocation: Burgers vector points in positive sense of line.

The right-hand and left-hand screw dislocations are said to be of opposite *sense*, and similarly the positive and negative edge dislocations are of opposite *sense*.

We now list some properties of the screw dislocation that help one understand its nature.

1. The Burgers vector is parallel to the dislocation line. Hence, specifying both the line and the *b* vector does not specify the glide plane as it does for an edge dislocation.

2. Glide motion causes the dislocation line to move at right angles to the glide direction. Therefore, the motion of the *line* is at right angles to both the stress vector and the slip produced by the stress.

3. There is no way to visualize this dislocation as an extra plane of atoms.

4. Passage of the dislocation line causes a movement of the atoms across the glide plane by one *b* vector relative to each other, analogous to Fig. 4.8 and illustrated in Fig. 4.12(*b*).

One of the most significant differences between an edge and screw dislocation is given in point 1 above. Figure 4.11(*a*) shows an edge

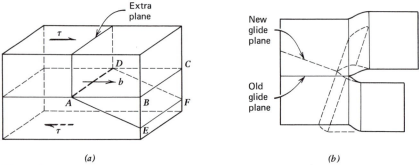

(a)   *(b)*

Figure 4.11 A change of glide plane for (*a*) an edge dislocation and (*b*) a screw dislocation.

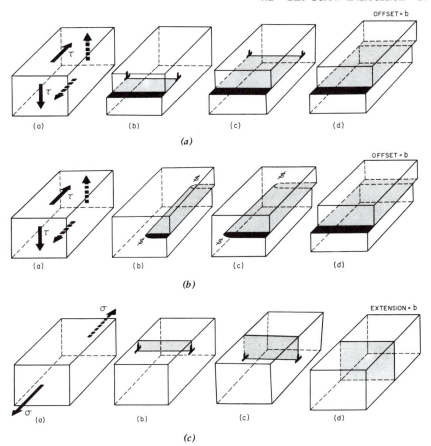

Figure 4.12 (*a*) Macroscopic deformation from edge dislocation glide. (*b*) Macroscopic deformation from screw dislocation glide. (*c*) Macroscopic deformation from edge dislocation climb. (From Ref. 9, used with permission of R. de Wit.)

dislocation moving to the right along glide plane *A-B-C-D*. Suppose now that this dislocation were to change its glide plane to *A-E-F-D*. Glide motion along this new glide plane would require the extra plane to rotate in order to become perpendicular to the new glide plane. This rotation would in turn require diffusion of atoms and therefore it would be slow and require a relatively high energy outlay. Consequently, the glide motion of edge dislocations is restricted to the glide plane defined by its line and *b* vector. Screw dislocations do not involve an extra plane of atoms and, consequently, there is no inherent restriction on the location of their glide planes. Figure 4.11(*b*) is an attempt to show what is involved in the changing of the glide plane for a screw dislocation. Suppose an edge dislocation moving along some glide plane becomes pinned. Increasing the shear stress may cause continued motion of the dislocation either by breaking through the pinning stress or by causing climb, both of which will require considerable energy. However, if the dislocation were a pure screw dislocation, continued glide motion could occur on an intersecting glide plane with perhaps a considerably less energy expenditure, since the motion of the screw dislocation is not restricted to its glide plane.

We are particularly interested in the macroscopic deformation produced by dislocation motion; this is illustrated very nicely in Fig. 4.12 taken from DeWit.[9] Both the shear stress and the final deformation in (a) and (b) are identical, but notice that the line moves in the direction of the stress couple in (a), and it moves at right angles to it in (b). Also notice that climb of edge dislocations produces a different type of deformation. There is no climb for screws.

4.3   MIXED DISLOCATIONS

If the b vector is perpendicular to the line, the dislocation is a *pure* edge dislocation; if the b vector is parallel to the line the dislocation is a *pure* screw dislocation. At other angles between the b vector and the line, the dislocation is a *mixed* dislocation. Figure 4.13(a) shows a block in which an offset is produced similar to Fig. 4.5. The offset has caused all atoms located above the area A-C-B to be shifted by the slip vector relative to the atoms below. The boundary of this shifting of atoms within the crystal is the curve running from A to C, which is shown best in Fig. 4.13(b). This curve, delineating the boundary of the shift, is the dislocation line. Since the shift vector (equivalent to the b vector) is everywhere the same, *the b*

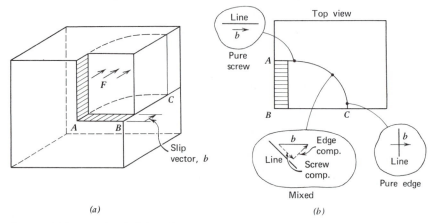

(a)   (b)

Figure 4.13   A mixed dislocation.

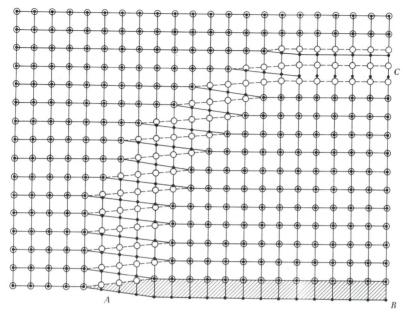

Figure 4.14   A top view of the mixed dislocation showing atom positions. Atoms above glide plane are open circles and atoms below are solid circles. (From Ref. 10, Copyright 1953, McGraw-Hill Book Co. Used with permission of McGraw-Hill Book Co.)

vector of the dislocation line is the same at all points along the line. Figure 4.13(b) shows that near A the dislocation is pure screw, near C it is pure edge, and everywhere else it is mixed. Notice that we may resolve the mixed dislocation into an edge component and a screw component, which are sometimes referred to as the edge character and screw character of the dislocation.

The geometry of mixed dislocations is quite difficult to visualize. Figure 4.14 presents a top view of the curved dislocation of Fig. 4.13 showing the atom locations.[10] The open circles represent the atoms just above the glide plane and the close circles represent atoms just below. At location C one sees a top view of a pure edge dislocation and at A the top view of a pure screw dislocation and the top view of a mixed dislocation at points in between.

4.4   TERMINOLOGY OF "CROOKED" DISLOCATIONS

Dislocations often contain a sharp break in their dislocation lines having a length of just a few atom spacings. These breaks are classified in two

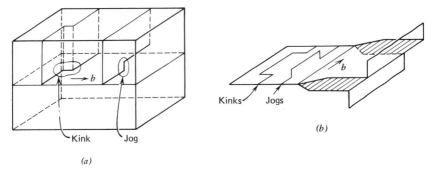

Figure 4.15   Illustrations of jogs and kinks in (a) edge dislocations and (b) screw dislocations.

categories as follows:

Jog: A sharp break in the dislocation line moving it out of the glide plane.

Kink: A sharp break in the dislocation line not moving it out of the glide plane.

Figure 4.15(a) shows the kink and jog in an edge dislocation. Notice that the kink segment is itself a screw dislocation. The jog segment remains an edge dislocation but with a new glide plane. The kink and jog in a screw dislocation are shown in Fig. 4.15(b); in this case both segments are pure edge in character. Notice that motion of these jog and kink segments back and forth across the crystal may cause the entire dislocation line to advance. For example, motion of the kinked segment in (a) toward the reader causes the edge dislocation to glide to the right while motion of the jogged segment toward the reader causes positive climb.

4.5   DISLOCATION LOOPS

A dislocation line may not terminate inside of a crystal. Hence, dislocation lines will either terminate at a free surface or an internal surface such as a grain boundary, or they will form loops. Figure 4.16(a) shows a circular dislocation loop and its glide plane. This dislocation is a mixed dislocation except at the two points where the *b* vector is perpendicular to the line and the two points where it is parallel to the line. If one applies a shear couple in the *b* vector direction as shown in Fig. 4.16(a), two very fascinating things happen: (1) The dislocation line expands outward in all directions normal to the line due to glide motion as shown by the arrows at right angles to the line. (2) After the dislocation has reached the outer edges of the crystal the net effect is a shift of the top of the crystal by one *b* vector

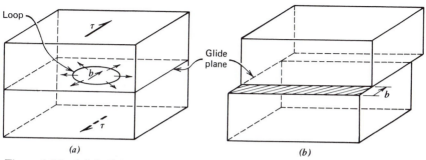

Figure 4.16 (a) A dislocation loop. (b) Offset produced by action of shear stress τ upon loop shown in (a).

relative to the bottom, Fig. 4.16(b). In order to understand what is going on here consider the square dislocation loop shown in top view on Fig. 4.17.

First we ask: What must happen in a perfect crystal in order to produce this loop? This loop could be generated by going into the perfect crystal and shifting all of the atoms above the glide plane and inside the square region by one b vector length relative to the atoms below the glide plane. This would push the top atoms within the loop up to the dashed curve of Fig. 4.17. The locus of points on the glide plane that delineates the boundary of this shift is the dislocation line, which in this case is a square loop. Now consider the effect of this shift on the atoms in the plane B-B, which is perpendicular to the glide plane. The shift will tend to "tear" this plane as shown in Fig. 4.18. The two dislocations produced at I and II are screw dislocations, and it should be clear from Fig. 4.18 that glide motion of these dislocations will cause their lines to move directly away from each other toward the outer edges of the crystal. We pick the positive sense of the dislocation line as shown in Fig. 4.17, and the Burgers circuits on both dislocations give the same b vector as shown in Fig. 4.18. Both dislocations must have the same b vector since they were both generated by the same

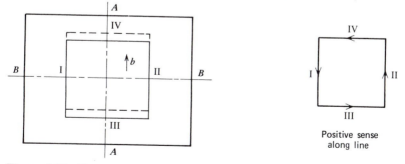

Figure 4.17 Top view of a square dislocation loop.

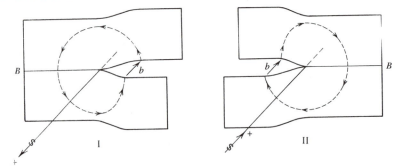

Figure 4.18 Effect of introducing dislocation loop upon the atom planes B–B.

slip vector. Notice that dislocation I is a right-handed screw and II a left-handed screw.

Now consider the effect of the lattice shift upon the atoms in the plane A-A. The shift produces an extra plane of atoms *above* the glide plane at IV and it leaves an extra plane of atoms *below* the glide plane at III; see Fig. 4.19 Hence these two edge dislocations are of opposite sense and the shear stress will cause them to move directly away from each other out to the edges of the crystal. Using the positive sense of the dislocation line chosen in Fig. 4.17, we see in Fig. 4.19 that both dislocations give the same b vector, as required. Notice that where dislocation III meets the crystal edge the extra plane comes out below the glide plane and where IV meets the crystal edge it comes out above the glide plane so that we obtain the offset shown in Fig. 4.16(b). The same shear stress has caused dislocations III and IV to move directly away from each other and, also, dislocations I

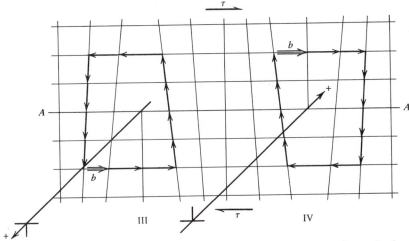

Figure 4.19 Effect of introducing dislocation loop upon the atom planes A–A.

and II to move directly away from each other because both pairs are of opposite sense. By analogy, then, the circular loop of Fig. 4.16(a) expands outward in all directions normal to the line under the same shear stress because the sense and character of the dislocation change as one proceeds around the loop.

4.6 MOBILE DISLOCATIONS IN REAL CRYSTALS

Consider a b vector in a bcc crystal running along the body diagonal from a corner atom to the center atom. The components of this b vector could be $\frac{1}{2}a, \frac{1}{2}a, \frac{1}{2}a$. It is customary to specify such a b vector as $b = a/2[111]$. This is a very useful convention because the direction of the b vector appears in its usual lowest integer form within the brackets.

In any real metal there could be a very large number of dislocations with different b vectors present after the metal solidifies from the melt. However, some limitations exist, which we will discuss more fully later on.

1. The b vectors of the stable dislocations extend from one atom to a nearest-neighbor atom. Such dislocations are called unit dislocations.

2. Unit dislocations tend to have lowest energy when their b vector extends along a close-packed direction.

3. In general, the lowest-energy dislocations have the greatest mobility.

Consequently, in fcc crystals we would expect to find the most mobile dislocations to have b vectors in the $\langle 110 \rangle$ directions with lengths of one-half of a face diagonal, that is, $b = a/2\langle 110 \rangle$. It is the glide motion of these most mobile dislocations which produces slip. The fact that slip in fcc crystals always occurs on $\{111\}$ planes means that the mobile dislocations have $\{111\}$ glide planes in fcc crystals. The dislocations that have been observed to be responsible for slip are listed in Table 4.1.

Table 4.1 Dislocations Producing Slip in the Three Important Crystal Lattices

		Mobile Dislocation	
Crystal	Slip System	b Vector	Glide Planes
fcc	$\langle 110 \rangle \{111\}$	$a/2\langle 110 \rangle$	$\{111\}$
bcc	$\langle 111 \rangle \{110\}$	$a/2\langle 111 \rangle$	$\{110\}$
bcc	$\langle 111 \rangle \{211\}$	$a/2\langle 111 \rangle$	$\{211\}$
bcc	$\langle 111 \rangle \{321\}$	$a/2\langle 111 \rangle$	$\{321\}$
hcp	$\langle 11\bar{2}0 \rangle (0001)$	$a/3\langle 11\bar{2}0 \rangle$	(0001)
hcp	$\langle 11\bar{2}0 \rangle \{10\bar{1}0\}$	$a/3\langle 11\bar{2}0 \rangle$	$\{10\bar{1}0\}$
hcp	$\langle 11\bar{2}0 \rangle \{10\bar{1}1\}$	$a/3\langle 11\bar{2}0 \rangle$	$\{10\bar{1}1\}$

4.7 OBSERVATION OF DISLOCATIONS

There are a number of techniques presently available for observing dislocations in metals. We will very briefly discuss the two methods that will be of the most use to us in later discussions; the student is referred to Chapter 2 of Ref. 7 for a more complete discussion.

We have stressed that a dislocation is a defect in a lattice. When a dislocation line intersects a surface the region of the lattice around this intersection will be distorted due to the dislocation. Hence, if one etches the surface with the proper etchant it is possible to remove atoms more quickly from the region where the dislocation intersects the surface; consequently, a small pit will form at the intersection. This technique is called the etch-pit technique and it reveals the locations of the intersections of dislocation lines with the surface.

The most versatile technique for observing dislocations is transmission electron microscopy of thin foils. The local bending of the lattice planes near the dislocations can cause a corresponding local diffraction of the electron beam if the beam makes the proper angle with the foil. The dislocation therefore appears as a dark line in the image due to the mechanism of diffraction contrast discussed in Section 2.2. Figure 4.20(a) shows a foil specimen containing a stacking fault in the back left-hand corner and three dislocations running from the top to the bottom of the foil along a common plane in the front right-hand corner. The dislocation images would appear in the transmitted image as shown on Fig. 4.20(b), and several actual dislocations may be seen on the TEM micrograph of Fig. 2.12. The stacking fault interacts with the electron beam in a manner that may produce a very striking pattern on the transmitted image, called a fringe pattern, as shown in Fig. 4.20(b). These fringe patterns are produced

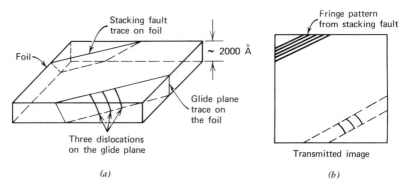

Figure 4.20 Images of dislocations and stacking faults produced in the transmission electron microscope.

from two-dimensional interfaces such as stacking faults, twin boundaries, and grain boundaries. Careful examination of the subboundaries of Fig. 2.12 shows that some of them display fringe patterns.

4.8 ELASTIC STRAIN ENERGY

In a perfect lattice all atoms are situated upon their equilibrium sites. If a tensile stress, σ, is now applied to the lattice the atoms will be moved slightly off of their equilibrium sites and this displacement is termed a strain. We call the energy introduced into the lattice by the action of the stress to produce a strain an elastic *strain energy*. Hence, whenever the atoms are displaced from their equilibrium lattice sites by some force a strain energy is introduced into the crystal. We now ask: How much energy do we introduce when we strain a metal? Consider the very simple case of a rod of length l and cross-sectional area A pulled in uniaxial tension as shown in Fig. 4.21(a). The elastic region of the stress–strain diagram for this material is shown in Fig. 4.21(b). If the applied force F increases the rod length by dl, the corresponding strain energy is simply $dE = F\,dl$. We take the stress as $\sigma = F/A$, strain as $d\varepsilon = dl/l$, and the volume of the rod as $V = A \cdot l$, and we obtain by direct substitution $dE = \sigma V\,dl/l$. Since the volume of the rod remains nearly constant we obtain

$$\int_0^{E/V} d\left(\frac{E}{V}\right) = \int_0^{\varepsilon} \sigma\,d\varepsilon \tag{4.1}$$

This shows that the strain energy per unit volume is simply the area under the stress–strain curve, Fig. 4.20b. Because Hooke's law applies in the elastic region, we may write $\sigma = \bar{E}\varepsilon$, and we find

$$\frac{E}{V} = \tfrac{1}{2}\sigma\varepsilon = \tfrac{1}{2}\bar{E}\varepsilon^2 \tag{4.2a}$$

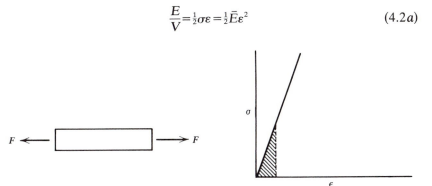

(a) (b)

Figure 4.21 Stress–strain relationship.

for uniaxial tensile or compressive stress and

$$\frac{E}{V}=\tfrac{1}{2}\tau\gamma=\tfrac{1}{2}G\gamma^2 \qquad (4.2b)$$

for simple shear, where $\tau = G\gamma$. In books on elasticity it is shown that a principle of superposition applies and the general relation is obtained as

$$\frac{E}{V}=\tfrac{1}{2}[\sigma_{xx}\varepsilon_{xx}+\sigma_{yy}\varepsilon_{yy}+\sigma_{zz}\varepsilon_{zz}+\tau_{xy}\gamma_{xy}+\tau_{xz}\gamma_{xz}+\tau_{yz}\gamma_{yz}] \qquad (4.3)$$

where the stress components are defined in Fig. 4.25 below. This result shows clearly that whenever a lattice contains a single strain component, ε, it will possess a strain energy per volume of simply $\tfrac{1}{2}\bar{E}\varepsilon^2$. It should be obvious that a dislocation will have a certain strain energy associated with it, since it distorts the lattice and, therefore, must introduce strain into its surroundings. Any defect that distorts the crystal lattice will introduce a strain field into the lattice, and as we will see later such strain fields exert a strong influence upon the motion of dislocations.

4.9   ENERGY OF DISLOCATIONS

A dislocation introduces atomic displacements into the lattice which are a maximum right at the dislocation line. Within some radius r_0 of the dislocation line the displacements are sufficiently large that Hooke's law does not apply. Hence, the energy of a dislocation is separated into two parts,

$$E = E_{\text{core}} + E_{\text{strain}}$$

where E_{core} is called the core energy and it refers to the energy within the region where the displacements are too large to apply to Hooke's law.

The screw dislocation involves only a single shear strain, and consequently it is possible to present a simple derivation for its strain energy. We perform the analysis on the cylindrical differential volume element shown in Fig. 4.22(a). This volume element is removed from the cylinder shown in Fig. 4.10(a) and it contains the offset b produced by the screw dislocation running down its center. To start we must first be able to determine the shear strain, γ, produced in the element by the displacement due to the screw dislocation. If we lay the element out flat as shown in Fig. 4.22(b) we see that the shear strain may be taken as $\gamma = b/2\pi r$. From Eqs. 4.2 we have $dE = \tfrac{1}{2}\tau\gamma\,dV$, and for the volume element we have $dV = 2\pi r\,dr \cdot L$. Taking $\tau = G\gamma$ we obtain by substitutions

$$\int_0^{E/L}\left(\frac{dE}{L}\right)=\int_{r_0}^{r}\frac{Gb^2}{4\pi}\frac{dr}{r} \qquad (4.4)$$

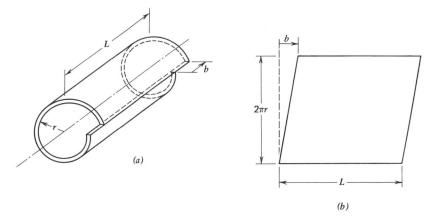

(a)

(b)

Figure 4.22 Effect of screw dislocation displacement upon a cylindrical differential volume element.

We take r_0 to be some small radius inside of which Hooke's law fails, and we account for this energy by a separate term as explained above:

$$\left(\frac{E}{L}\right)_{sc}=\frac{Gb^2}{4\pi}\ln\frac{r}{r_0}+\left(\frac{E}{L}\right)_{core} \tag{4.5}$$

Notice that the energy is evaluated per unit length of the dislocation line and it has units of ergs/cm. This simply reflects the fact that dislocations are line defects.

The stress field around an edge dislocation is more complex and it is necessary to use formal elasticity theory to determine its strain energy. One obtains[5,6] the following expression:

$$\left(\frac{E}{L}\right)_{edge}=\frac{Gb^2}{4\pi(1-\nu)}\ln\frac{r}{r_0}+\left(\frac{E}{L}\right)_{core} \tag{4.6}$$

where ν is Poisson's ratio. We may draw a number of conclusions based on these equations.

1. Since the strain energy of a dislocation is proportional to $\ln r$, it increases slowly with radius. Consequently dislocations possess a *long-range* strain field. Estimates of the core energy show that it is less than one-fifth the total energy for $r>10^{-4}$ cm in Eqs. 4.5 and 4.6, and so the core energy is often neglected.

2. The strain energy is proportional to the square of the Burgers vector. Consequently, the stable dislocations in metal crystals will be those having the lowest b vector because they will also have the lowest strain energy.

3.   Why does a bubble in a glass of beer always form as a sphere? Because the spherical geometry produces the minimum area for the fixed volume, and by minimizing its area the bubble minimizes its surface energy. The dislocation has an energy per length that we could call a line energy by analogy to surfaces. Consequently, a straight dislocation between two points will possess less energy than a curved dislocation and so straight dislocations are more stable. Later we will define a line tension that is analogous to surface tension.

4.10   FORCES UPON DISLOCATIONS

We found in discussing dislocation loops that an external shear stress causes the dislocation line to move outward in a direction normal to the line. One could imagine then that the applied shear stress has produced a force at right angles to the dislocation line that causes it to move. We may define this force that the dislocation feels, F, as the work required to move it a unit length, $F = dW/dl$. Suppose you move a block across the floor by applying a shear stress τ as shown in Fig. 4.23(a). The work required to move the block through a distance X may be written as $W = \tau \cdot$ (area of block) $\cdot X$.

Now suppose that an applied shear stress causes the dislocation segment of length ds to glide by an amount dl as in Fig. 4.23(b). The motion of this segment of the line causes the top of the crystal above the area dA to be shifted by one b vector, b, relative to the bottom. Hence, by analogy with the block we may write for the work required to move the segment, $dW = \tau(dA) \cdot b$, where τ is the shear stress component in the slip vector direction. (It is interesting to note that τ does not necessarily lie in the direction of motion of the dislocation line. It only does so for pure edge dislocations.) The factor dA is the area shifted and b is the magnitude of the shift. Since $dA = dl \cdot ds$, we may write

$$\frac{dW}{dl} = \tau \cdot ds \cdot b = F \tag{4.7}$$

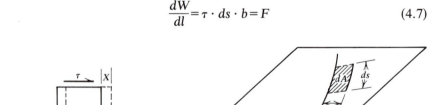

(a)   (b)

Figure 4.23   (a) Shear stress moving a block over a surface. (b) Motion of the dislocation line segment ds through distance dl.

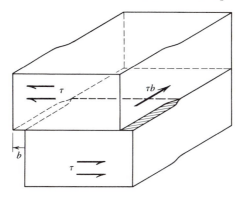

Figure 4.24   Geometric relationship of the force experienced by a pure screw dislocation line and the shear stress causing the force.

It is more convenient to consider the force per unit length of the dislocation line, so we write

$$\frac{F}{ds} = F_d = \tau b \qquad (4.8)$$

where F_d is the force per unit length. Equation 4.8 gives us the force per unit length that a dislocation feels due to a shear stress component, τ, in the b-vector direction. This force always acts at right angles to the dislocation line in the glide plane, irrespective of the angle the line makes with the shear stress vector. For example, in a screw dislocation the force experienced by the line, τb, acts at right angles to the shear stress direction as shown in Fig. 4.24. It may be shown that the force per unit length causing climb in edge dislocations is given by σb, where σ is the applied tensile or compressive stress normal to the dislocation line.

We would like to be able to describe the forces between dislocations in a more general way. It is shown in courses in engineering mechanics that in order to describe the state of stress in a material we must specify three tensile and three shear stress components. Figure 4.25 shows the stress components in both rectangular and cylindrical coordinates. The first subscript on the shear components denotes the direction of the normal to the plane in which the shear stress acts and the second subscript denotes the direction of the shear stress. Consider a dislocation line aligned in the X direction having a mixed b vector aligned in the X-Y plane as shown in Fig. 4.26. We now ask: What is the force experienced by this dislocation due to the stress components shown in Fig. 4.25(a)? We know that this force will act at right angles to the line, so it will have only $\bar{j}$ and $\bar{k}$ components. Since the dislocation is mixed we resolve it into its edge and screw components

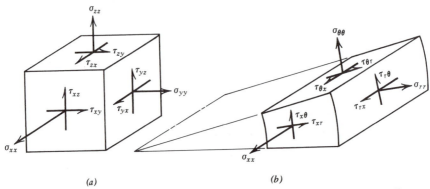

(a)   *(b)*

Figure 4.25 Definition of the stress components in (a) rectangular coordinates and (b) cylindrical coordinates.

and write

$$\bar{F}_d = [(F_y)_{edge} + (F_y)_{sc}]\bar{j} + [(F_z)_{edge} + (F_z)_{sc}]\bar{k} \tag{4.9}$$

where $(F_y)_{edge}$ is the force on the edge component in the Y direction and the other terms are similarly defined. The positive direction of the dislocation line is taken as the positive direction of the X axis. First consider the edge dislocation part, b_y. It moves in the Y direction by glide. Since its slip vector (b vector) lies in the Y direction, the shear stress causing the glide motion must also lie in the Y direction. By comparing the dislocation configuration of Fig. 4.26 to the stress components of Fig. 4.25(a) it is seen that the shear stress causing glide motion of the edge component must be

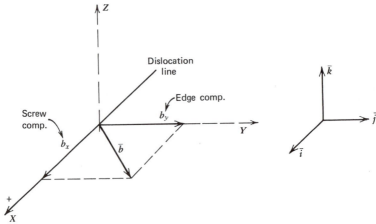

Figure 4.26 The dislocation configuration arbitrarily selected for derivation of Eq. 4.10.

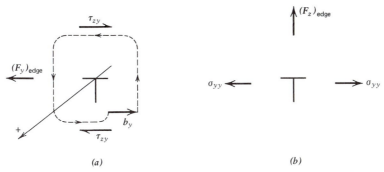

(a)   *(b)*

Figure 4.27   Determining the correct sense of the edge component in Fig. 4.26.

either τ_{xy} or τ_{zy}. The glide plane of the edge is the x-y plane. Since the shear stress couple of τ_{zy} acts across this plane but that of τ_{xy} does not, we conclude that stress τ_{zy} is the component causing glide of the edge component. This dislocation may be positive or negative depending on which sense gives the b vector in the plus Y direction as required by the geometry of Fig. 4.26. By doing a Burgers circuit as in Fig. 4.27(a) we find that the sense must be negative as shown. Therefore, the shear stress τ_{zy} causes motion in the negative Y direction. Therefore, the force must be negative and since τ_{zy} and b_y are both positive we write $(F_y)_{edge} = -\tau_{zy}b_y$. The edge part would move in the Z direction by climb, and the tensile stress causing this motion would be σ_{yy} as shown in Fig. 4.27(b). Since this stress causes motion in the positive direction, we write $(F_z)_{edge} = \sigma_{yy}b_y$.

The shear stresses causing motion of the screw component must be in the b-direction of the screw component, that is, the X direction. Hence, we must consider τ_{zx} and τ_{yx}. First consider the shear couple τ_{zx}. Since the stress couple acts across the x-y plane, this plane must be the glide plane of the screw dislocation affected by τ_{zx}. Therefore, the component τ_{zx} will move the screw component either in the $+Y$ or $-Y$ direction depending on its sense. Figure 4.28(a) shows that it must move to the left to give the proper sign of b_x so we have $(F_y)_{sc} = -\tau_{zx}b_x$. The other shear stress in the X direction, τ_{yx}, will move the screw component in either the $+Z$ or $-Z$ directions and we find as in Fig. 4.28(b) that it moves in the $+Z$ direction; therefore, $(F_z)_{sc} = \tau_{yx}b_x$. Substituting these expressions into Eq. 4.8 we obtain

$$\bar{F} = -[\tau_{zy}b_y + \tau_{zx}b_x]\bar{j} + [\sigma_{yy}b_y + \tau_{yx}b_x]\bar{k} \tag{4.10}$$

which is a limited form of an equation commonly called the Peach–Koehler equation. Equation 4.10 requires the dislocation of interest to be parallel to the X direction. This limitation may be relaxed by using a vector form[6] but for our purposes Eq. 4.10 is sufficient.

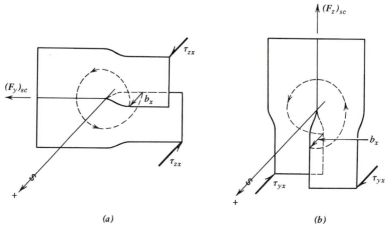

Figure 4.28   Determining the correct sense of the screw component in Fig. 4.26.

The strength and ductility of metals are controlled by the motion of dislocations, and the motion of dislocations is controlled by the forces acting upon them. Consequently, Eq. 4.10 is an important fundamental equation. We can draw an interesting analogy from physics. The force on a particle of charge, q, may be written

$$\bar{F} = q\bar{E} = q[E_x\bar{i} + E_y\bar{j} + E_z\bar{k}] \qquad (4.11)$$

where E_x, E_y, and E_z are the components of the electric field vector $\bar{E}$. Hence an electric *field* produces a force on a particle of charge q as given by Eq. 4.11. Similarly, a stress *field* produces a force on a dislocation of "strength b" as given by Eq. 4.10. Consider a single dislocation in a perfect crystal. This dislocation will experience zero force because the stress field of a perfect crystal is zero. If we generate a stress field in the crystal by any means, the dislocation will experience a force that may be determined by substituting the generated stress components at the point of the dislocation into Eq. 4.10. We may generate a stress field at the point of the dislocation by either applying an external stress or introducing defects into the lattice. Such defects as precipitate particles, solute atoms, and, in particular, other dislocations will generate a stress field at the dislocation of interest and hence these defects will exert a force on this dislocation.

4.11   THE STRESS FIELD PRODUCED BY DISLOCATIONS

The stress field (or displacement field) produced by a screw dislocation is very easy to calculate because the screw dislocation involves only a simple shear. Consider a screw dislocation lying in the X direction of Fig. 4.25(b).

The *b* vector would be aligned in the *X* direction and the shear stress producing slip could be $\tau_{\theta x}$. Hence we have

$$\tau_{\theta x} = G\gamma_{\theta x} = G\frac{b}{2\pi r} \tag{4.12}$$

where shear stress was previously shown to be $b/2\pi r$ (p. 96) and the remaining stress components of Fig. 4.25(b) are zero.

From elasticity theory we may derive analytical expressions for the stress field around an edge dislocation. As above, the analysis assumes an isolated straight dislocation in an isotropic crystal. For a positive edge dislocation aligned in the *X* direction of Fig. 4.25(b) we find

$$\sigma_{rr} = \sigma_{\theta\theta} = \frac{G}{2\pi(1-\nu)}\frac{b\sin\theta}{r}$$

$$\tau_{r\theta} = \tau_{\theta r} = \frac{-G}{2\pi(1-\nu)}\frac{b\cos\theta}{r} \tag{4.13}$$

where θ is measured up from the horizontal glide plane and the remaining stress components are all zero. Equations 4.13 assume that the positive direction of the dislocation line lies in the positive *X* direction. As illustrated in Fig. 4.27(a), our convention then requires the extra half-plane of the edge dislocation to lie *below* the glide plane. Now notice from Eqs. 4.13 that for θ between 0° and 180° both σ_{rr} and $\sigma_{\theta\theta}$ are positive, while for θ between 180° and 360° they are negative. This means that above the glide plane the stresses are tensile and below they are compressive. Hence, compressive stresses are produced in the region of the extra plane and tensile stresses are produced above this region. One expects this result upon considering the type of displacements produced in a lattice when an extra plane of atoms is inserted.

It is frequently useful to use the stress fields around dislocations in rectangular coordinates. These equations are presented here for reference. The stresses around a straight dislocation aligned in the *X* direction are as follows.

1. Screw dislocation:

$$\tau_{yx} = \tau_{xy} = \frac{Gb}{2\pi}\frac{z}{y^2+z^2}$$

$$\tau_{zx} = \tau_{xz} = \frac{-Gb}{2\pi}\frac{y}{y^2+z^2} \tag{4.14}$$

$$\sigma_{xx} = \sigma_{yy} = \sigma_{zz} = 0; \qquad \tau_{yz} = \tau_{zy} = 0$$

2. Edge dislocation:

$$\sigma_{xx} = \nu(\sigma_{yy} + \sigma_{zz})$$

$$\sigma_{yy} = \frac{Gb}{2\pi(1-\nu)} \frac{z(3y^2+z^2)}{(y^2+z^2)^2}$$

$$\sigma_{zz} = \frac{-Gb}{2\pi(1-\nu)} \frac{z(y^2-z^2)}{(y^2+z^2)^2} \qquad (4.15)$$

$$\tau_{yz} = \tau_{zy} = \frac{-Gb}{2\pi(1-\nu)} \frac{y(y^2-z^2)}{(y^2+z^2)^2}$$

$$\tau_{yx} = \tau_{xy} = \tau_{zx} = \tau_{xz} = 0$$

Below are two problems that you are advised to work at this point in order to establish a firmer grasp of the above concepts.

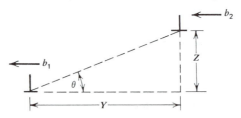

1. Two parallel edge dislocations of the same sense lie on parallel glide planes as shown above. From the stress field around an edge dislocation given above, and the Peach–Koehler relation, determine:

a. The force that the two dislocations exert upon each other in the *slip direction.*

b. Show that the value of this force goes to zero at $\theta = 45°$ and $90°$.

c. For values of (1) $0° < \theta < 45°$ and (2) $45° < \theta < 90°$ will the above force tend to pull the dislocations toward one another or push them apart?

d. What is the answer to part c if dislocation 2 has the opposite sense of dislocation 1?

2. For the case where the above dislocations are pure screw of the same sense, use the Peach–Koehler relation and the above equations for the elastic stresses produced by a screw dislocation and show that the force acting along the line between the dislocations is

$$|\bar{F}| = \frac{Gb_1 b_2}{2\pi r}$$

where r is the distance between the two dislocations.

In order to work problem 1 consider dislocation (2) to be placed in a perfect crystal. The force on it will be zero. Now introduce dislocation (1) and ask what force dislocation (2) feels from the stress field produced by dislocation (1). Based on your answer to problem 1 consider how you would expect dislocations to position themselves in crystals in order to minimize the forces they feel.

4.12   LINE TENSION

Similar to the surface tension, a line tension of a dislocation may be defined as the work required to generate a unit length of dislocation line. The line tension of a straight dislocation would then be given essentially by Eqs. 4.5 and 4.6. For a curved dislocation the line tension has the form

$$T = \frac{Gb^2}{4\pi K} \left[\ln \frac{R}{r_0} + \text{const} \right] \approx k Gb^2 \tag{4.16}$$

where R is the radius of curvature, $K = 1$ for a screw and $(1 - \nu)$ for an edge, and const depends on the shape of the line. Taking $k = \frac{1}{2}$ gives a good approximation of the line tension. The line tension has units of energy per length, which is the same as force. Hence, the line tension may be thought of as a force in the direction of the line similar to the force acting along the axis of a stretched rubber band.

Consider a segment of dislocation of length dS having a radius of curvature R as shown in Fig. 4.29. The line tension acting along the dislocation line tends to keep it straight, so that one must apply a force normal to the line in order to curve the line. This normal force per length of

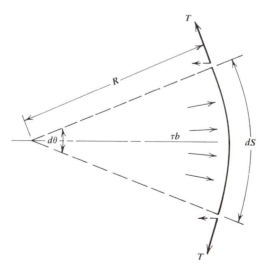

Figure 4.29   Forces acting on a segment of dislocation line.

line would be $F_d = \tau b$ and it would bend the dislocation line until it was balanced by components of the line tension force. The total force to the right on segment dS would be $\tau b\, dS$. The balancing force to the left due to the line tension components may be shown by a little geometry to be $T\, d\theta$ in the limit of small $d\theta$. Recognizing that $dS = R\, d\theta$ we obtain

$$\tau b = \frac{T}{R} \approx \frac{Gb^2}{2R} \tag{4.17}$$

This result shows that if a shear stress produces a force τb on a dislocation that is not free to move (pinned or partially restrained), then the dislocation will bend out to a radius of curvature, R, given by Eq. 4.17. This result is of use in understanding the motion of pinned dislocations.

4.13 EXTENDED DISLOCATIONS

Consider the very simple model of an edge dislocation where the "extra plane of atoms" consists of *two* half-planes of atoms in a {100} orientation as shown on the left of Fig. 4.30. If the two extra half-planes repel each other along the glide plane it is probable that they would separate to form two dislocations as shown at the right, b_2 and b_3, each with a Burgers vector of only one-half of the original dislocation. We may call this separation a dislocation reaction, which can be represented by analogy to chemical reactions as

$$b_1 \rightleftharpoons b_2 + b_3 \tag{4.18}$$

As with chemical reactions, whether this reaction will proceed to the right

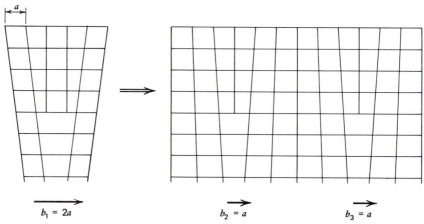

$b_1 = 2a$ $b_2 = a$ $b_3 = a$

Figure 4.30 Dissociation of an edge dislocation containing two adjacent extra planes of atoms.

or remain on the left depends on which configuration has the lower Gibbs free energy. We write the free-energy change in going from left to right according to convention as $\Delta G = G_2 + G_3 - G_1$ and if ΔG is negative the separated dislocations will have the lower free energy and, hence, be the stable configuration. We can, to a good approximation, equate the free energy of a dislocation with its strain energy. Therefore, we have from the expression for strain energy per length of dislocation line,

$$\Delta G \approx k b_2^2 + k b_3^2 - k b_1^2 = k[b_2^2 + b_3^2 - b_1^2] \tag{4.19}$$

For the above example we find $\Delta G = k[a^2 + a^2 - 4a^2] = -k2a^2$; that is, the separated dislocations are the stable configuration since they have a lower strain energy per length. Hence, we except the dislocation composed of two extra half-planes to dissociate into two dislocations with single extra half-planes.

By extension of the above arguments one can see that any dislocation having two or more planes of atoms as the "extra plane of atoms" would be unstable and should decompose into a series of dislocations each having one single extra half-plane of atoms. A dislocation with a Burgers vector of $n \cdot a$ would decompose into n dislocations each having Burgers vectors a. Extending this argument further one might expect that the dislocation having one extra plane of atoms and a Burgers vector a would further decompose somehow into two dislocations each with a Burgers vector of less than a, say $\frac{1}{2}a$. By analogy to the above arguments the combined strain energy of the two resultant dislocations would be less than the parent dislocation. An additional complication will arise in this case, however. We have seen earlier that the glide movement of a dislocation produces a relative motion of the crystal across the glide plane by a magnitude and direction given by the b vector. Consider the glide motion of the dislocation in Fig. 4.30 having b vector $n \cdot a$. After the dislocation passes a given point, the top of the crystal is displaced by $n \cdot a$ relative to the bottom. It is apparent that if n is not an integer, the passage of the dislocation will produce a planar fault in the crystal structure right on the glide plane. For example, if n were $\frac{1}{2}$, the passage of the dislocation would shift the top of the crystal by $\frac{1}{2}a$ relative to the bottom. The fault so produced on the glide plane is shown in Fig. 4.31.

Any dislocation whose glide motion produces a fault in the crystal is called a *partial dislocation* as opposed to a perfect dislocation. It can be seen that for a perfect dislocation the Burgers vector must extend between some two atoms of the crystal, whereas this rule does not hold for a partial dislocation. Now consider the decomposition of the perfect dislocation, $b_1 = a$, into two partial dislocations, $b_2 = \frac{1}{2}a$, and $b_3 = \frac{1}{2}a$. The effect of this reaction upon the lattice is shown in Fig. 4.32. If you imagine b_2 as

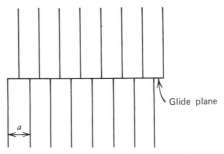

Figure 4.31   Fault produced on the glide plane by the glide motion of a dislocation with Burgers vector $\frac{1}{2}a$.

stationary, then as b_3 glides away from b_2 it generates a faulted region along the glide plane. We can represent this reaction schematically as in Fig. 4.33 where the faulted region between the partials is represented by the cross hatching. This combination of the two partial dislocations plus the fault is called an *extended dislocation*. In terms of an equation we may write,

$$b_1 \rightleftharpoons b_2 + b_3 + \text{fault} \tag{4.20}$$

To determine which way this reaction will prefer to go one must consider the free-energy contribution of the fault. Again the dislocation free energy contribution of the fault. Again the dislocation free energy will simply be taken as the strain energy. Referring to Fig. 4.32 note that in the faulted region the lattice is not strained, because the atoms lie in metastable equilibrium sites midway between the lower atoms. The energy of the faulted region is *not* the result of a straining of the lattice; rather, it results from different bond energies produced by the different atom configuration along the faulted plane. It is the same type of energy as produced by a stacking fault. Let the energy per unit area of the fault be given as E_f. Then, if the distance between the partial dislocations is given as r_e we find for the free-energy change of the above reaction per length of line

$$\Delta G = kb_2^2 + kb_3^2 + E_f r_e - kb_1^2$$
$$= k[b_2^2 + b_3^2 - b_1^2] + E_f r_e \tag{4.21}$$

For the example considered here we have $\Delta G = -2ka^2 + E_f r_e$. Hence the dislocation will only dissociate into partial dislocations if $E_f r_e < 2ka^2$. In other words if the faulted region has a low fault energy then the stable dislocations will be partial dislocations. One may conclude, then, that if partial dislocations do not form in a crystal, the stable dislocations will be those perfect dislocations that have the smallest possible Burgers vector.

In close-packed crystals partial dislocations do form, and hence they will be discussed in more detail. In particular we will now discuss how the

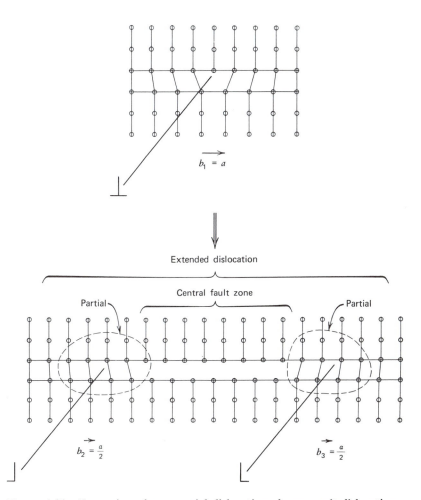

$\overrightarrow{b_1} = a$

Extended dislocation

Central fault zone

Partial ⌐ ⌐ Partial

$\overrightarrow{b_2} = \dfrac{a}{2}$ $\overrightarrow{b_3} = \dfrac{a}{2}$

Figure 4.32 Formation of two partial dislocations from a unit dislocation.

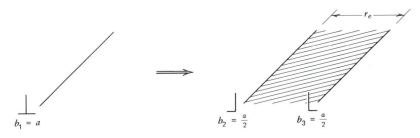

$b_1 = a$ $b_2 = \dfrac{a}{2}$ $b_3 = \dfrac{a}{2}$

r_e

Figure 4.33 Schematic representation of the dissociation of unit dislocation into partial dislocations.

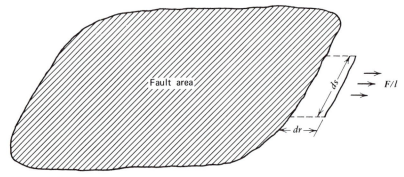

Figure 4.34 Extension of a segment of a two-dimensional (area) fault.

distance of separation, r_e, may be determined. It may be shown that partial dislocations on the same glide plane repel each other due to the interaction of their stress fields with a force per unit length

$$F_{\text{rep}} \approx \frac{G(\bar{b}_2 \cdot \bar{b}_3)}{2\pi r} \tag{4.22}$$

(see Problem 4.6) where $\bar{b}_2 \cdot \bar{b}_3$ is the dot product of the Burgers vectors of the two dislocations. This repulsive force is counter-balanced by an attractive force that results from the faulted region generated between the two partial dislocations as they move apart. Suppose we let the faulted region have a surface tension of γ in units of dynes/cm. The surface tension is defined as the work necessary to generate a unit area of the fault plane, $\gamma = dw/dA$. Consider a length ds along the boundary of the faulted area. The work needed to extend this length outward by distance dr, as shown in Fig. 4.34, is

$$dW = \frac{F}{l} ds \cdot dr$$

where F/l is the force per unit length exerted on the boundary to move it outward. Hence one has

$$\frac{F}{l} = \frac{dw}{ds \cdot dr} = \gamma$$

by definition of γ. Hence, the surface tension γ may be thought of as a force per unit length acting inward along the boundary, so that in order to extend the boundary outward one must exert an outward force per length of boundary that is greater than γ. It can be shown that the surface tension of an area is approximately the same as energy per unit area,[11] $\gamma = E_f$ (see p. 182).

We may now return to the problem of calculating r_e. It can be seen from Eq. 4.22 that the force per unit length repulsing the partial dislocations drops as the dislocations move apart. When the force drops below E_f it will no longer be strong enough to generate more faulted region. Hence the dislocations will separate until the repulsive force is balanced by the surface tension:

$$\gamma = E_f = \frac{G(\bar{b}_2 \cdot \bar{b}_3)}{2\pi r_e} \qquad (4.23)$$

The equilibrium spacing of the partial dislocations may then be determined as

$$r_e = \frac{G(\bar{b}_2 \cdot \bar{b}_3)}{2\pi E_f} \qquad (4.24)$$

Suppose that the dislocations in two metals having the same crystal structure separate into partials, and that the modulus G of the two metals is nearly the same so that the repulsive force between the partials (Eq. 4.22) is approximately the same. However, the fault energy of the region between the partials is much higher in metal 2 than in metal 1. One then expects the equilibrium separation distance to be less in metal 2 since the repulsive force will not be able to spread the dislocations as far due to the large surface tension of the fault in metal 2. This is shown graphically in Fig. 4.35(a). If we apply a shear stress of the proper orientation we can cause the partial dislocations to come together again. For reasons that will become obvious later we would like to know the relative amounts of energy necessary to recombine the partial dislocations in metals 1 and 2. Consider the partials in metal 1. When they are separated by some distance

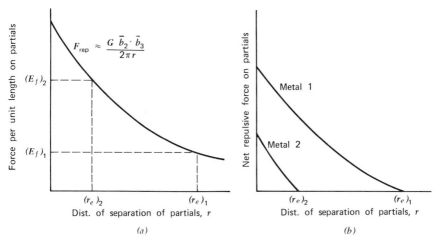

Figure 4.35 (a) Diagram illustrating equilibrium spacing of two partial dislocations. (b) Net force on partials as a function of separation distance.

r less than their equilibrium spacing, $(r_e)_1$, there will be a net repulsive force on the partials given as $[(G_1\bar{b}_2 \cdot \bar{b}_3/2\pi r)-(E_f)_1]$. This net repulsive force is shown as a function of separation distance r in Fig. 4.35(b). To collapse two partial dislocations into the parent dislocation, the externally applied shear stress must overcome this net repulsive force. Hence, we may write the energy required to collapse the extended dislocation as

$$E_{\text{collapse}} = -\int_{r_e}^{0} (\text{Net repulsive force}) \, dr \qquad (4.25)$$

From this equation it is seen that the energy to collapse the extended dislocations is simply the area under the two curves in Fig. 4.35(b). It is obvious then that it takes less energy to collapse the more narrow extended dislocation in metal 2 than the wider dislocation in metal 1. Hence, lower fault energies will tend to produce more stable partial dislocations because they will be harder to collapse back into their perfect form.

4.14 DISLOCATIONS IN FCC METALS

For simplicity we will restrict ourselves to a discussion of dislocations in fcc metals since these metals have been well studied and have, perhaps, the most interesting dislocation arrangements. As mentioned earlier, in these crystals the mobile dislocations have a b vector of $a/2\langle 10\bar{1}\rangle$. Figure 4.36($a$) shows the locations of the atoms on the section of the (111) plane contained in the unit cell. Similar to Fig. 1.11 the atoms on the (111) plane are labeled B atoms, the atoms in the next close packed plane, $(\frac{1}{2}\frac{1}{2}\frac{1}{2})$, are

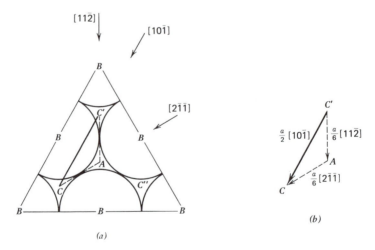

(a)

(b)

Figure 4.36 Dislocations in fcc crystals.

labeled C atoms, and the corner atoms are A atoms. Consider a dislocation gliding along the plane between B and C having a b vector extending from atom C' to C. This b vector would be $a/2[10\bar{1}]$. The glide motion of this dislocation causes all atoms above the glide plane to be shifted by the b vector relative to the lower atoms, so that the atom at C' will move to C. In addition to a direct path from C' to C one can see from Fig. 4.36(a) that the motion could be accomplished by a two-step path going from C' to A and then from A to C. If the motion occurs by this two-step path the perfect dislocation could degenerate into the two partial dislocations shown in Fig. 4.36(b).

$$\frac{a}{2}[10\bar{1}] \rightarrow \frac{a}{6}[11\bar{2}] + \frac{a}{6}[2\bar{1}\bar{1}] + \text{fault} \qquad (4.26)$$

The two partial dislocations are commonly called Shockley partial dislocations because they were first proposed by Heidenreich and Shockley in 1948.[12] Together they form an extended dislocation.

Suppose a Shockley partial dislocation $a/6[11\bar{2}]$ glides along the plane between B and C of Fig. 4.36(a). This motion shifts the *entire* crystal above the glide plane by the vector $a/6[11\bar{2}]$. Hence, the atom at C' moves to A and all such atoms at C sites move to A sites. Go back to Fig. 1.11(b), which shows an edge view of the B and C planes. We redraw these planes horizontally in Fig. 4.37 and then we allow all atoms above the glide plane to shift $a/6[11\bar{2}]$. Notice that this shifts *all* atoms above the glide plane producing displacements from C to A, from A to B, and from B to C.

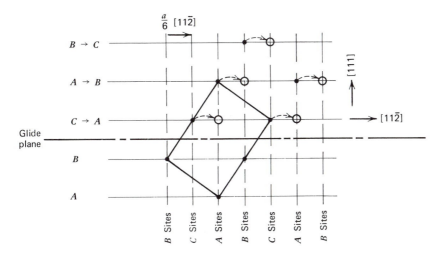

Figure 4.37 The atomic shift caused by glide of the Shockley partial dislocation $[11\bar{2}]a/6$. (See Figs. 1.11 and 1.12 for construction of this diagram.)

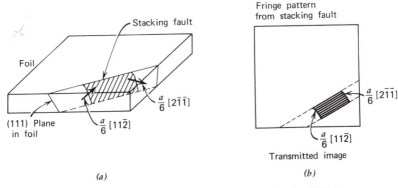

Fringe pattern
from stacking fault

(a)   (b)

Figure 4.38   Appearance of extended dislocation in the TEM.

Hence, motion of this Shockley partial dislocation produced a one-layer stacking fault, $-A\text{-}B\text{-}C\text{-}A\text{-}B\downarrow A\text{-}B\text{-}C\text{-}$.

Whether or not the extended dislocation on the right-hand side of Eq. 4.26 will exist depends on the energy of the faulted region. In this case the fault is very simple, a stacking fault, and consequently it has a relatively low energy. The stacking fault energy in Al is around 200 ergs/cm² and in Au around 10 ergs/cm². From Eq. 4.24 we calculate the equilibrium spacing, r_e, for Al as 1–2 atom diameters and for Au as around 30 atom diameters. Therefore, we would expect the perfect dislocations to form extended dislocations in Au but not in Al. It is apparent that the stacking fault energy controls whether or not extended dislocations form in any given fcc metal.

Extended dislocations can be dramatically observed in transmission electron microscopes and the expected behavior is observed. Figure 4.38 shows a foil containing the extended dislocations of Eq. 4.26 upon a (111) plane of the foil. The stacking fault between the extended dislocations appears upon the transmitted electron beam image as a fringe pattern. Application of a stress is observed to cause the two partial dislocations to move across the glide plane as a coupled pair with the fault in between.

Whether or not the dislocations of an fcc metal are extended can have a strong influence on its plastic behavior. Slip occurs by the motion of many dislocations on a given slip plane. Suppose that slip was occurring on the (111) plane in the [10$\bar{1}$] direction by the motion of screw dislocations and that this slip became blocked somehow. If additional stress caused slip to continue in the [10$\bar{1}$] direction but on the intersecting slip plane, (1$\bar{1}$1), we have what is called *cross slip*. Since screw dislocations are involved, this cross slip could occur fairly easily. However, if edge dislocations were involved, the cross slip would be much more difficult because they cannot easily change glide planes. But, suppose the screw dislocations had disassociated into extended dislocations as shown in Fig. 4.39. Notice that neither of the partial dislocations are pure screw. It should be clear that there is no way that both Shockley partial dislocations can ever be pure screw. Hence, cross slip requires that the extended dislocations first contract into perfect dislocations and then cross slip onto the new plane. It was shown above that the wider the separation of partials the greater the energy required to contract them. Consequently, we see that cross slip should be more difficult if we can lower the stacking fault energy since this increases the separation of the partials, r_e.

To summarize:

1. Cross slip requires pure screw dislocations.

2. The partials must first be contracted into the unextended form to permit cross slip.

3. The wider the extension the more difficult is cross slip.

Dislocations are often classified according to their mobility as follows:

Glissile: Dislocations able to move by pure glide.

Sessile: Dislocations not able to move by glide. Some sessile dislocations can move by climb.

The motion of partial dislocations is restricted to the plane of their fault. Motion off their fault plane does not occur because it requires so much energy that ordinarily other dislocations will move and relieve the stress before the partial moves off the fault plane. Hence, the Shockley partial dislocation would be classed as a glissile dislocation because its fault plane is also its glide plane. However, there exists another important partial dislocation in fcc metals that is a sessile dislocation. Figure 4.40(a) shows an edge view of the close-packed planes in an fcc crystal. Imagine that we remove atoms from the C plane, thus forming a disk of vacancies on this plane which in an edge view appears as in Fig. 4.40(a). If the diameter of

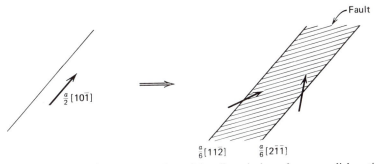

Figure 4.39 Schematic representation of the dissociation of a screw dislocation in an fcc lattice into Shockley partials.

the vacancy disk becomes more than a few atom distances, the surrounding lattice cannot support the void space so produced and it collapses onto the disk area in a manner similar to Fig. 4.40(*b*); thus a stacking fault -*A-B-C-A-B↓A-B-C-* is formed across the vacancy disk area. Notice that two edge dislocations now appear at the outer disk boundary on Fig. 4.40(*b*). These two edge dislocations are really just one dislocation called a *Frank partial* dislocation, which forms a loop running around the outer boundary of the vacancy disk region. The *b* vector must be perpendicular to the (111) planes and its length is the distance between (111) planes. By inspection of Fig. 1.11(*b*) this distance is found to be one-third of the body diagonal and so the *b* vectors of Frank partials are $a/3\langle 111 \rangle$ and these dislocations are pure edge. Since this partial dislocation is restricted to motion on its fault plane, it can only move by changing the size of its extra plane of atoms, that is, by climb. Consequently, Frank partials are sessile dislocations.

When two parallel dislocations meet in a crystal they combine to form a third dislocation and we have a dislocation reaction. The dislocations and the dislocation reactions in fcc crystals may be well illustrated using any of the tetrahedrons surrounding the tetrahedral voids that were discussed in Chapter 1. One of these is shown in Fig. 4.41 with the four apexes labeled *A*, *B*, *C*, and *D*. The center of the face opposite apex *A* is labeled α, the center opposite *B* is β, and so on. Notice the following:

1. AB = unit dislocation = $a/2[10\bar{1}]$
2. $A\delta$ = Shockley partial dislocation = $a/6[11\bar{2}]$
   δB = Shockley partial dislocation = $a/6[2\bar{1}\bar{1}]$
3. $A\alpha$ = Frank partial dislocation = $a/3[11\bar{1}]$

It should be clear that all possible orientations of the above three types of dislocations are represented on the tetrahedron, which is referred to as the Thompson reference tetrahedron.

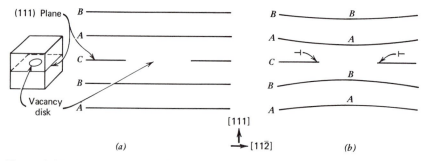

Figure 4.40   Formation of the Frank partial dislocation on {111} planes in an fcc lattice.

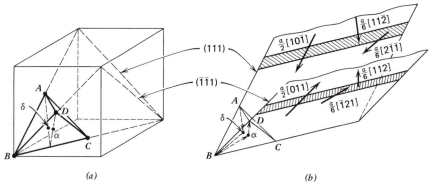

Figure 4.41   (a) A Thomson reference tetrahedron. (b) Dislocations involved in the formation of the Cottrell–Lomer lock.

Dislocation reactions are easily described with the Thompson reference tetrahedron. For instance it may be seen from Fig. 4.41 that the vector sum of $A\delta$ plus δB gives AB, which is the dislocation reaction of Eq. 4.26. Consider the reaction: unit disl, AB + Shockley part, $B\alpha$ → Frank part, $A\alpha$, which would be written

$$\frac{a}{2}[10\bar{1}] + \frac{a}{6}[\bar{1}21] \rightarrow \frac{a}{3}[11\bar{1}] \tag{4.27}$$

This reaction illustrates the use of the reference tetrahedron to describe dislocation reactions. It also illustrates that two glissile dislocations may react to form a sessile dislocation. Such reactions are important because they represent a mechanism of causing dislocation motion to become more difficult, which is therefore a strengthening mechanism.

A somewhat similar type of dislocation reaction, first considered by Lomer, is the combination of two parallel unit dislocations of b vectors AB and BD to give AD,

$$\frac{a}{2}[10\bar{1}] + \frac{a}{2}[011] \rightarrow \frac{a}{2}[110] \tag{4.28}$$

Dislocation AB would be gliding along the (111) plane and dislocation BD along the ($\bar{1}\bar{1}$1) plane both having their lines parallel to the intersection of their glide planes, B-C. Hence the resulting $a/2[110]$ dislocation would be pure edge because its b vector is perpendicular to B-C. This dislocation must have a (001) glide plane, and since the mobile dislocations in fcc crystals must have (111) glide planes this dislocation is a sessile dislocation. Hence, the Lomer reaction provides a means of locking the motion of the mobile unit dislocations of fcc crystals at the intersections of the slip planes.

Later Cottrell reexamined the Lomer reaction and considered what would happen if the two dislocations AB and BD were extended, as is shown schematically in Fig. 4.41(b). If these two extended dislocations move toward each other on their glide planes, they meet at the intersection B-C just as in the Lomer reaction. Only now the leading partial dislocations, δB and $B\alpha$ will react and form a new partial dislocation $\delta\alpha$ according to the equation

$$\delta B + B\alpha \rightarrow \delta\alpha \qquad (4.29)$$

$$\frac{a}{6}[2\bar{1}\bar{1}] + \frac{a}{6}[\bar{1}21] \rightarrow \frac{a}{6}[110]$$

and you may show yourself that this reaction reduces the strain energy. Hence, we end up with three partial dislocations as shown on the right of the complete reaction,

$$AB + BD \rightarrow A\delta + \delta\alpha + \alpha D$$

$$\frac{a}{2}[10\bar{1}] + \frac{a}{2}[011] \rightarrow \frac{a}{6}[11\bar{2}] + \frac{a}{6}[110] + \frac{a}{6}[112] \qquad (4.30)$$

The three partial dislocations are called the Cottrell–Lomer dislocations. The $\delta\alpha$ dislocation line lies along the intersection BC, and the lines of the two Shockley partials, $A\delta$ and αD, are separated from it by the two stacking faults. The $\delta\alpha$ dislocation line appears to hold the faults together much as a stair-rod holds a rug onto a staircase. Hence, the $\delta\alpha$ dislocation has been termed a stair-rod dislocation. This dislocation, $a/6[110]$, is an edge dislocation similar to the Lomer dislocation and it has an (001) glide plane. Consequently, the three partial dislocations each have different glide planes and so the Cottrell–Lomer dislocation is sometimes called a supersessile dislocation because it cannot move without disassociating. The Cottrell–Lomer dislocation is also often referred to as the Cottrell–Lomer lock because it blocks the motion of dislocations on either of the two intersecting {111} glide planes. Cottrell–Lomer dislocations have been observed in fcc crystals and they illustrate one mechanism for blocking dislocation motion and thus producing strain hardening in fcc metals.

4.15   FRANK–READ GENERATOR

The density of dislocations in a crystal is characterized by giving the number of dislocations passing through a unit area. Annealed metals contain dislocation densities on the order of 10^7 per square centimeter and cold-worked metals on the order of 10^{11} per square centimeter. It is apparent that mechanical working greatly increases the number of dislocations in a metal. A number of mechanisms have been proposed for

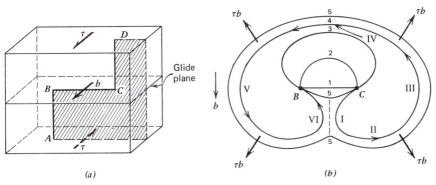

Figure 4.42   The Frank–Read generator.

dislocation multiplication within crystals,[7] and here we will discuss the best known of these, the Frank–Read generator. Consider the dislocation line A-B-C-D, shown in Fig. 4.42(a). Suppose that only the portion from B to C lies upon a glide plane as shown, so that the portions A-B and C-D are sessile. If we apply the shear stress τ parallel to the b vector, the dislocation feels a normal force τb and the glissile portion B-C begins to move out upon the glide plane but remains pinned at points B and C. Figure 4.42(b) shows five successive positions of the segment B-C as it moves out under the shear stress τ. Notice the interesting fact that in position 5 the dislocation has broken into two parts, one still connected to points B and C, and the other that has become a surrounding closed loop. The closed loop will grow out to the crystal extremities as described earlier. The portion still connected to B and C will return to position 1 and then continue out through positions 2 through 5 generating more and more loops as this cycle is repeated. This action may be understood by noting the following points.

1.   Irrespective of where you are along the dislocation line the b vector is always the same. It is invariant.

2.   The force on the line, τb, always acts at right angles to the line as shown for position 5.

3.   The positive direction of the line is shown on position 4, consistent with the b vector and the location of the extra plane shown on Fig. 4.42(a). Notice that:

a.   At points IV and II the dislocation is pure edge with the extra plane below the glide plane at IV and above it at II. Hence, these two segments are of opposite sense and will move away from each other under the shear stress τ.

b.   At points I, III, V, and VI the dislocation is pure screw. The screw is left-handed at I and V and right-handed at III and VI. Therefore, since I and VI are of opposite sense they move in opposite directions toward each other and when they meet they annihilate along their intersection causing the line to break into two parts as shown at the dotted portion of position 5.

It is quite fascinating that by the time the line has bowed around to position 4, the edge segment at point II has its extra plane above the glide plane whereas the original edge dislocation had its extra plane below the glide plane. It is difficult to visualize how this atomic motion is produced, but a good discussion of this point may be found in Ref. 10.

Earlier we showed that a shear stress τ will bend a dislocation to a radius of curvature R given by

$$\tau \approx \frac{Gb}{2R} \tag{4.31}$$

Let the distance from B to C be called X. Then at position 1 the radius of curvature is infinite. At position 2, $R = \frac{1}{2}X$ and at position 3 the mean radius of curvature is greater than $\frac{1}{2}X$. The mean radius of curvature starts at infinity, reaches a *minimum value* of $\frac{1}{2}X$ at the semicircular position, 2, and then increases in going to position 3. Consequently, the semicircular position, 2, is a critical position because it produces the minimum radius of curvature, and, hence, requires the maximum shear stress. One sees from Eq. 4.31 that in order to push the dislocation from position 1 to 2 the stress must increase from 0 to Gb/X, and the stress to push from 2 to 3 will be less than Gb/X. Therefore, one finds a critical shear stress, $\tau_{CR} = Gb/X$, required to "operate" the Frank–Read source. If $\tau < \tau_{CR}$ the dislocation line will bow out to a position somewhere between positions 1 and 2. If $\tau > \tau_{CR}$ the dislocation passes through position 2 and is then able to proceed through all successive positions thereby producing loops.

Beautiful electron microscope pictures of operating Frank–Read sources are available (Fig. 8.7, Ref. 7) in semiconductors and salts. However, in metals these sources have been observed only very rarely. It appears that in metals the majority of the dislocations are generated at grain boundaries.

4.16 INTERPRETATION OF PLASTIC FLOW IN TERMS OF DISLOCATION MOTION

In Chapter 3 we examined the plastic flow of single crystals from a macroscopic approach. We will now reexamine the stress–strain curve of the D specimen shown in Fig. 3.8(b) and consider how its behavior is controlled by dislocation motion.

Stage I. As the stress is increased, the critical resolved shear stress is obtained on the active slip system. At this point the shear stress acting on the dislocations of the active slip system is able to overcome the Peierls–Nabarro force and any additional force holding the dislocations, so that the dislocations begin to move, causing slip. Sufficient numbers of new dislocations are generated by source mechanisms such as the Frank–Read source to produce the relatively large strain of stage I. This generation and motion of dislocations during stage I occurs with very little hinderance and so the stress does not rise much.

Stage II. We saw earlier that stage II begins when the crystal has rotated sufficiently to produce slip on an intersecting system. This means that the dislocations on the intersecting slip systems will begin to intersect with each other as they move. This interaction causes many of the dislocations to become pinned, and so the metal begins to harden because dislocation motion is becoming more difficult. A pinned dislocation prevents other dislocations of the same sense from moving past it on the same glide plane because the stress field of the pinned dislocation produces a repulsive force on the approaching dislocations, as may be calculated from the Peach–Koehler equation. This repulsive force from the pinned dislocation will eventually cause the source that produced the pinned dislocation to become inoperative as shown schematically in Fig. 4.43(a). Possible pinning mechanisms may be such things as

1. Formation of Cottrell–Lomer locks at the intersecting slip planes.
2. Formation of jogs in the moving dislocations, which reduce the dislocation mobility (see Problem 4.1).
3. Formation of dislocation tangles.

Stage III. Eventually the stress becomes high enough to cause the pinned dislocations to move and the work hardening rate drops a bit as stage III begins. It appears that cross slip is the mechanism primarily responsible for motion of the pinned dislocations. The process of double cross slip is illustrated in Fig. 4.43(b). Here the dislocations are able to continue moving on the same slip system by simply cross slipping over to a parallel glide plane. A more detailed discussion of these ideas may be found in Refs. 13 and 14.

Figure 4.43 (a) Dislocation pile-up behind a pinned dislocation. (b) Continued dislocation motion via double cross slip.

Hopefully, the above discussion illustrates how the strength and plasticity of metals are controlled by the mobility of dislocations. In polycrystalline alloys the situation is considerably more complex due to the presence of so many grains of different orientations and to the requirement of coherency at the grain boundaries; the same principles apply, however. One may find a discussion of polycrystalline metals in Refs. 13 and 14. It was pointed out on p. 56 that there are four primary mechanisms of plastic flow in metals. Dislocation motion plays a strong, but not as well-defined, role in the three other mechanisms besides slip. Hence, any strengthening mechanism that one may hope to develop will at some level be related to an interaction with dislocations, and this interaction controls dislocation motion. Consequently, it behooves us to understand the nature of dislocations and their motion, even though in most practical systems the dislocation arrangements and interactions are too complex to be quantitatively described.

In the present chapter the importance of dislocations in controlling the mechanical properties of metals has been stressed. It should be realized, however, that the influence of dislocations upon the electromagnetic properties of metals and semiconductors may also be very important. For example, the best commercially available hard superconducting wires make use of high dislocation densities for flux pinning, and control of dislocation densities can be very critical in the fabrication of semiconductor devices.

REFERENCES

1. G. I. Taylor, Proc. Roy. Soc. **145A,** 362 (1934).

2. E. Orowon, Z. Phys. **89** 605, 614, 634 (1934).

3. M. Polanyi, Z. Phys. **89** 660 (1934).

4. E. Orowon, *Dislocations in Metals*, M. Cohen, Ed., American Institute Metallurgical Engineers, New York, N.Y., 1954.

5. A. H. Cottrell, *Dislocations and Plastic Flow in Crystals*, Oxford Univ. Pr., Oxford, 1952.

6. J. Weertman and J. R. Weertman, *Elementary Dislocation Theory*, Macmillan, New York, 1964.

7. D. Hull, *Introduction to Dislocations*, Pergamon, New York, 1965.

8. J. M. Burgers, Proc. Kon. Ned. Akad. Wet. **42,** 293, 378 (1939).

9. R. de Wit, *Theory of Dislocations: An Elementary Introduction*, NBS Monograph 59, pp. 13–34, U.S. Dept. of Commerce, 1963.

10. W. T. Read, Jr., *Dislocations in Crystals*, McGraw-Hill, New York, 1953.

11. R. A. Swalin, *Thermodynamics of Solids*, Wiley, New York, 1962, Chapter 12.

12. R. D. Heidenreich and W. Shockley, *Bristol Conference*, Physical Society, London, 1948.

13. R. E. Smallman, *Modern Physical Metallurgy*, Butterworth, London, 1962.

14. R. W. K. Honeycombe, *The Plastic Deformation of Metals*, St. Martins Press, New York, 1968.

PROBLEMS

4.1

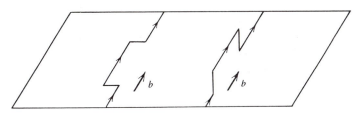

Shown above are two pure screw dislocations, one of which contains a kink and the other a jog. Take the positive direction of the lines as shown by the arrows. Both the kinked segments and the jogged segments are pure edge dislocations.

(a) Redraw the above kinked dislocation and show on your drawing the location of the extra plane of atoms for both kinked segments. Be careful to get the sense of both segments correct.

(b) Repeat (a) for the jogged dislocation.

(c) Suppose the glide plane shown above were a (111) plane in a fcc crystal. Which of the two pairs of segments would you think could be removed by gliding motion of themselves more easily? Explain.

(d) Explain how the glide motion of screw dislocations containing jogs could produce vacancies.

4.2 A metal crystal is made up of atoms A and B randomly distributed on a bcc lattice. With proper heat treatment the atoms arrange themselves in an ordered array on the lattice having A atoms at 0, 0, 0 and B atoms at $\frac{1}{2}, \frac{1}{2}, \frac{1}{2}$. The dislocations will move in $\langle 111 \rangle$ directions on the $\{1\,1\,0\}$ planes.

(a) Make a sketch of the planes of the lattice that are perpendicular to the (101) plane and contain the $[\bar{1}11]$ direction. Let A atoms on your sketch be represented as dots and B atoms as small circles.

(b) Sketch the above plane for the crystal containing a pure edge dislocation gliding on the (101) plane in the $[\bar{1}11]$ direction.

(c) Movement of this dislocation by glide creates a disruption of the perfect ordering of the lattice. On a sketch of the planes in part (a), show the disrupted area so produced. These areas are called antiphase boundaries (APB).

(d) It has been observed that the dislocations in these ordered alloys move in pairs, that is, two closely parallel dislocations. Why would you imagine this pair configuration to be stable?

4.3 Suppose an edge dislocation has a Burgers vector in the $[0\bar{1}0]$ direction and glides in the (100) plane.

(a) Will this dislocation be kinked or jogged if another edge dislocation having a b vector direction of $[010]$ moving in the (001) plane passes through it?

(b) Will it be kinked or jogged if a screw dislocation having a b vector in the $[100]$ direction gliding in the (001) plane passes through it?

4.4 LiF forms an ionic crystal of the NaCl structure. The Li atoms are at the corners and face centers of the unit cell and the F atoms are at the middle of each of the edges of the unit cell and one F atom is at the body center position. Hence there are eight atoms per unit cell.

It is known that the slip planes of LiF crystals are the {110} planes. Give the Burgers vector of the dislocation you would expect to be responsible for this slip. Remember, this is an ionic crystal and there will be a strong repulsion between like-type atoms. Explain your reasoning fully.

4.5 If a total dislocation in a fcc lattice has a Burgers vector $[\bar{1}10]a/2$ and lies in a {111} plane, indicate the b vectors of all of the pairs of Shockley partial dislocations into which it may dissociate.

4.6 (a) The dislocation $[10\bar{1}]a/2$ dissociated into two Shockley partial dislocations, $[11\bar{2}]a/6$ and $[2\bar{1}\bar{1}]a/6$. Using the Peach–Koehler relation and the equations for the state of stress around dislocations calculate the force acting in the glide plane between the two partial dislocations for the following cases.

1. $[10\bar{1}]a/2$ is pure screw; answer, $\dfrac{Gb^2}{8\pi y}\left[\dfrac{2-3\nu}{1-\nu}\right]$

2. $[10\bar{1}]a/2$ is pure edge; answer, $\dfrac{Gb^2}{8\pi y}\left[\dfrac{2+\nu}{1-\nu}\right]$

where b is the length of the partial dislocations and y is their distance of separation. The stress around a mixed dislocation is given by the sum of the stresses due to the edge components and screw component.

(b) If the stacking fault energy is 10 ergs/cm^2, the shear modulus 7×10^{11} dynes/cm^2, the lattice parameter 3.0 Å and Poisson's ratio $\frac{1}{3}$, determine the equilibrium spacing between the partial dislocations for the above two orientations.

(c) Calculate the energy per unit length of dislocation line required to collapse the above two dislocations from their equilibrium spacing to a spacing of 1 Å.

4.7 Into what partial dislocations would you expect the total dislocation of Burgers vector $[11\bar{2}0]a/3$ to dissociate? Give your answer in vector notation, and assume slip occurs in the basal plane.

4.8 The dislocation of Burgers vector $[\bar{1}01]a/2$ moves along the $(1\bar{1}1)$ plane of an fcc crystal and interacts with a parallel dislocation on the $(11\bar{1})$ plane having Burgers vector $[1\bar{1}0]a/2$. The interaction forms a Cottrell–Lomer dislocation.

Determine the three partial dislocations making up the Cottrell–Lomer dislocation. Which of these three dislocations is the stair rod dislocation? Note: You may find it useful to use the Thomson reference tetrahedron for this problem.

4.9 A unit dislocation in the $[\bar{1}01]$ direction of an fcc metal can combine with a Shockley partial dislocation in the $[12\bar{1}]$ direction to form a Frank partial dislocation.

(a) What is the Burgers vector of the resultant Frank partial?

(b) Why is this dislocation called sessile?

(c) Determine whether the reaction is energetically favorable.

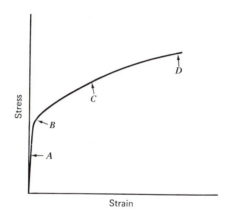

4.10 Shown below is a stress strain diagram for a *polycrystalline* metal.

(a) Suppose the dislocation density of the unstrained metal were 10^7 dislocations per cm^2. Make a qualitative estimate of dislocation density at the Hooke's law stress, *A*, at the yield stress, *B*, and at two flow stresses, *C* and *D*.

(b) According to dislocation theory what happens in the metal as the stress goes past the yield point *B*?

(c) As the stress increases from point *B* to *C* to *D* the metal becomes stronger. This phenomena is called work hardening. How does dislocation theory account for this increased strength?

(d) Aluminum alloys are strengthened by the precipitation of a second phase. How do you explain this source of strength in terms of dislocation theory?

(e) Can all strengthening mechanisms be explained in terms of dislocations? That is, are there plastic flow mechanisms that do not involve dislocations?

(f) Could you determine the critical resolved shear stress from the above stress–strain diagram for a polycrystalline metal? If so, explain how; if not, explain why not.

4.11 Suppose the active slip system in an fcc crystal is $[\bar{1}10](11\bar{1})$.

(a) Give the Burgers vector for the unit dislocation causing slip and explain how you know this.

(b) If slip occurs by the motion of pure unit edge dislocations give the direction of the dislocation line.

(c) If slip occurs by the motion of pure unit screw dislocations give the direction of the dislocation line.

(d) Give the direction in which the dislocation line would move during slip in (b) and (c) above.

(e) Suppose a shear stress of 100 psi acts on the $(11\bar{1})$ plane in the $[\bar{1}10]$ direction. Calculate the direction and magnitude of the force per unit length which the shear stress exerts on (1) a pure unit edge dislocation and (2) a pure unit screw dislocation. Take the lattice parameter, a, as 2 Å.

4.12

Shown above are two pinned edge dislocations, A-B and C-D, both of which are of the same length, X, and have the same sense and Burgers vector. Each dislocation segment will act as a Frank–Read generator.

(a) Show the dislocations at successive positions and indicate what happens when the dislocations interact with each other.

(b) Show by your diagrams whether the interactions of the two expanding loops will cause the dislocation motion to become pinned, or whether they will form one large Frank–Read source.

(c) If one large source is produced, what is the critical shear stress to operate it? Take the distance from A to D as 5X.

(d) Answer part (b) for the case where edge dislocation A-B has the opposite sense from edge dislocation C-D.

CHAPTER 5
VACANCIES

Very often in the study of physical metallurgy we are interested in whether or not certain physical processes will occur. In the last chapter we wanted to know if a dislocation with two extra planes would separate into two dislocations each with one extra plane. In this chapter we want to know if a metal containing a large number of lattice vacancies will dispel these vacancies to become a more nearly perfect lattice of atoms. To answer these questions properly we make use of some principles of thermodynamics. Suppose we are considering a physical process such as either of the above, and we wish to know if the process will occur spontaneously. The second law of thermodynamics tells us that the entropy change for any spontaneous process plus the entropy change in those surroundings affected by the process must be positive. Hence, our process will occur spontaneously if $\Delta S_{process} + \Delta S_{surroundings} > 0$. This criterion is awkward to use; instead we use the free-energy function to determine if a process should occur spontaneously. The Gibbs free energy is defined as $G = E + PV - TS$. By combining the definition of entropy with the first law of thermodynamics we obtain $dG = V\,dP - S\,dT$. It may be shown from this relation and the second law that:

For any spontaneous process in which the products and reactants are at constant temperature and pressure and in which no work is done other than against pressure, the Gibbs free energy will decrease, $\Delta G < 0$.

(See for example Ref. 1, p. 187 or Ref. 2, p. 80.) Hence, thermodynamics tells us that if $\Delta G > 0$ for such a process, the process will not occur; and, if $\Delta G < 0$ the process should occur spontaneously. It is important to realize that the fact of $\Delta G < 0$ for some process does not guarantee that the process will spontaneously occur. For example, when a metal is quenched to a very low temperature the mobility of the atoms becomes very small so

127

that even though $\Delta G < 0$ for some preferred atom movement, the movement occurs so slowly that the process may not take place in our lifetime. It should be noted that thermodynamics provides us with other functions that may be used as stability criteria. For example, the Helmholtz free energy, $A = E - TS$, for spontaneous processes at constant temperature and volume is negative. The Gibbs free energy is generally more convenient because many processes occur isothermally at atmospheric pressure (constant T, P).

Whether or not a process can occur spontaneously depends on the function $\Delta G = \Delta E - P \Delta V - T \Delta S$ for the process. In most processes involving solids and liquids at atmospheric pressures the term $P \Delta V$ is negligibly small, so that we have

$$\Delta G \approx \Delta E - T \Delta S \qquad (5.1)$$

Suppose you were considering some process that caused a large decrease in energy. Intuitively you might suspect that such a process would occur spontaneously because the system could thereby appreciably lower its energy, such as a ball rolling down a hill or a wave damping out on a lake. However, Eq. 5.1 shows that we must consider the entropy term $T \Delta S$. If the large decrease in energy is accompanied by a corresponding decrease in entropy, then the process may not be able to occur spontaneously.

In studies of statistical mechanics it is shown that the entropy may be expressed as

$$S = k \ln \Omega \qquad (5.2)$$

where k is Boltzman's constant and Ω is the number of microstates per macrostate. The macrostate is simply the observable state of the system. The number of microstates may mean different things. Consider two examples: (a) the number of microstates may mean the number of ways the quantum states of the energy levels may be filled in a metal, (b) the number of ways the carbon atoms in steel may be distributed over their interstitial sites. The first example is concerned with a distribution in energies, and the second example involves a geometrical distribution. The entropy is frequently separated into two parts related to these two types of distributions, and they are termed vibrational entropy and configurational entropy, respectively. If the "number of ways" the distribution may be filled is large, we say the system has a large *randomness* and if it is small the system has a large *order* (nonrandomness).

Hence, we have the following conclusions: (1) The entropy term favors a process if it is positive. (This contrasts with the energy change, which is negative for a favored process.) (2) Processes that introduce order will decrease randomness and therefore have a negative (unfavorable) entropy change. Consequently, *if a process introduces order the entropy change acts to prevent the process.* (3) If the ΔE for a process is negative the process will always have a negative ΔG if order is destroyed. However, if the process is introducing order, the ΔG may be positive in spite of the reduction in energy, because of the entropy term. There are many processes in which the entropy term, $T\Delta S$, is relatively small and may be neglected, such as the dislocation reaction mentioned above. However, one must first make some estimate of the magnitude of ΔS in order to know this. In this chapter we consider a process, vacancy formation, in which the entropy term is very important.

5.1 VACANCY FORMATION

In a perfect crystal all lattice sites are occupied by atoms, so that no vacancies are present. If we now introduce a number of vacancies into the lattice by removing atoms, will these vacancies remain in the lattice as a stable defect or will they spontaneously migrate to the surface restoring the perfection of the lattice? We analyze this problem by considering the effect of the vacancies upon the Gibbs free energy of the lattice. Given a lattice of N lattice sites, how does its free energy change as a function of the number of vacancies, n, introduced into it? If the introduction of n vacancies causes G to decrease, then the vacancies will be *thermodynamically stable* and vice versa.

Let the free-energy change due to the introduction of n vacancies be $\Delta G = G - G(\text{perfect})$ where $G(\text{perfect})$ is the free energy of the perfect lattice. We write

$$\Delta G = G - G(\text{perfect}) \approx \Delta E - T\Delta S \qquad (5.3)$$

Now let

E_v = energy per vacancy

S_v = vibrational entropy per vacancy

S_c = configurational entropy of the entire crystal

Hence we have

$$\Delta G = n\,\Delta E_v - T(n\,\Delta S_v + \Delta S_c) \qquad (5.4)$$

We will consider each of the terms in Eq. 5.4 beginning with the configurational entropy, ΔS_c. The process we are describing is the introduction of n vacancies into a lattice of N sites. For this process there is a configurational entropy, because there are many different geometrical ways in which the vacancies may arrange on the lattice. The value of ΔS_c is determined from Eq. 5.2 where we have

Macrostate: A crystal with n vacancies
Microstate: One particular arrangement of the n vacancies on the N lattice sites

Hence Ω is simply the number of distinguishably different ways that n vacancies may be arranged on N lattice sites.

The calculation of Ω is therefore simply a problem in permutations and combinations. To determine Ω we consider an analogous problem: *Given 100 lattice sites and the three atoms Fe, Co, and Ni, in how many distinguishably different ways can we put these three atoms on the 100 lattice points?* Suppose we place the atoms on the lattice in the sequence: first the Fe atom, then Ni, and last Co. There are 100 different sites on which we may place the Fe atom. *For each* one of these 100 choices we may place the Ni atom on any of 99 remaining sites, and *for each* of these 99 choices we may place the Co atoms on any of 98 sites. Hence, we have:

Fe atom may go on in 100 ways
For each of these Ni atom may go on in 99 ways
For each of these Co atom may go on in 98 ways

so that the answer to our question is simply

$$100 \cdot 99 \cdot 98 = \frac{100!}{97!} = \frac{100!}{(100-3)!} = 970,200$$

What is the answer to the above question if all three atoms are Fe atoms? Consider two of the 970,200 arrangements for the case of the three different atoms,

. . . Fe Ni . . Co . . → 100

. . . Ni Fe . . Co . . → 100

where the three atoms happen to be arranged on the same set of three positions as shown here. Notice that if all three atoms were Fe we could not distinguish between these two arrangements. We therefore ask: For each set of three sites, such as shown above, how many distinguishably different ways can we arrange the Fe, Ni, and Co atoms? The answer is 3! because

Fe atoms may go on in 3 ways
For each of these the Ni atoms may go on in 2 ways
For each of these the Co atoms may go on in 1 way

Hence, for each set of three sites we have counted 3! permutations of the three atoms. If the three atoms are the same, these 3! permutations will not be distinguishable. Therefore, since we have counted too large by 3! *for each set* of three sites we must *divide* by 3!. Hence, for the case where all three atoms are Fe (indistinguishable) the answer is

$$\frac{100!}{97!\,3!} = \frac{100!}{(100-3)!\,3!} = 161,700*$$

Vacancies are indistinguishable from one another, so the calculation of Ω is analogous to the latter of the above two examples. By analogy we have,

$$\Omega_n = \frac{N!}{(N-n)!\,n!}$$

where N is the total number of lattice sites and n is the number of vacancies. Hence, we have for the configurational entropy

$$\Delta S_C = S(n \text{ vacancies}) - S(0 \text{ vacancies})$$

$$= k \ln \Omega_n - k \ln \Omega_0$$

$$\Delta S_C = k \ln \frac{N!}{(N-n)!\,n!} \qquad (5.5)$$

Since values of N and n are quite large we may apply a mathematical approximation known as Stirling's approximation,

$$\ln X! = X \cdot \ln X - X \qquad \text{for } X \text{ large}$$

After some algebraic manipulation we obtain

$$\Delta S_C = -k\left[N \ln \frac{N-n}{N} + n \ln \frac{n}{N-n} \right] \qquad (5.6)$$

Notice the following,

$$\frac{N-n}{N} = \frac{\text{No. atoms on lattice}}{\text{No. lattice sites}} < 1$$

$$\frac{n}{N-n} = \frac{\text{No. vacancies on lattice}}{\text{No. atoms on lattice}} < 1$$

Hence, from Eq. 5.6 we see that ΔS_C must be positive, which means that the configurational entropy change for our process of introducing vacancies into a perfect lattice is *positive*, that is, it favors the process.

In addition to the configurational entropy we must consider the vibrational entropy, which is related to the ways in which the energy levels in the

---

* Alternately, one may see this as follows. The 970,200 permutations counts each set of 3 sites 3! times. If we let X equal the total number of 3 site sets, then we must have $X \cdot 3! = 970,200 = 100!/97!$.

solid are occupied. This is a much more formidable problem and it will simply be stated that one may show from a study of statistical mechanics starting basically from Eq. 5.2 that an approximate form of the vibrational entropy per atom is given by

$$\Delta S_V \approx 3k \ln \left(\frac{\nu}{\nu'}\right) \qquad (5.7)$$

where ν' is the final frequency of the atoms around the vacancy and ν is their original frequency. The vacancy tends to increase the vibrational amplitude and lower ν so that $\nu/\nu' > 1$ and ΔS_V is positive. Therefore the total entropy change associated with introducing vacancies, $\Delta S_C + n \Delta S_V$, must be positive.

We may visualize adding a vacancy as a process of removing an atom to the surface or to an interface such as a grain boundary. This process will require energy so that ΔE_V must be positive. We may now evaluate each term in Eq. 5.4. Figure 5.1 is a plot of the terms in Eq. 5.4 as a function of the number of vacancies, n. It may be seen that the energy term raises the free energy but the total free energy actually drops at first due to the increased entropy upon introducing vacancies. The free-energy function has a minimum in it. A maximum decrease in free energy is obtained at this minimum, so this point represents the equilibrium condition, and the value of n at this minimum is the equilibrium number of vacancies, n_e. We determine n_e by setting the derivative of the free-energy function equal to zero,

$$\frac{d \Delta G}{dn} = 0 = \Delta E_V - T \Delta S_V - T \frac{d \Delta S_C}{dn} \qquad (5.8)$$

The derivative of ΔS_C is evaluated from Eq. 5.6 and we find

$$\frac{n_e}{N} = \exp \left(-\frac{\Delta E_V}{kT} + \frac{\Delta S_V}{k}\right) \qquad (5.9)$$

In order to determine the order of magnitude of n_e we must obtain values for ΔE_V and ΔS_V. These numbers may be obtained from two types of experiments that are described in detail in Ref. 3. The values of ΔE_V range from 20 to 30 kcal/mole (80 to 160 kJ/mole) and $\Delta S_V/k$ from 1.0 to 2.0. Taking Cu for a specific example ($\Delta S_V/k = 1.5$, $\Delta E_V = 113$ kJ/mole) we determine n_e/N from Eq. 5.9 and we approximate the number of atoms per volume from the density as 5×10^{22} sites/cm^3. The number of vacancies per cubic centimeter is then found to be

Temperature	Vacancies/cm^3
1000°C	5×10^{18}
22°C	2×10^3

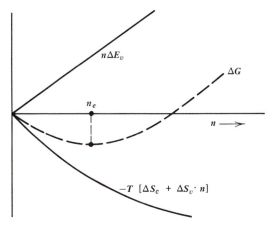

Figure 5.1 A plot of the terms of the free-energy change for vacancy formation as a function of the number of vacancies formed.

This result illustrates two interesting things. (1) A perfect metal is *not* thermodynamically stable. The free energy is lowered when vacancies are added, and the equilibrium number of vacancies is quite large, although it is still a small fraction of the total number of lattice sites. (2) The number of vacancies is a very strong function of temperature. As we will see later this latter fact is a major factor in the success of many heat-treating operations.

In general a pure metal will contain numerous defects in its crystal structure; we systematically list them as follows:

1. A foreign atom substitutionally placed.
2. A foreign atom interstitially placed.
3. Parent atoms interstitially placed.
4. Vacancies.
5. Twins or stacking faults.
6. Grain boundaries.
7. Dislocations.
8. Voids.
9. Inclusions.

It has become customary to class these defects as point defects, line defects, and plane defects. The first four are *point defects*, the dislocation is the only *line defect*, and 5 and 6 are *plane defects*. Defects 8 and 9 are so large relative to the lattice that they could probably be called bulk defects.

For each of these defects we may carry out an analysis similar to above to determine if they are thermodynamically stable. For the point defects we obtain an equation similar to Eq. (5.9) showing that all point defects have an equilibrium concentration whose magnitude is determined primarily by the ΔE of formation of the defect. Hence, we have the interesting result that a pure metal is really thermodynamically unstable relative to impurity atoms.

We will briefly consider defect 3. An interstitial Cu atom in pure Cu would be an example of this type of defect. Intuitively we do not expect many of these type of defects because an interstitial Cu atom in its own lattice would be a relatively large defect and, hence, would produce a severe strain in the surrounding lattice. Approximate quantum-mechanical calculations give $\Delta E(\text{interstitial}) = 7 \cdot \Delta E(\text{vacancy})$ in Cu. One then calculates from Eq. 5.9 for the two defects that the fraction of interstitials is $10^{-35400/T}$ times the fraction of vacancies. Hence, at 1000°C one would have only one interstitial per 10^{10} grams of Cu. Therefore, such defects are generally not important.

If one carries out this same kind of a calculation for dislocations it is also found[4] that dislocations would not be expected to occur in crystals. As mentioned previously it is extremely difficult to produce a metal with less than 10^4 dislocations per cm^2. Hence, even though we could lower the free energy of a metal by removing all dislocations, this cannot be done. Apparently, this is because even an extremely small stress generates dislocations in metals and no way has been found to crystallize metals or anneal them without introducing significant numbers of dislocations. It has been possible, however, to obtain semiconductor materials such as Ge and Si with essentially no dislocations.

REFERENCES

1. L. S. Darken and R. W. Gurry, *Physical Chemistry of Metals*, McGraw-Hill, New York, 1953.

2. J. Mackowiak, *Physical Chemistry for Metallurgists*, Allen and Unwin Ltd., London, 1965.

3. P. G. Shewmon, *Diffusion in Solids*, McGraw-Hill, New York, 1963.

4. A. H. Cottrell, *Dislocations and Plastic Flow in Crystals*, Oxford Univ. Pr., Oxford, 1953.

PROBLEMS

5.1   In a previous problem we considered an ordered bcc alloy in which all of the A atoms were on the cube corners, and all of the B atoms were on the cube centers. When the atoms become ordered in many binary alloys they lower the energy of the system so that ΔE_{order} is negative. However, alloys never exhibit ordering except at low temperatures. Question: If the ΔE for ordering is negative, why does the metal not become ordered at all temperatures? Please explain your answer in some detail.

5.2   Take the energy to form a mole of vacancies in Cu as 20,000 calories, and the vibrational entropy as $1.0k$, where k is Boltzman's constant. Compute the number of vacancies per cubic centimeter of Cu at 20°C and at its melting point. Take the density of Cu as 8.94 g/cm³.

5.3   Suppose you are carrying out a heat-treating process on a piece of steel by annealing at 850°C and then quenching to room temperature. If the energy per mole of vacancies in iron is 25,000 calolries, calculate how many times the number of vacancies would be increased after heating from 20°C to 850°C.

In a couple of sentences explain what you think might happen to these "extra" vacancies after quenching to room temperature.

CHAPTER 6
DIFFUSION

Consider the simple experiment shown in Fig. 6.1. A vertical glass tube contains a column of clear water above separated from a column of inked water below by a thin diaphragm. At time zero we gently remove the diaphragm in a manner such that the water remains stagnant. We now ask: As time passes what happens to the ink? One finds that the dark color of the ink slowly migrates upward so that after 5 hours the column might appear as shown in Fig. 6.1. Somehow the molecules producing the dark color of the ink have migrated upward. The water has remained stagnant so that this motion must have been accomplished by an actual preferred migration of the ink molecules; that is, the movement results from molecular motion. This is a form of *mass transport* that is called *diffusion.* Mass transport in gases and liquids generally occurs by a combination of convection (fluid motion) and diffusion. In solids, convection does not occur and consequently diffusion is generally the only available mass transport mechanism; therefore, it is an important mechanism controlling the rate of many physical processes of interest to us.

We can study diffusion from two general approaches.

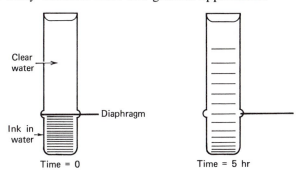

Figure 6.1 Diffusion of ink in water.

137

1. Phenomenological approach. Here we ask: How can we describe the rate and, hence, the amount of mass transport that occurs in terms of parameters we can measure? This approach is important to our ability to control such processes as carburizing, nitriding, tempering, homogenization of castings, and the like.

2. Atomic approach. Here we ask: What is the atomic mechanism by which atoms move? This approach is important to our understanding of how diffusion mechanisms affect such processes as precipitation hardening. We will discuss diffusion under these two above categories.

6.1 PHENOMENOLOGICAL APPROACH

A simple unidirectional diffusion experiment involving solids is shown in Fig. 6.2. A rod of pure iron is butt welded to a steel rod containing 1 wt% carbon and this "diffusion couple" is heated to 700°C to allow diffusion to occur at a significant rate. After some time at temperature the diffusion couple is quenched to room temperature and the composition of carbon along the rod is determined by chemical analysis. The composition profile might look as shown by the dashed curve labeled $t = t$ in Fig. 6.2. After an infinite time the composition will become constant at the average value. Our problem is to describe the rate at which the carbon atoms move to the right. This type of a diffusion problem was studied by Adolf Fick in a paper published in 1855.[1] He found that the *flux* of atoms was proportional to the

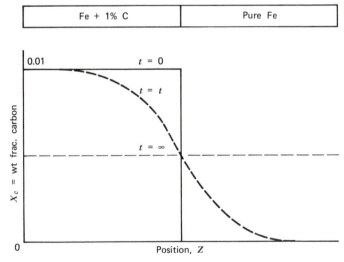

***Figure* 6.2** A diffusion couple showing variation of composition with position and time.

volume concentration gradient, so that one has

$$J_1 = -D_1 \frac{dC_1}{dZ} \qquad (6.1)$$

This equation is called Fick's first law of diffusion. The following points should be noted.

1. J_1 is defined as the flux of atoms 1 (in this case carbon atoms), and it has units of either g/cm²-sec or atoms/cm²-sec. The flux may be thought of as the rate at which atoms cross a unit area, that is, atoms per sec/cm².

2. D_1 is simply the proportionality constant and it is called the diffusion coefficient. Its units are cm²/sec.

3. C_1 is a volume concentration of component 1, either g/cm³ or atoms/cm³. Notice that a chemist does not directly measure *volume* concentration but rather he measures weight percent carbon, which is a *fractional* concentration that we will term X_1. The relation between these two concentrations is simply $C_1 = X_1 \cdot$ (density), where the density is either a mass density or an atom density depending on the units being used for volume concentration. (In this text we will use the terms concentration and composition interchangeably.)

4. The minus sign is required because the atoms flow toward the lower concentrations. Notice in Fig. 6.2 that the atoms move to the right, which is the positive direction. However, the concentration gradient that causes them to migrate is negative; see the curve for $t = t$. Hence, to make the flux positive in Eq. 6.1 we must use a minus sign to compensate for the negative gradient.

From the above we conclude that whenever a concentration gradient is present in a metal, a diffusion flux will occur. Later we will see that this is generally, but not always, true. Our problem now is: How do we actually determine D? By considering the experiment of Fig. 6.2 you will see that one cannot measure either J_1 or D_1 directly. What we can measure is the composition profile after various times so that we measure composition as a function of Z and t. The concentration gradient of Eq. 6.1 will vary with both position and time, and consequently so will the flux. Therefore, we must determine a differential equation for this diffusion process. To do this we simply perform a mass balance upon a differential volume element perpendicular to the mass flow direction as in Fig. 6.3. We may write for carbon transport on this element

$$\text{Mass in} - \text{mass out} = \text{Accumulation} \qquad (6.2)$$

By considering some time interval we have

$$\text{Rate in} - \text{rate out} = \text{Rate accumulation} \qquad (6.3)$$

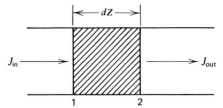

Figure 6.3 A differential volume element for unidirectional diffusion.

All of the mass comes into the volume element through plane 1 so that the rate of mass in is simply the flux at 1 times the area at 1,

$$\text{Rate mass in} = (JA)_1 \tag{6.4}$$

Since dZ is a differential length we obtain the rate out of the volume element at plane 2 by adding the change in rate going across the volume element as

$$\text{Rate mass out} = (JA)_1 + \frac{\partial(JA)}{\partial Z}\, dZ \tag{6.5}$$

The rate of accumulation is now written in terms of the volume concentration as

$$\text{Rate accumulation} = \frac{\partial[CA \cdot dZ]}{\partial t} = A\, dZ \cdot \frac{\partial C}{\partial t} \tag{6.6}$$

We now substitute Eqs. 6.4, 6.5, and 6.6 into Eq. 6.3, cancel terms, and obtain

$$-\frac{\partial J}{\partial Z} = \frac{\partial C}{\partial t} \tag{6.7}$$

where we have assumed that the area A is constant in the Z direction. This is a very important equation, called the continuity equation, which is limited by our derivation to unidirectional transport. Notice that our treatment here assumed that mass transport occurred in only one direction. For the general three-dimensional case the derivation is similar and we obtain Eq. 6.7 with $\partial/\partial Z$ replaced by the del operator. The equation holds for all material flow processes when no material is generated within the volume element, for example, flow of heat, neutrons, electrons, and so on. Our flow process is mass diffusion and by substitution of the diffusion flux equation, Eq. 6.1, we obtain for one-dimensional diffusion

$$\frac{\partial[D_1\, \partial C_1/\partial Z]}{\partial Z} = \frac{\partial C_1}{\partial t} \tag{6.8}$$

This equation is sometimes called Fick's second law. It is a partial

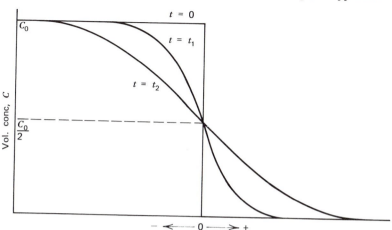

Figure 6.4 Composition profiles in an iron–steel diffusion couple.

differential equation with C_1 as the dependent variable and Z and t as the two independent variables. Hence, solutions to this equation will give C_1 as a function of Z, t, and D_1. Since our experimental data give us $C_1 =$ function (Z, t) as shown on Fig. 6.2, we may fit these data to the mathematical solution of Eq. 6.8 for various values of D_1 and then determine D_1 as that value giving the best fit.* An example solution will now be presented to illustrate this technique.

We consider the Fe–C example mentioned earlier and redraw the composition profiles as shown in Fig. 6.4. The volume concentration C is plotted here versus distance for various times. We assume that D is a constant, so that Eq. 6.8 becomes a linear partial differential equation,

$$D\frac{\partial^2 C}{\partial Z^2} = \frac{\partial C}{\partial t} \tag{6.9}$$

We may solve this problem fairly easily using the Laplace transformation if we make two additional assumptions, which turn out to be very realistic. Assume:

1. At all times $t > 0$, the concentration at the interface, $Z = 0$, remains at $C_0/2$. This assumption requires that the atom velocities do not depend on concentration so that the decay on the left is symmetric to the buildup on the right.

2. The bar is sufficiently long that the concentrations at either end are

---

* Because of the complexity of the solutions it is generally not possible to determine D_1 as an explicit function of C_1, Z, and t and, therefore, directly solve for D_1.

unaffected by the diffusion process. To solve the problem we consider only the portion of the rod where $Z>0$ and we write

$$\text{Boundary conditions:} \quad C(Z=0, t) = \frac{C_0}{2}$$

$$C(Z=\infty, t) = 0 \qquad (6.10)$$

$$\text{Initial condition:} \quad C(Z, 0) = 0$$

The Laplace transformation using t as the independent variable is applied to Eq. 6.9 and an ordinary differential equation is obtained that may be easily solved for the boundary conditions. The transformed solution is found to be

$$c(Z, s) = \frac{C_0}{2}\left[\frac{1}{s} e^{-Z\sqrt{s/D}}\right] \qquad (6.11)$$

Taking the inverse transform we find as the solution of Eq. 6.9 under conditions 6.10

$$C(Z, t) = \frac{C_0}{2}\left[1 - \frac{2}{\sqrt{\pi}}\int_0^{Z/2\sqrt{Dt}} e^{-y^2}\, dy\right] \qquad (6.12)$$

This equation is not nearly as messy as it looks at first encounter because the function e^{-y^2} decays rapidly from one to zero. The integral function is called the *error function* and it is defined as,

$$\text{Error function of } \beta = \text{Erf}[\beta] = \frac{2}{\sqrt{\pi}}\int_0^\beta e^{-y^2}\, dy \qquad (6.13)$$

Consequently we may write the solution as

$$C(Z, t) = \frac{C_0}{2}\left[1 - \text{erf}\frac{Z}{2\sqrt{Dt}}\right] \qquad (6.14)$$

The values of the error function are tabulated as shown in Table 6.1, so that it is no more difficult to evaluate an equation like 6.14 than an equation containing a sine or a cosine function. Since assumpion 1 requires symmetry about $Z=0$ we obtain by inspection for $Z<0$, $C(Z, t) = C_0 - [1 + \text{erf } Z/2\sqrt{Dt}] \cdot C_0/2$, which is identical to Eq. 6.14 upon simplification. [Note: The plus sign on the erf term is required here because $\text{erf}(-\beta) = -\text{erf}(\beta)$, and β is negative for $Z<0$.] Equation 6.14 is frequently called the Grube solution after G. Grube.[2] In our solution we have assumed the original composition of carbon on the right-hand side to be identically zero. If we let the composition on the right-hand side have some finite value C_1, which is less than C_0, the Grube solution is simply

$$C(Z, t) = C_1 + \frac{C_0 - C_1}{2}\left[1 - \text{erf}\frac{Z}{2\sqrt{Dt}}\right] \qquad (6.15)$$

Table 6.1 Tabulation of the Error Function, erf [β], For Various Values of β From 0 to 2.7

β	0	1	2	3	4	5	6	7	8	9
0.0	0.0000	0.0113	0.0226	0.0338	0.0451	0.0564	0.0676	0.0789	0.0901	0.1013
0.1	0.1125	0.1236	0.1348	0.1459	0.1569	0.1680	0.1790	0.1900	0.2009	0.2118
0.2	0.2227	0.2335	0.2443	0.2550	0.2657	0.2763	0.2869	0.2974	0.3079	0.3183
0.3	0.3286	0.3389	0.3491	0.3593	0.3694	0.3794	0.3893	0.3992	0.4090	0.4187
0.4	0.4284	0.4380	0.4475	0.4569	0.4662	0.4755	0.4847	0.4937	0.5027	0.5117
0.5	0.5205	0.5292	0.5379	0.5465	0.5549	0.5633	0.5716	0.5798	0.5879	0.5959
0.6	0.6039	0.6117	0.6194	0.6270	0.6346	0.6420	0.6494	0.6566	0.6638	0.6708
0.7	0.6778	0.6847	0.6914	0.6981	0.7047	0.7112	0.7175	0.7238	0.7300	0.7361
0.8	0.7421	0.7480	0.7538	0.7595	0.7651	0.7707	0.7761	0.7814	0.7867	0.7918
0.9	0.7969	0.8019	0.8068	0.8116	0.8163	0.8209	0.8254	0.8299	0.8342	0.8385
1.0	0.8427	0.8468	0.8508	0.8548	0.8586	0.8624	0.8661	0.8698	0.8733	0.8768
1.1	0.8802	0.8835	0.8868	0.8900	0.8931	0.8961	0.8991	0.9020	0.9048	0.9076
1.2	0.9103	0.9130	0.9155	0.9181	0.9205	0.9229	0.9252	0.9275	0.9297	0.9319
1.3	0.9340	0.9361	0.9381	0.9400	0.9419	0.9438	0.9456	0.9473	0.9490	0.9507
1.4	0.9523	0.9539	0.9554	0.9569	0.9583	0.9597	0.9611	0.9624	0.9637	0.9649
1.5	0.9661	0.9673	0.9684	0.9695	0.9706	0.9716	0.9726	0.9736	0.9745	0.9755

1.55	1.6	1.65	1.7	1.75	1.8	1.9	2.0	2.2	2.7
0.9716	0.9763	0.9804	0.9838	0.9867	0.9891	0.9928	0.9953	0.9981	0.9999

A.　CARBURIZING

Perhaps the most important application of the principles of diffusion in metallurgy involves the carburization of steel. Suppose a rod of pure iron has one end packed against graphite as shown in Fig. 6.5 and it is heated to 700°C. Within a few minutes of achieving temperature a *local equilibrium* will be established at the graphite–iron interface. This means that the compositions of the two phases touching each other at the interface are given by the equilibrium phase diagram at 700°C. From Fig. 6.5(*b*) we see that at 700°C α iron is in equilibrium with the carbide phase, Fe$_3$C, called cementite. (Actually, Fe$_3$C is a metastable phase, but it nevertheless usually forms). In order to establish the local equilibrium it is necessary to mix the graphite with a catalyst or to use certain gas atmospheres, but we will not discuss this difficulty here. A carbide layer forms on the surface of the iron as a result of the *local equilibrium* and the carbon composition in the iron right at this interface is determined from the phase diagram as C_s. Physically, this means that at the left surface of the iron bar the composition jumps to a value of C_s at time zero and remains there. This causes a very large carbon concentration gradient to be generated at the left end of the iron rod and so carbon diffuses into the rod at a high rate producing composition profiles in the rod that vary with time as shown, for example, on Fig. 6.5(*a*). To determine these concentration profiles we must solve Fick's second law for the boundary conditions of this example. We may write these conditions as

$$\text{Boundary conditions:} \quad C(Z=0, t)=C_s$$
$$C(Z=\infty, t)=0 \qquad (6.16)$$
$$\text{Initial condition:} \quad C(Z, 0)=0$$

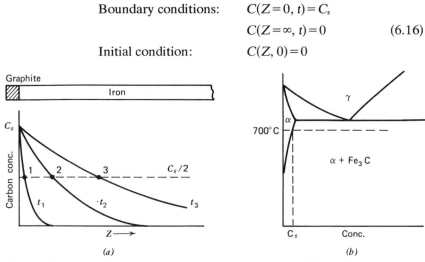

Figure 6.5 (*a*) Composition profiles for carburizing iron. (*b*) Pertinent portion of iron–carbon phase diagram.

where we have assumed the bar sufficiently long so that no carbon diffuses to the right-hand end during the 700°C anneal. Notice that these conditions are identical to conditions 6.10 with $C_0/2$ replaced by C_s. Hence the solution to our carburizing problem is

$$C(Z, t) = C_s[1 - \text{erf } Z/2\sqrt{Dt}] \qquad (6.17)$$

This equation is restricted to carburization of a rod that contains no carbon. If the rod to be carburized already contains a uniform composition of carbon less than C_s, call it C_i, then you may show yourself from Eq. 6.15 that the solution is

$$C(Z, t) = C_s\left[1 - \left(1 - \frac{C_i}{C_s}\right)\text{erf } Z/2\sqrt{Dt}\right] \qquad (6.18)$$

Suppose we define the *case depth* as the depth into the bar that contains a carbon concentration above some arbitrary value C_c. The case depths at the three times t_1, t_2, and t_3 of Fig. 6.5(a) are shown on the figure for the condition of $C_c = C_s/2$. Suppose we ask: How far does the case depth extend into the bar as a function of time for the condition $C_c = C_s/2$. Assuming the bar originally contained no carbon we write Eq. 6.17 as

$$\frac{C_s}{2} = C_s\left[1 - \text{erf}\left(\frac{Z_{0.5}}{2\sqrt{Dt}}\right)\right] \qquad (6.19)$$

where $Z_{0.5}$ is the case depth for $C_c = 0.5C_s$. Rearranging we have,

$$\tfrac{1}{2} = \text{erf}\left(\frac{Z_{0.5}}{2\sqrt{Dt}}\right) \qquad (6.20)$$

From Table 6.1 we find $\text{erf } 0.477 = \tfrac{1}{2}$ so that we obtain

$$Z_{0.5} = 0.954\sqrt{Dt} \qquad (6.21)$$

If we had taken $C_c = 0.25C_s$ we would have found the same result with the constant 0.954 replaced by 1.6. The general result may be written as

$$Z_{C_c} = \text{const} \cdot \sqrt{Dt} \qquad (6.22)$$

This is a very significant result because it shows that the thickness of a "case depth" is proportional to $\sqrt{Dt}$. Often in considering annealing processes it is desirable to be able to estimate how far atoms will move by diffusion in a given time. In the absence of a solution to Fick's second law for the problem at hand one may estimate the diffusion distance simply as $\sqrt{Dt}$ for a reasonable first-order approximation.

Suppose one had heated the iron rod of Fig. 6.5 to a temperature T_1

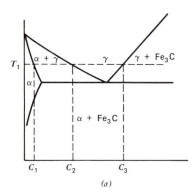
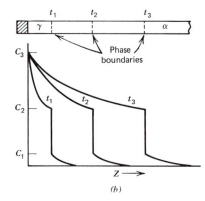

(a) (b)

Figure 6.6 (a) Pertinent portion of iron–carbon phase diagram. (b) Composition profiles for carburizing iron above the eutectoid temperature.

above the eutectoid temperature as shown on Fig. 6.6(a). In this case a very interesting fact is observed when the rod is examined. A sharp phase boundary is found to move down the bar as diffusion proceeds. The local equilibrium at the graphite–iron interface fixed the composition at the left end of the bar at C_3. At temperature T_1 iron with a carbon content between C_3 and C_2 must be γ iron (fcc) and iron with $C < C_1$ must be α iron (bcc). Since the carbon content at the surface is fixed at C_3 we must have γ iron at the iron–graphite interface and we must have α iron far from the interface where $C < C_1$. Notice on the phase diagram that at temperature T_1 it is not possible to have iron containing a composition between C_1 and C_2. C_1 is the maximum amount of carbon the α iron will dissolve and C_2 is the minimum amount γ iron will hold. However, the iron bar could have an *average* composition between C_1 and C_2 if it consisted of a mixture of α iron ($C = C_1$) and γ iron ($C = C_2$). Hence, on a plot of composition versus distance as shown in Fig. 6.6(b) one could only have compositions between C_1 and C_2 if a two-phase mixture were to occur. However, it is found that: *Two phase regions never form in diffusion couples.* Consequently, one always observes sharp phase boundaries in diffusion couples. This is illustrated for you in the photograph of Problem 6.7. These phase boundaries are points of *local equilibrium*, since the two phases contacting each other are essentially in equilibrium with each other at the temperature of the experiment. The phase boundary moves down the rod, and one may calculate the rate of this motion from a solution of Fick's second law for this problem.[3] This solution is quite complex and will not be discussed here. Later we will discuss why two phase regions do not form in diffusion couples. The student is urged to work Problems 6.1 and 6.2 to appreciate better the application of the principles of diffusion to this important practical problem of carburizing.

B. SUBSTITUTIONAL DIFFUSION

In the example of Figs. 6.2 and 6.4 we considered the diffusion of C in Fe, which is an example involving the diffusion of an interstitial solute. In this case we made no mention of a diffusive motion of the Fe atoms because any such motion is negligible compared to the diffusive motion of the smaller and more mobile C atoms. Suppose, however, the diffusion couple were made of Cu and Ni as shown in Fig. 6.7(a). These atoms are nearly the same size and so they dissolve in each other as substitutional solutes and one expects their mobility to be about the same order of magnitude. Consequently, we must consider both the diffusion of Cu to the right and the diffusion of Ni to the left. In general, substitutional solutes do *not* diffuse into each other at equal and opposite rates. Suppose that the Ni atoms diffuse to the left faster than the Cu atoms diffuse to the right. To assist us in determining what effect this relative motion has upon the diffusion couple we place inert markers on the weld interface. These markers could be an inert material such as Mo or Ta wires, or oxide particles, or even the small voids generated by the welding process at the interface. After diffusion has occurred for a number of hours we will have produced a net transport of atoms from the right of the markers to their left because the Ni atoms are moving faster. The extra atoms that arrive at the left-hand side of the markers will cause the lattice to expand on the left, whereas the loss of atoms from the right-hand side will cause the lattice on the right to shrink. Consequently, the entire center section of the bar will shift to the right as shown in Fig. 6.7(b) as diffusion causes atoms to be deposited on the left and removed from the right. Hence, if the atoms move at different rates one expects to see a shift of the markers relative to the ends of the bar as shown in Fig. 6.7(b). Such a shift does occur. It was first reported in metals by Kirkendall and it has come to be called the Kirkendall effect. The occurrence of this shift means that the entire crystal lattice is actually moving with respect to the observer during the diffusion process. This is a type of bulk motion similar to convective motion in liquids and we must take this into account in analyzing the diffusion process occurring here. Such an analysis was first done for alloys in 1948 by L. Darken.[4]

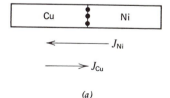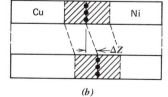

(a) (b)

Figure 6.7 The Kirkendall shift.

Let

$$v_B = \text{bulk velocity of the lattice}$$
$$= \text{velocity of the markers} = v_m$$
$$v_D = \text{velocity due to diffusion alone}$$
$$= \text{velocity of atoms relative to markers}$$

Then we may write $v_{total} = v_B + v_D = v_m + v_D$. Earlier we defined a flux of atoms i as the rate of transport of i per unit area in atoms per sec/cm^2. One may show that *if the atoms i have a volume concentration, C_i, and move at velocity v their flux may be written as $C_i v$*. This is a very useful result that we will have occasion to use many times. Hence, the total flux of atoms 1 with respect to an observer is simply $C_1[v_m + (v_D)_1]$. But $C_1(v_D)_1$ is simply the diffusive flux of atoms 1, which we may write as $-D_1\, dC_1/dZ$. We have for both components

$$(J_1)_T = C_1 v_m - D_1 \frac{dC_1}{dZ}$$

$$(J_2)_T = C_2 v_m - D_2 \frac{dC_2}{dZ} \tag{6.23}$$

Darken now assumed that the molar density (atoms/cm^3) remained constant, which turns out to be a good assumption here. This requires $(J_1)_T = -(J_2)_T$ since the diffusion process may not change the number of atoms per unit volume. By equating Eqs. 6.23, recognizing that, (a) volume concentration i equals molar density times atom fraction i, and (b) molar density is constant, we obtain

$$v_m = (D_1 - D_2) \frac{dx_1}{dZ} \tag{6.24}$$

where x_1 is the mole (or atom) fraction. Substituting this equation back into Eqs. 6.23 we obtain

$$(J_1)_T = -(D_1 x_2 + D_2 x_1) \frac{dC_1}{dZ} = -\bar{D} \frac{dC_1}{dZ}$$

$$(J_2)_T = -(D_1 x_2 + D_2 x_1) \frac{dC_2}{dZ} = -\bar{D} \frac{dC_2}{dZ} \tag{6.25}$$

This result shows that we may analyze the diffusion process of Fig. 6.7 with Fick's first law even though a bulk motion of the rod occurs. However, the quantity we measure as $\bar{D}$ is not a simple diffusion coefficient but is related to the simple diffusion coefficients as shown in Eqs. 6.25. The quantity $\bar{D}$ is called the *mutual diffusion coefficient*. By measuring the velocity of the markers and $\bar{D}$ one may calculate the *intrinsic diffusion coefficients* D_1 and D_2 from Eqs. 6.24 and 6.25.

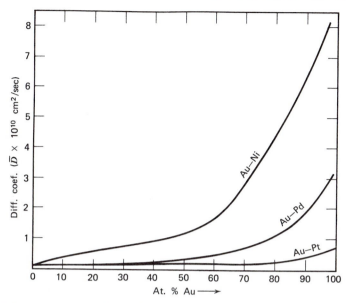

Figure 6.8 Concentration dependence of diffusion coefficient in some gold alloys. (From Ref. 5, used with permission of Springer-Verlag.)

Many experiments have been done and a great deal has been written on the Kirkendall effect, most of which is mainly of theoretical interest.[4,5] From a practical point of view one most often wants to know $\bar{D}$ since this gives us the total flux relative to the observer. When one does a Grube analysis on substitutional solute diffusion, the measured diffusion coefficient is actually $\bar{D}$. It is possible to measure $\bar{D}$ by a method called the Matano interface technique,[3–6] which allows one to relax the assumption that D is independent of concentration. This technique is generally used in studies on substitutional diffusion; it allows one to determine the concentration dependence of the diffusion coefficient. Summaries of experimental results on the concentration dependence of $\bar{D}$ in many alloys may be found in Chapter 5 of Ref. 3 and Chapter 6 of Ref. 5. Figure 6.8 shows a typical result for the variation of $\bar{D}$ with concentration.[5] These results show that if diffusion is occurring over a wide concentration range one must be very careful about assuming $\bar{D}$ to be constant in solving Fick's second law.

C. DRIVING FORCE FOR DIFFUSION

Fick's first law of diffusion was formulated on an empirical basis. We would like to consider the diffusion process from a more fundamental approach, but in order to do so we must first determine an expression for the force

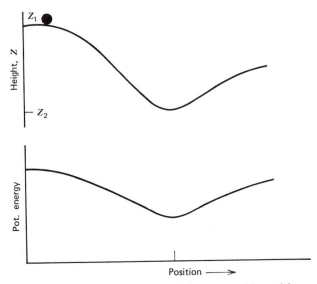

Figure 6.9 Variation of potential energy with position.

that causes diffusion to occur. We are used to thinking of simple mechanical forces or of field forces such as the electrical, magnetic, or gravitational force. Here, however, we are dealing with what might be called a chemical force. It arises as a result of the interatomic forces in a very complex way, and it may not be determined from a simple force equation as the above field forces may. Consider the following analogy. The upper part of Fig. 6.9 shows a ball on a hill. We know that the ball experiences a force from the gravitational field, and from physics we know the force function to be

$$F_{\text{down}} = \frac{kM_{\text{earth}} \cdot m_{\text{ball}}}{Z^2} = \frac{\text{constant}}{Z^2} \tag{6.26}$$

The potential energy is now determined as a function of height, Z, by simply integrating this function as

$$\text{P.E.} = \int_{\infty}^{Z} F_{\text{down}} \, dZ = -\frac{\text{constant}}{Z} \tag{6.27}$$

From Eq. 6.27 we may now determine the potential energy at any position along the hill and so we are able to construct a map of the potential energy as shown in the lower part of Fig. 6.9.

Now suppose that the lower part of Fig. 6.9 was all the information that you had, and you desired to know the downward force, F_{down}, upon the ball.

It should be clear from above that you could write

$$F_{\text{down}} = \frac{d(\text{P.E.})}{dZ} \qquad (6.28)$$

From Fig. 6.9 you could determine that P.E. $= -\text{const}/Z$ and from Eq. 6.28 you could then determine the force function by differentiation. We want to emphasize two points about this example. (1) The ball will seek the position of lowest potential energy and (2) the force on the ball is given by the derivative of the potential-energy function, Eq. 6.28.

From thermodynamics we know that systems at constant temperature and pressure seek the lowest Gibbs free energy of the system. To consider the free energy per atom of a certain type in the system we use the partial molal Gibbs free energy, which is often called the chemical potential and is defined as

$$\text{Chemical potential of element } i \equiv \left(\frac{\partial G}{\partial n_i}\right)_{T,P,n_j} = \mu_i \qquad (6.29)$$

where G is the free energy of the system considered, n_i is the number of i atoms, and n_j is the number of other atoms. The definition of chemical potentials, μ_i, appears quite formal but it may be shown that, physically, μ_i is simply the Gibbs free energy of an i atom when it is in solution in the alloy.

By analogy to the above example we now have

$$\text{Chemical force per } i \text{ atom in } Z \text{ direction} = -\frac{\partial \mu_i}{\partial Z} \qquad (6.30)$$

The minus sign did not appear in Eq. 6.28 because the integral of Eq. 6.27 is taken by convention from infinity to Z. Equation 6.30 is a fundamental result showing that whenever there is a gradient in the chemical potential of atoms i in an alloy, these atoms will experience a force that may cause them to move.

D. MOBILITY AND DIFFUSION COEFFICIENT

In discussing diffusion it is quite useful to introduce the concept of mobility. Consider the force balance upon a parachute as shown in Fig. 6.10(a). The man of mass m is in a gravitational field and so he experiences a downward gravitational force, F_g, proportional to his mass, $F_g = mg$, where g is the gravitational acceleration. Initially after his jump there is zero force holding the man up. However, as his downward velocity increases the air colliding with his parachute produces an upward drag force that will be proportional to his velocity, $F_d = kv$, so that we have for

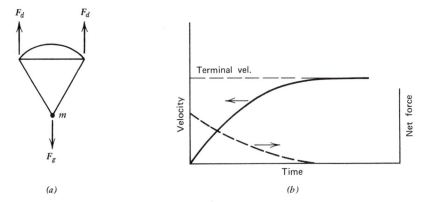

Figure 6.10 Attainment of a terminal velocity.

the net force

$$F_{net} = F_g - F_d = mg - kv = ma \tag{6.31}$$

At sufficiently high velocity the drag force equals the gravitational force. At this point the net force is zero and, hence, the acceleration is zero. Consequently, as shown on Fig. 6.10(b) a terminal velocity is reached at this point of force balance. The *mobility* is defined as

$$\text{Mobility} \equiv \frac{\text{Terminal velocity}}{\text{Unit applied force}} = B \tag{6.32}$$

As an example consider an electron moving through a metal under an applied electric field E. One may consider the electron to move with a terminal velocity that is achieved when the drag force of the lattice balances the applied force of the electric field. The resistivity, ρ, gives a measure of the drag force and the applied force is simply eE, where e is the electron charge. As a simple exercise you may show yourself using Ohm's law,

$$B = \frac{1}{\rho n e^2} \tag{6.33}$$

where n is the volume concentration of electrons.

Equation 6.33 relates the mobility of the electron to the resistivity. Similarly, we may consider the mobility of an atom under a chemical force and relate this mobility to the diffusion coefficient. As previously mentioned we may write the flux of component i as a product of volume concentration and velocity, $J_i = C_i v_i$. From the definition of mobility we have for the velocity $v_i = B_i F_i$ where F_i is the force on the atoms i. Using Eq.

6.30 we obtain

$$J_i = -C_i B_i \frac{\partial \mu_i}{\partial Z} \qquad (6.34)$$

In writing this equation we have assumed that there is no net force exerted on the atoms i by the flow of additional solutes in the alloy.[4] This assumption is not required in binary alloys and it is generally a good approximation in higher-order systems. From thermodynamics we have for the chemical potential,

$$d\mu_i = kT \, d \ln a_i \qquad (6.35)$$

where a_i is the activity of atoms i. Substituting into Eq. 6.34 and equating to Fick's first law we have

$$J_i = -C_i B_i kT \frac{d \ln a_i}{dZ} = -D_i \frac{dC_i}{dZ} \qquad (6.36)$$

By algebraic simplification we determine the relation between the diffusion coefficient, D_i, and the mobility, B_i, as

$$D_i = B_i kT \frac{d \ln a_i}{d \ln C_i} \qquad (6.37)$$

The activity is usually related to the atom fraction concentration, x_i, as $a_i = \gamma_i x_i$, where γ_i is called an activity coefficient. Assuming constant molar density, Eq. 6.37 becomes

$$D_i = B_i kT \left[1 + \frac{d \ln \gamma_i}{d \ln x_i} \right] \qquad (6.38)$$

In ideal solutions or in dilute solutions, γ_i is constant, so that

$$[D_i = B_i kT]_{\text{ideal or dilute soln}} \qquad (6.39)$$

These equations show that there exists a direct relationship between mobility and diffusion coefficient. If an atom has a high mobility it has a high diffusion coefficient.

The absence of two phase zones in a diffusion couple may be understood quite readily from Eq. 6.34. At any average composition within the two-phase boundaries on the phase diagram the alloy will be composed of two phases each of which has a constant composition independent of the average composition. Consequently, in a two-phase region on a phase diagram the chemical potential, μ_i, is constant. Equation 6.34 shows that if a two-phase region did form, the flux through it would be zero because the chemical potential gradient would be zero in the region. It is a simple matter then to show yourself that if a two-phase region did form in a diffusion couple, diffusion into and out of its boundaries would cause it to disappear because of the fact that diffusion within the region is zero; see Problem 6.5.

E. TEMPERATURE DEPENDENCE

The diffusion coefficient is a very strong function of temperature; it virtually always may be expressed as

$$D = D_0 \exp\left[-\frac{Q}{RT}\right] \tag{6.40}$$

where D_0 is a constant and Q is a constant called the activation energy. The values of D are almost always quoted in cgs units and the units of D are cm^2/sec. Approximate values for the temperature dependence of D are shown in Fig. 6.11. Note the following: (1) The diffusion coefficients of interstitial solutes are significantly higher than for substitutional solutes. (2) The diffusion coefficient in the solid at the melting point (along solidus) is nearly the same for different alloys, and similarly, the diffusion coefficient in the liquid at the freezing temperature (along liquidus) is nearly the same for different alloys; see Fig. 6.11 for typical D values. (3) The temperature dependence is very strong, with high-melting metals having the higher room-temperature D values shown on Fig. 6.11 and low-melting metals having the lower D values shown. Extensive data on the temperature dependence of D may be found in Chapter 5 of Ref. 3, and in Chapter 4 of Ref. 5.

F. INTERFACE DIFFUSION

Figure 6.12(a) shows that in polycrystalline metals diffusion may occur along the grain boundaries and the surface as well as through the volume of

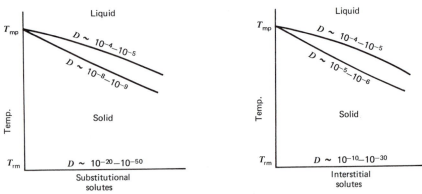

Figure 6.11 Range of diffusion coefficient values near liquidus, solidus, and room temperatures.

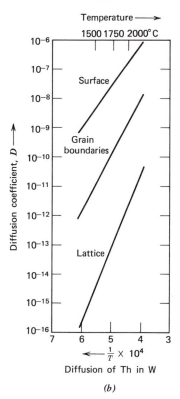

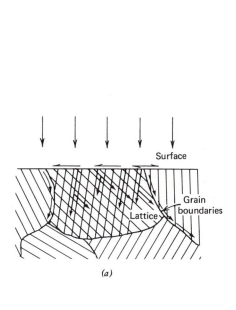

(a)

(b)

Figure 6.12 Lattice, grain boundary, and surface diffusion. (From Ref. 5, used with permission of Springer-Verlag.)

the grains. In recent years many experimental measurements have been conducted on surface diffusion and grain-boundary diffusion. One expects the mobility of an atom along a grain boundary or a surface to be higher than through the crystal volume because these interfaces have a more open structure and, hence, should offer less resistance to atom motion. Consequently, one expects interface diffusion coefficients to be higher than volume diffusion coefficients since the diffusion coefficient is directly related to mobility, Eq. 6.39. Experimental results bear this out and Fig. 6.12(b) presents some typical experimental data. Grain-boundary diffusion makes a significant contribution to the total diffusion only when the grain size is quite small.

6.2 ATOMISTIC APPROACH

As we will see later the atoms of a metal are not fixed on their lattice sites but rather they continually move about. Consequently, if a concentration

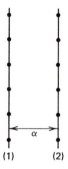

(1)   (2)

***Figure* 6.13**   Lattice plane spacing for diffusion model.

gradient exists in a metal the random motion of the atoms will eventually cause the gradient to disappear. Therefore, we may relate the diffusion coefficient of an atom to its jumping properties in the crystal. Consider two neighboring planes in a set of occupied (hkl) planes as shown in Fig. 6.13 where the spacing between the planes is called α. We now define the following three terms,

(a) Γ = jump frequency
  = No. times per second a given atom jumps to a neighboring position
(b) p = probability that any jump of an average atom on plane 1 will carry it to plane 2
  = fraction of jumps going from plane 1 to plane 2
(c) n_1, n_2 = No. atoms/cm^2 on planes 1 and 2.

An individual atom will jump from plane 1 to plane 2 $p\Gamma$ times per second. Consequently we may write

No. atoms/cm^2 jumping $1 \rightarrow 2$ in time $\delta t = n_1 \cdot (p\Gamma) \cdot \delta t$

No. atoms/cm^2 jumping $2 \rightarrow 1$ in time $\delta t = n_2 \cdot (p\Gamma) \cdot \delta t$

Net No. atoms/cm^2 jumping $1 \rightarrow 2$ in time $\delta t = (n_1 - n_2) \cdot (p\Gamma) \cdot \delta t$

If J is the flux of atoms from 1 to 2, in atom/sec-cm^2, we have

$$J \, \delta t = (n_1 - n_2) \cdot p \cdot \Gamma \cdot \delta t \qquad (6.41)$$

The volume concentration at plane 2, C_2, may be written as

$$C_2 = C_1 + \frac{\partial C}{\partial Z} \cdot \alpha \qquad (6.42)$$

where Z is the distance normal to the planes shown in Fig. 6.13. The volume concentration, C, may be related to the area concentration, n, as,

$C = n/\alpha$. Substituting this relation we obtain

$$n_2 - n_1 = \alpha^2 \frac{\partial C}{\partial Z} \tag{6.43}$$

Combining Eqs. 6.41 and 6.43 we obtain the following expression for the flux:

$$J = -\alpha^2 \cdot p \cdot \Gamma \cdot \frac{\partial C}{\partial Z} \tag{6.44}$$

Comparing this equation with Fick's first law the final result is obtained,

$$D = \alpha^2 \cdot p \cdot \Gamma \tag{6.45}$$

To determine the physical significance of Eq. 6.45 we will consider the diffusion of carbon in iron. It is given as Problem 6.4 to show that for an interstitial atom diffusing in an fcc lattice the diffusion coefficient is

$$D = \frac{a^2 \Gamma}{12} \tag{6.46}$$

where a is the lattice parameter. Taking D for carbon in austenite from the data in Problem 6.1 we obtain at 925°C

$$\Gamma_{925} = 1.7 \times 10^9 \text{ jumps/sec}$$

At room temperature we find the following value of Γ for carbon in retained austenite:

$$\Gamma_{20°C} = 2.1 \times 10^{-9} \text{ jumps/sec}$$

This result points up two very interesting things: (1) At high temperatures the interstitial carbon atoms are changing their lattice positions at a fantastic rate, on the order of a billion times a second, and (2) the jump frequency is extremely sensitive to temperature. It is apparent from the above analysis that the basic mechanism of the diffusion process is closely related to the jumping characteristics of the atoms involved. Consequently, if one carries out a statistical analysis relating the net motion of an atom to its individual jumping characteristics it is possible to gain further insight into the diffusion problem. Consider an atom located at some position on a crystal lattice called position zero. Now,

1. Allow the atom to make jumps of length r only.
2. Assume jumps in any direction are equally probable, that is, each jump independent of preceding jump.
3. Let R_n be the net displacement of the atom from position zero after n jumps.

We now ask: How does R_n vary with n? Since the jumps are random, one might expect R_n to approach zero as n becomes sufficiently large. However, statistical analysis shows that this is not true and one finds[4]

$$\overline{R_n^2} = nr^2 \qquad (6.47)$$

where $\sqrt{\overline{R_n^2}}$ is the root-mean-square displacement. Hence, we see that $R_n = \sqrt{n} \cdot r$, which is considerably different from zero. This problem is called the random-walk problem and it is discussed in detail in Ref. 4, p. 47.

Consider the diffusion of carbon atoms in a rod of γ iron. In this case the jump distance, r, is related to α as $r^2 = 2\alpha^2$, and p is $\frac{1}{3}$ so that we obtain from Eq. 6.45 and Eq. 6.47

$$r^2 = \frac{6D}{\Gamma} = \frac{\overline{R_n^2}}{n} \qquad (6.48)$$

The ratio n/Γ is the number of jumps divided by the jumps per sec which is simply the time, t, so that we may write Eq. 6.48 as

$$\overline{R_n^2} = 6Dt, \qquad (R_n)_{R.M.} = 2.45\sqrt{Dt} \qquad (6.49)$$

This result shows that the root mean displacement is proportional to $\sqrt{Dt}$. Note the similarity of this result with the result obtained earlier for the problem of the case depth upon carburizing: both go as the $\sqrt{Dt}$. It is quite interesting to examine the physical significance of Eq. 6.49. Suppose we ask: What is the root mean displacement of a carbon atom in γ iron after 4 hours. The result is given at 925°C and at room temperature in Table 6.2. The total distance that the carbon atom moved may be calculated by multiplying the jump distance by the total number of jumps, and these results are also included in Table 6.2. Two results are apparent from the data of this table: (1) The root mean displacement is quite sensitive to temperature, and (2) the atoms must travel a tremendous total distance in order to obtain a significant root-mean-square displacement.

Table 6.2 Migration Distances of Carbon Atoms in γ Iron for 4 Hours

Temp. (°C)	$(R_n)_{R.M.}$	Total Distance Moved
925	1.3 mm	3.9 miles (6.3 km)
20	1.4×10^{-9} mm	zero

A.   DIFFUSION MECHANISMS

There are a number of different diffusion mechanisms that have been postulated.[4] The two most important mechanisms, interstitial and vacancy diffusion will be discussed here. In interstitial diffusion the atoms simply jump through the interstitial voids in the lattice. This mechanism generally occurs for small atoms in metals such as C, O, N, and H. In vacancy diffusion the atoms migrate by jumping into near-neighbor vacancies. This mechanism occurs predominately for substitutional solutes and also for self-diffusion. The self-diffusion of metal A is measured by observing the migration of an isotope of element A in a crystal of metal A. Self-diffusion lends itself more easily to theoretical interpretation since one does not have to worry about chemical interactions between solute and solvent atoms. Note that in vacancy diffusion one has a vacancy flux equal and opposite to the atom flux. It was pointed out earlier that a Kirkendall shift occurs when the fluxes of the two substitutional species are not equal. Since these fluxes occur by a vacancy mechanism, there must exist a net vacancy flux in a direction opposite to the net atom flux. In Fig. 6.7(a), for example, a net vacancy flux would occur to the right, toward Ni. This requires that vacancies be continually generated on the left in the Cu and annihilated on the right in the Ni. This is a very interesting conclusion, and one may find a more extensive discussion of it in Ref. 4. One of the mechanisms proposed for the continual generation of vacancies is a Frank–Read generation of edge dislocations where the edge dislocations are always moving by climb.[4]

B.   RATE DATA AND THE ARRHENIUS EQUATION

Quite often in physical metallurgy it is necessary to be able to describe the rate of some process in an alloy as a function of temperature. For example, in heat-treating operations it is necessary to have a knowledge of the rate of grain growth as a function of temperature in order to avoid excessive grain growth; the rate of creep at high temperatures in alloys used in turbine blades is a critical property in the development of useful alloys for this purpose; the rate of diffusion as a function of temperature is a critical factor in the control of the carburizing process, and so on. It is almost always found that if one plots the log of the rate concerned as an inverse function of temperature a straight line is observed as shown in Fig. 6.14. The equation for this straight-line function is called the Arrhenius equation and it is written as

$$R = Ae^{-Q/RT} \tag{6.50}$$

where A is the intercept and Q is called the activation energy. This

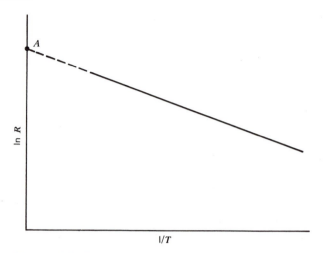

Figure 6.14   The Arrhenius plot.

empirical equation will frequently be encountered when considering the temperature dependence of solid-state processes.

From a comparison of Eqs. 6.40 and 6.50 it is apparent that the diffusion coefficient follows an Arrhenius equation. Using the atomistic model to describe the diffusion process, one can arrive at a physical interpretation for the activation energy, which is determined from the Arrhenius plot. We will consider two cases depending on whether the diffusing atoms are interstitial or substitutional.

C.   INTERSTITIAL DIFFUSION

In this case the solute is an interstitial atom that migrates by jumping between the interstitial voids, for example, carbon in iron. The first question we ask ourselves is how may we determine an expression for Γ, the rate at which an atom changes its position in the lattice. Figure 6.15 shows a plot of the free energy of a solute atom as a function of its position in the lattice. At the two lattice positions shown, 1 and 2, its free energy is a minimum. We now define the following terms:

1. f = fraction of atoms at any time having sufficient energy to change position, that is, having $G > G_2$.

2. Z = number of nearest-neighbor interstitial voids around each solute atom. We assume they are all unoccupied.

3. ν = frequency of vibration toward *each* of these Z voids.

Our problem is first to calculate Γ, the rate at which any given solute atom changes position. If a solute atom were to change positions on every vibration we would have $\Gamma = \nu \cdot Z$. However, an atom will change position only if it has sufficient energy to make the jump. We will assume that the probability that an atom has sufficient energy to jump, $G > G_2$, is given by f. Hence we have

$$\Gamma = \nu \cdot Z \cdot f \tag{6.51}$$

Again, in order to proceed it is necessary to call upon studies of statistical mechanics, which show that the distribution of free energies over the atoms follows a Maxwell–Boltzman law. Accordingly we have for the fraction of atoms with free energy $G > G_i$

$$\frac{n(G > G_i)}{N} = e^{-G_i/kT}$$

where N refers to the total number of atoms. Therefore, we may write

$$\frac{n(G > G_2)}{n(G > G_1)} = e^{-(G_2 - G_1)/kT} = e^{-\Delta G/kT} \tag{6.52}$$

where G_2 and G_1 are defined in Fig. 6.15. G_1 is the free energy of the atom when it lies directly on the lattice site, and hence it is the minimum free energy. Therefore all atoms have a free energy $G > G_1$. Consequently Eq. 6.52 gives us the value of f,

$$f = \frac{n(G > G_2)}{N} = e^{-\Delta G/kT} \tag{6.53}$$

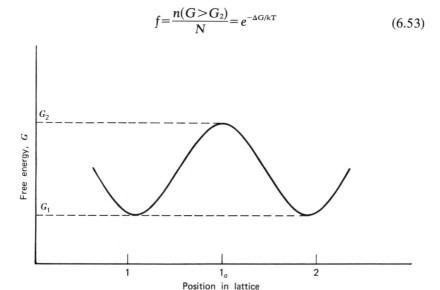

Figure 6.15 Variation of free energy of an atom with position in the lattice.

Now by combining Eqs. 6.45, 6.51, and 6.53 we obtain for D

$$D = \alpha^2 p(Zve^{\Delta S/k})e^{-\Delta E/kT} \tag{6.54}$$

where we have taken $\Delta G \approx \Delta E - T\,\Delta S$. In general, ΔS is not very temperature dependent so that the only temperature-sensitive term in Eq. 6.54 is the final term. Hence, by comparing Eq. 6.54 to Eq. 6.50 one sees that the activation energy Q is identical to ΔE. The above analysis shows that ΔE is the energy difference for an atom placed at position 1a and position 1 of Fig. 6.15. In order to move from position 1 to 2 the atom must pass through the maximum energy position at 1a. The increase in energy required to move the atom to this position is called the activation energy, for obvious reasons. Hence in this case Q is the true activation energy, ΔE, for the process that we are considering, atom migration. In general, the Q of an Arrhenius equation will not correspond directly to the true activation energy for the rate process being considered. This may be demonstrated for our second case.

D.   SUBSTITUTIONAL DIFFUSION

In this case the solute atom is a substitutional solute and, hence, is constrained to move on the lattice sites. We now let Z be the number of nearest-neighbor lattice sites. A significant difference occurs in this case because in order for a jump to occur to a near-neighbor site the site must be vacant. We take the number of near-neighbor sites that are vacant as Z times the fraction of all sites that are vacant. This latter expression is given by Eq. 5.9. Hence for this case Eq. 6.51 becomes

$$\Gamma = v \cdot Z \cdot e^{-\Delta E_v/kT} \cdot e^{\Delta S_v/k} \cdot f \tag{6.55}$$

where ΔE_v is the energy per vacancy and ΔS_v is the vibrational entropy per vacancy. In this case then the expression for the diffusion coefficient becomes

$$D = \alpha^2 p(Zve^{(\Delta S + \Delta S_v)/k})e^{-(\Delta E + \Delta E_v)/kT} \tag{6.56}$$

Consequently we obtain $Q = \Delta E + \Delta E_v$. Therefore Q is not simply an activation energy but in this case it is the sum of the true activation energy and the energy per vacancy. Q is a true activation energy only for very simple processes, and in general it is best thought of as an empirical constant. The above discussion shows that a statistical interpretation of the diffusion process provides one with a physical insight into the actual atomic processes giving rise to the mass transport produced by diffusion.

REFERENCES

   1.   A. Fick, Poggendorff's Annalen **94,** 59 (1855).

   2.   G. Grube and A. Jedele, Z. Elektrochem. **38,** 799 (1932).

3. W. Jost, *Diffusion in Solids, Liquids, Gases,* Academic, New York, 1952.

4. P. Shewmon, *Diffusion in Solids,* McGraw-Hill, New York, 1963.

5. W. Seith and T. Heumann, *Diffusion in Metals: Exchange Reactions,* Springer-Verlag, New York, 1955. Translation from the German, Office of Technical Services, Dept. of Commerce, Washington 25 D.C., 1962, AEC-TR-4506.

6. R. E. Reed Hill, *Principles of Physical Metallurgy,* Van Nostrand, New York, 1964.

PROBLEMS

6.1 You have a plate of 1010 steel (0.1 wt % carbon) that must act as a bearing surface. To achieve the necessary hardness you decide to carburize the surface and then heat treat. You would like to achieve an as-quenched hardness of at least 60 Rockwell C in the outer 1 mm layer. The as-quenched hardness varies with carbon content as shown below. The hardness falls at high carbon contents due to retained

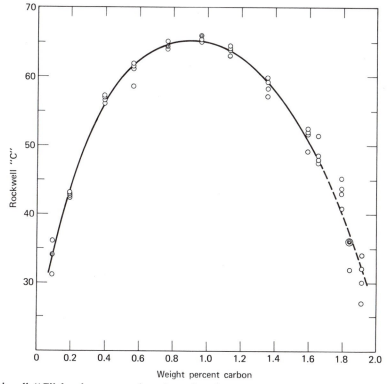

Rockwell "C" hardness as a function of carbon content for as-quenched iron–carbon alloys. (From A. Litwinchuk, "Densities, Microstructures, and Hardness of As-Quenched Iron Carbon Alloys," M.S. thesis, Iowa State Univ. Library, 1973.)

austenite. The diffusion coefficient of C in γ iron goes as $D = 0.12 \exp(-32000/RT)$ where $R = 1.987$ cal/mole-°K and D is in units of cm^2/sec.

(a) You pack carburize (as discussed on p. 144) for 2.4 hours at 1050°C. Using the appropriate equation and the phase diagram of Fig. 9.64 (p. 306), calculate the composition as a function of position in the steel surface region. Now plot the composition at 0.02-cm intervals from the surface to a depth of 1 mm. Below this, plot the as-quenched hardness that would be obtained on heat treatment.

(b) You will notice that the hardness is lower than desired at the surface due to retained austenite. To overcome this problem it is necessary to reduce the carbon content at the surface during carburization. As an alternative to pack carburizing you may gas carburize in an atmosphere of methane or carbon monoxide. By controlling the CH_4/H_2 or CO/CO_2 ratio you can achieve various surface concentrations in gas carburizing below the saturation value shown on the phase diagram.* Suppose you pick a ratio to give a surface concentration of 0.8% C. Now design a carburizing treatment to achieve your objective of a minimum 60 Rockwell C as-quenched hardness in the outer 1 mm layer (that is, specify temperature and time). Note: Do not carburize much above 912°C as excessive grain growth will occur.

(c) If you were to choose a temperature between 727 and 912°C you could not answer part (b) from equations developed in this chapter. Explain why not. If you were to choose a temperature below 727°C you could not achieve your objective. Explain why not.

6.2

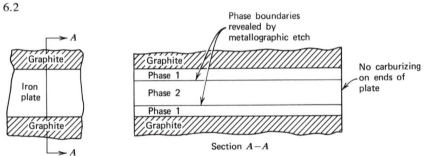

Section $A-A$

An iron plate is packed in graphite as shown above and heated to 740°C for 4 hours. A metallographic examination revealed the above two-phase picture.

(a) Make a plot of how the composition of carbon (in wt %) might vary across the section A-A. Label the phase regions and those compositions you can determine from the phase diagram. (Fig. 9.64, p. 306).

(b) The graphite is now removed from the plate and it is again heated to 740°C. Assume that no decarburization occurs at the surface. On your diagram for (a) above, show how the composition profile would look after many weeks at 740°C (equilibrium). Explain.

(c) If the temperature were now increased to 800°C show how the composition profile would look after equilibrium was again reached.

* It is also possible to reduce the surface concentration below the equilibrium saturation value in pack carburizing, apparently by control of the catalyst.

6.3   Equation 6.8 was derived for the case of unidirectional diffusion along a bar. Suppose a round bar were packed in graphite and diffusion occurred in the radial direction of the bar. As one moves in the radial direction the cross-sectional area of a volume element in cylindrical coordinates is not a constant. Hence, the derivation of Fick's second law (Eq. 6.8) does not apply directly to the case of radial diffusion. Your problem is to derive Fick's second law for the case of radial diffusion; that is, your independent variable will now be the radius, r, rather than distance Z.

6.4   A derivation is presented on pp. 156–157 for the equation $D = \alpha^2 \cdot p \cdot \Gamma$, which relates the diffusion coefficient to a spacing between planes, α, a jump frequency, Γ, and the probability, p, that a jump of an average atom will carry it from plane 1 to plane 2.

(a) Calculate p for a carbon atom moving between octahedral sites in (1) fcc iron, and (2) bcc iron. Note that in bcc crystals the octahedral sites are located at both the face centers and the edge centers of the unit cell.

(b) From your answer show that in terms of their respective lattice parameters,

$$D_{bcc} = \frac{a_{bcc}^2 \Gamma}{24}$$

$$D_{fcc} = \frac{a_{fcc}^2 \Gamma}{12}$$

6.5   It was stated on p. 146 that a two-phase region never appears on a diffusion couple. If a two phase region did appear it might look like the sketch shown here.

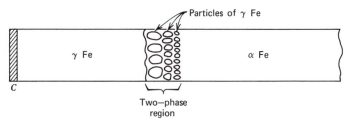

Your problem is to show that the two-phase region will disappear. Make a sketch of the above. Directly underneath this sketch make a plot of how the chemical potential of carbon varies from the left-hand end to the right-hand end of the bar. Remember, the chemical potential is proportional to carbon content and is *constant* in a two-phase region. Now from an analysis of this plot determine how the carbon will diffuse with respect to the two-phase region and explain why the two-phase region will slowly disappear as diffusion continues.

6.6

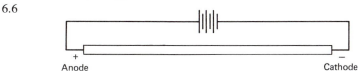

When a dc electric current passes through a rod of steel it causes the carbon atoms to move relative to the iron atoms. This phenomenon is called electrotransport or electromigration. The mobility of carbon atoms due to the electric current has been measured and reported in terms of velocity per unit electric field:

$$\text{Mobility} = U = \frac{\text{velocity}}{\text{elect. field}} = \frac{v}{E}$$

(a) Consider a bar of 1080 steel that is heated to 1000°C and has a dc current, j, passing through it as shown above ($E = j\rho$, where ρ is the resistivity and j is the current density). The dc current causes the carbon atoms to migrate toward the cathode. Draw a graph of carbon concentration versus distance along the rod showing qualitatively how you think the concentration distribution would look after (a) 2 hours and (b) 100 hours.

(b) Write an equation for the flux of carbon atoms through any cross section of the bar at any time. Hint: Your equation should include a term for each mechanism of mass transport, that is, electrotransport and diffusion transport. Your electrotransport term will involve the mobility U.

(c) After a very long period the concentration distribution along the rod will no longer change with time. It will have reached a "steady state." Can you explain why this should happen? That is, why is the electric current (or field) no longer able to push more and more carbon atoms to the cathode end of the bar after a very long time?

(d) From your results of (b) and (c) write a differential equation for this system at steady state.

$$\text{Answer:} \qquad \frac{dC}{C} - \frac{UE}{D}\, dZ = 0$$

6.7 ∎

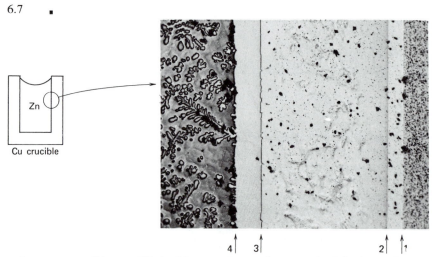

A copper crucible was filled with pure zinc as shown at the left above. It was heated to 500°C, held for 15 hours, and then furnace cooled; it was sectioned in half longitudinally and the surface was polished and etched. The interface region between the Cu and Zn was examined in an optical microscope and a photomicrograph at 33 × from this region is shown above.

Four distinct interfaces were observed. To the right of interface 1 the metal had the color of Cu, between interfaces 1 and 2 the color of brass, and the other regions were colorless. All of the regions are single phase except that to the left of interface 4. (The small black specks on the three phases present to the right of interface 3 are etch pits.)

(a) Using the Cu-Zn phase diagram given on p. 321 identify the phases present in the five regions of the above photomicrograph. Explain your reasoning.

(b) Make a plot of the variation of the zinc concentration from the left end to the right end of the above photomicrograph. Label all concentrations that you can quantitatively specify from the phase diagram.

CHAPTER 7
INTERFACES

Virtually all of the metals with which one comes into daily contact are polycrystalline. Metallurgists refer to the small crystalline regions as grains, and the average grain size in most metal objects ranges from 0.015 to 0.24 mm. As we will see in the later chapters, grains are frequently not perfect single crystals but are divided into smaller subgrains that are nearly perfect single crystals. The grain boundaries and subgrain boundaries are *interfaces* between crystals of different orientation. Since the physical properties of metals are dependent on the physical properties of these interfaces, the study of the nature of metal interfaces is very important in physical metallurgy.

7.1 CLASSIFICATION, GEOMETRY, AND ENERGY OF INTERFACES

In order to describe the geometry of grain boundaries we begin by considering two dimensional lattices. Figure 7.1(*a*) shows two lattices that are oriented at some angle to each other; that angle is indicated as θ. When these two lattices are brought together a grain boundary is formed between them. Figure 7.1(*b*) shows two different ways in which the boundary might form, depending on the angle between the boundary and a plane in one of the lattices. It should be clear from Fig. 7.1 that specifying the angle between the lattices of these two-dimensional crystals, θ, is not sufficient to specify the boundary. To completely define the boundary one must specify:

1. The orientation of one lattice with respect to the other, θ.
2. The orientation of the boundary with respect to a lattice, ϕ.

Since the boundary can be specified with two angles it could be called a two-degree-of-freedom boundary.

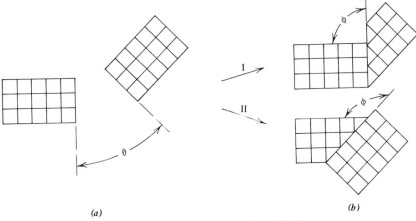

(a)

(b)

Figure 7.1   A grain boundary in a two-dimensional lattice.

To describe a boundary between three-dimensional crystals one must specify both the orientation of the crystals with respect to each other and the orientation of the boundary relative to one of the crystals. Consider the crystal shown in Fig. 7.2(a). Imagine that this crystal is sectioned on the X-Z plane and then the right section is rotated about the X axis to form an orientation mismatch between the two crystals as shown in Fig. 7.2(b). In general one could rotate about each of the three axes, X, Y, and Z so that in order to specify the orientation of two crystals one must specify three angles. Now consider the boundary between two crystals of fixed orientation as shown in Fig. 7.2(b). In this case the boundary lies in the X-Z plane. The boundary plane could be changed by rotating it about either X or Z but not by rotation about Y. Hence, to specify a boundary orientation between two crystals we must specify two angles. Therefore we conclude: The *general grain boundary has five degrees of freedom; three degrees specify*

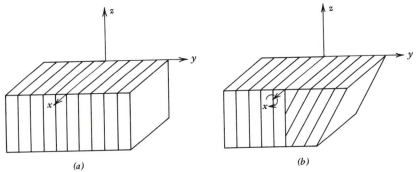

(a)

(b)

Figure 7.2   A grain boundary in a three-dimensional lattice.

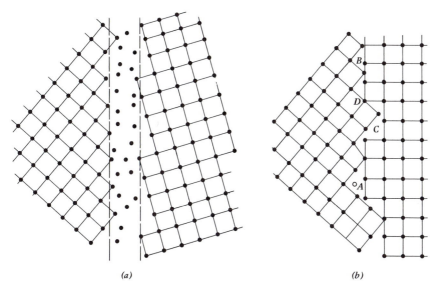

Figure 7.3   Grain-boundary models.

the orientation of one grain relative to the other and two degrees specify the orientation of the boundary relative to one of the grains.

The interfaces that form in crystals because of orientation mismatches between different regions of the crystal are termed grain boundaries. We are also interested in other interfaces that are not grain boundaries such as free surfaces, stacking faults, and antiphase domain boundaries.

Figure 7.3 shows two different models for a grain boundary. Figure 7.3(a) considers that the two nearly perfect crystals extend up to a thin layer of atoms that has an amorphous structure and acts almost as a liquid boundary layer separating the crystals. This is an old model for grain boundaries, stemming from the ideas of Beilby early this century.[1] Recent experimental studies indicate that grain boundaries are more closely represented by the model of Fig. 7.3(b). In this model the nearly perfect crystals extend up to each other and touch at irregular points. The boundary contains atoms that belong to both crystals, D, and atoms belonging to neither crystal, A; it contains compression zones, B, and tensile zones, C. In general, the width of the grain boundary is thought to be quite narrow, only a few angstroms.

A.   SMALL-ANGLE BOUNDARIES

If the orientation mismatch between two crystals is quite small the boundary between the crystals is called a small-angle boundary. Very

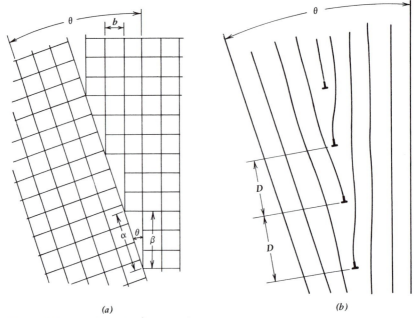

(a)   *(b)*

Figure 7.4 Small-angle tile boundary.

simple small-angle boundaries can be described by dislocation arrangements. Figure 7.4(a) shows two grains brought into contact with each other. Only one of the three degrees of orientation mismatch, θ, is allowed. The structure of Fig. 7.4(a) is an idealized picture of the grain-boundary region. Actually, this region will relax under the high strains due to the mismatch to produce a row of edge dislocations spaced at a distance D as shown in Fig. 7.4(b). From Fig. 7.4(a) one sees that the following relations must hold

$$\sin\theta = \frac{b}{\alpha}, \qquad \tan\theta = \frac{b}{\beta} \tag{7.1}$$

At small values of θ we have $\alpha \to \beta \to D$ and also from Eqs. 7.1, $\theta \to b/\alpha$ and $\theta \to b/\beta$. Therefore we have

$$\theta = \frac{b}{D} \tag{7.2}$$

where b is the Burgers vector of the dislocations. This simple boundary is called a *tilt boundary* and Eq. 7.2 has been well verified by experiment. The tilt boundary is a one-degree-of-freedom boundary.

Figure 7.5 shows an array of edge dislocations lined up to form a tilt boundary. It was shown in Chapter 6 that in this array the dislocations do

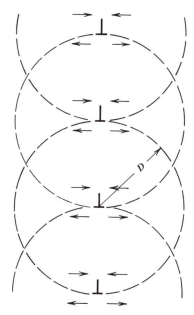

Figure 7.5   Strain field along a tilt boundary.

not exert a force upon each other in the slip vector direction, so one expects it to be stable. It is indicated schematically on Fig. 7.5 that a compressive stress exists above each dislocation and a tensile stress below each dislocation. Alternating tensile and compressive regions exist along the boundary and these stresses will tend to cancel each other. It may be shown[2] that the stress fields of the dislocations essentially cancel each other outside of a circle of radius D. Consequently, there is no long-range stress field produced by the tilt boundary, so that the crystals around the boundary may be stress free beyond distance D.

We now consider the energy of a tilt boundary. Figure 7.6 shows three edge dislocations of a tilt boundary spaced at distance D running in and out of the page. We must determine the energy per unit area of this boundary. The shaded area shown on Fig. 7.6 may be thought of as a "unit cell" area for the boundary because the entire boundary may be generated by stacking such areas together. Therefore, the energy per area of the boundary may be determined from this area. From Chapter 4 we have for the energy per length of the dislocations

$$E = \frac{Gb^2}{4\pi(1-\nu)} \ln \frac{r}{r_0} + C \tag{7.3}$$

The energy of the unit cell area is simply $E \cdot 1$, where E is the energy per

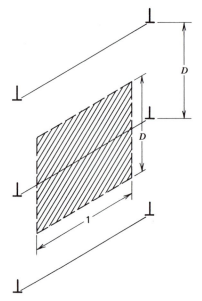

Figure 7.6   Edge dislocations aligned in a tilt boundary.

length of the dislocation which threads this area. Therefore, the energy per area of the boundary is

$$E_B = \frac{E \cdot 1}{1 \cdot D} = \frac{E}{D} \tag{7.4}$$

Combining Eqs. 7.2 and 7.3 with Eq. 7.4 we have

$$E_B = \frac{Gb\theta}{4\pi(1-\nu)} \ln \frac{r}{r_0} + C\frac{\theta}{b} \tag{7.5}$$

Because the stress fields cancel beyond distance D we take $r = D$, and we somewhat arbitrarily choose $r_0 = b$. We then obtain by algebra

$$E_B = E_0\theta[A - \ln \theta] \tag{7.6}$$

where $E_0 = Gb/4\pi(1-\nu)$ and $A = C4\pi(1-\nu)/Gb^2$. A more satisfying derivation of this equation may be found in Ref. 3. This equation is plotted schematically in Fig. 7.7. We would expect this equation to hold only at small values of θ (2°–3°) because (a) θ was assumed small to take the sine as its argument and (b) at large θ the values of D become small, see Eq. 7.2, so that Eq. 7.3 cannot be applied because r approaches r_0.

There is a second type of simple grain boundary which can be described with dislocations. Figure 7.8(a) shows a single crystal rod that has been cut

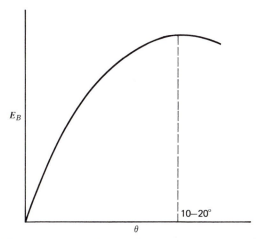

Figure 7.7 Predicted grain boundary energy versus angle of tilt.

along one face. The lower half of the rod is now rotated about the Z axis by an amount θ thereby forming a grain boundary as shown in Fig. 7.8(b). This boundary is called a *twist boundary* and it also is a one-degree-of-freedom boundary. One can show that the twist boundary may be described with a dislocation model consisting of a square grid of crossed screw dislocations.[2,3] The twist boundary is the simplest screw dislocation grain boundary, whereas the tilt boundary is the simplest edge dislocation grain boundary.

Examination of the tilt boundary in Fig. 7.4 reveals that the boundary is a mirror plane between the crystals. If the plane of the boundary is simply

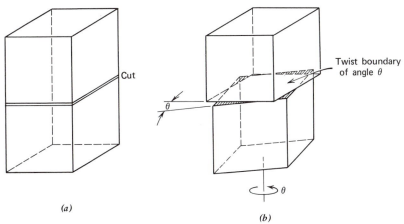

Figure 7.8 The small-angle twist boundary.

rotated off this symmetry position one forms a two-degree-of-freedom tilt boundary. Such a boundary can still be described by a dislocation array.[2,3] However, if the boundary becomes much more complex it is extremely difficult to describe with dislocation models and one generally refers to these more complex boundaries with a nonspecific model such as Fig. 7.3(*b*). Grain boundaries are often classified by a scheme which is given roughly as follows:

Small-angle boundaries	$\theta = 0° \rightarrow 3°$ to $10°$
Medium-angle boundaries	$\theta = 3°$ to $10° \rightarrow 15°$
Large-angle boundaries	$\theta = 15° \rightarrow$

where we are referring to a one-degree-of-freedom boundary, and θ is the mismatch angle.

A number of methods exist for the measurement of grain-boundary energies. These methods are described in Ref. 1 and will not be discussed here. The results of such studies are shown in Fig. 7.9 for a one-degree-of-freedom boundary. Three points seem worth emphasizing about this result.

1. The small-angle formula, Eq. 7.6, gives the proper form (shape) of the E_B versus θ curve all the way up to $15°$–$20°$. However, it does not give the correct absolute values of E_0 and A when fitted to the data beyond $5°$–$6°$.

2. The energies of large-angle grain boundaries are approximately constant at around 500–600 ergs/cm^2.

3. In polycrystalline metals over 90% of all the grain boundaries are high-angle boundaries because the probability that all three orientation angles are low is very small. Consequently, the grain boundary energy in

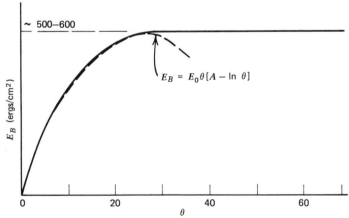

Figure 7.9 Experimental variation of grain-boundary energy with tilt angle.

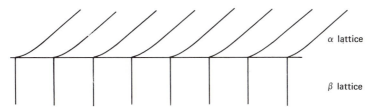

Figure 7.10   A coherent grain boundary.

polycrystalline metals may be taken as constant at around 500–600 ergs/cm^2.

B.   COHERENT BOUNDARIES

This is the name for a special type of boundary having to do with the degree of fit of the lattices across the boundary. Figure 7.10 illustrates a *fully coherent* boundary. There is a one-to-one matching of the lattice planes across the boundary. This generally produces lattice strains around the boundary where the lattice planes must be "bent" to give this one-to-one matching. In an *incoherent boundary* there is no regularity of lattice-plane matching across the boundary. High-angle boundaries are incoherent boundaries. A *partially coherent or semicoherent boundary* is shown in Fig. 7.11. Shown here are two cubic lattices touching on their respective (001) planes. The lattice parameters are taken so that $a_\alpha > a_\beta$. Consequently, an effect similar to a vernier caliper is produced. For the example shown in Fig. 7.11 a perfect lattice matching occurs every six spacings of the β lattice. Notice also that every six spacings one obtains a β plane situated directly between the two lower α planes. In real crystals the boundaries relax under the forces between atoms into the dashed positions of Fig. 7.11, and this leaves an edge dislocation at the point where the β plane is located symmetrically between two α planes. Hence edge dislocations are obtained periodically with a spacing D, where for this example, $D = 6$. Consequently, we may consider a semicoherent boundary as consisting of regions of coherency, A, and regions of disregistry, B, as shown in

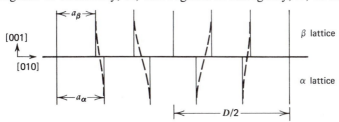

Figure 7.11   A partially coherent grain boundary.

Figure 7.12   Regions of coherency A and noncoherency B in a partially coherent boundary.

Fig. 7.12. A term called the disregistry, δ, may be defined as

$$\delta \equiv \frac{a_\alpha - a_\beta}{a_\alpha}, \qquad a_\alpha > a_\beta \tag{7.7}$$

It is a simple geometrical exercise to show yourself that the spacing between the dislocations, D, is related to the disregistry as

$$D = \frac{a_\beta}{\delta} \tag{7.8}$$

When D becomes sufficiently small one obtains complete disregistry across the boundary and hence an incoherent boundary.

C.   TWIN BOUNDARIES

Twin boundaries are perhaps the simplest of all grain boundaries. One may classify twin boundaries as either coherent or partially coherent. Figure 7.13(a) illustrates a coherent twin boundary. Notice that the twin boundary is a symmetry plane, a mirror plane. Complete coherency at the boundary is obtained without any straining of the lattices because a perfect registry of the lattices is naturally obtained at these special boundaries. If the twin boundary plane rotates off the symmetry plane as shown in Fig. 7.13(b) one obtains a partially coherent twin boundary (usually called a noncoherent twin). It may be seen from Fig. 7.13(b) that the disregistry of the boundary is given by $\delta = (a_1 - a_2)/a_1$, which is a function of the angle ϕ. This boundary is an example of a two-degree-of-freedom boundary.

D.   ENERGY OF INTERFACES

The interface energy is a measure of the energy per area of the interface region minus the energy per area of that region without an interface; that is the energy is defined relative to a perfect lattice. It is frequently useful to consider the interface energy to be partitioned into two forms, strain energy and chemical energy. Figure 7.14 shows a hypothetical energy-versus-distance curve for an atom within a crystal lattice, where positions S are the lattice sites. When an atom is displaced from its lattice site by some force (stress), a strain energy is produced as shown as B on Fig. 7.14. It was

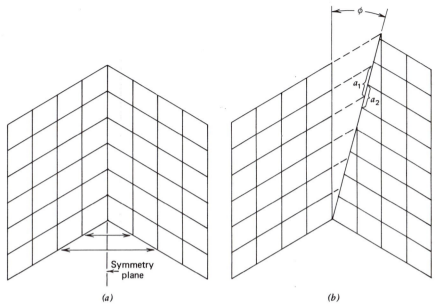

Figure 7.13 (a) A coherent twin boundary. (b) A partially coherent twin boundary.

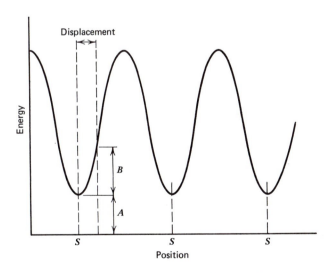

Figure 7.14 Partitioning of energy into chemical energy A and strain energy B.

Table 7.1   Partitioning of Energy for Various Types
of Grain Boundaries

Type of Interface	Strain E	Chem. E
Coherent grain boundary	High	Low
Stacking fault and coherent twin boundary	0	Low
High-angle boundaries	Low	High

shown in Chapter 4, Eq. 4.2, that this energy is given per unit volume as $\frac{1}{2}\sigma\varepsilon$ for uniaxial stress. The chemical energy is shown on Fig. 7.14 as A. This energy may be thought of as arising from the unstrained chemical bonds, and its magnitude depends on the number and strength of these bonds. Consider now the semicoherent interface of Fig. 7.12. In regions A, atom displacements are required in order to produce coherency, see Fig. 7.11, so that here the interface energy is mainly strain energy. However, at regions B the number and strength of the chemical bonds relative to a perfect lattice are affected at the dislocation cores so that here the energy is mainly chemical. One may summarize the partitioning of energies at interfaces as in Table 7.1. The strain energy of the coherent boundary is large because of the atomic displacements required for the one-to-one lattice registry. The chemical energy of the high-angle boundary is large because the number and strength of the chemical bonds to a boundary atom are much different than to an atom not on a boundary.

One may now ask: Would a high-angle boundary or a coherent boundary be more effective at stopping dislocation motion? To stop a dislocation the dislocation must feel a force retarding its motion. In Chapter 4 we showed that stress fields in the crystal produce forces upon dislocations, Eq. 4.10. Since the strain energy of the coherent boundary is large, the corresponding stress field could be effective in stopping dislocations. However, the strain energy of high-angle boundaries is very small, and consequently these boundaries will exert little force on dislocations gliding into them.

7.2   SURFACE TENSION AND SURFACE FREE ENERGY OF INTERFACES

There are three related quantities that one may define for any interface, the surface tension, the surface free energy, and the surface stress. We will now discuss each of these in turn.

1. *Surface tension,* γ: *Work required to* <u>*form*</u> *(create) a unit area of new surface at constant T, V, and* μ_i,

$$\gamma = \left(\frac{dW}{dA}\right)_{T,V,\mu_i} \qquad (7.9)$$

From this definition of the surface tension it is apparent that the surface tension is related to the energy associated with the chemical bonds that must be broken to generate the surface. As an approximation we may write

$$\gamma = \left[\frac{\text{No. of chemical bonds broken}}{\text{per unit area of surface formed}}\right] \cdot [\text{energy/bond}]$$

As an example, take a simple cubic lattice and consider only the bonds between nearest-neighbor atoms. Figure 7.15 shows the location of an (001) and an (011) surface in this crystal. It can be seen that the distance between bonds along the (001) surface is a, while it is only $0.707 \cdot a$ along the (011) surface. Hence, the number of bonds per unit area is higher along the (011) plane so we have $\gamma_{011} > \gamma_{001}$. There are two important conclusions that follow from this discussion:

1. The surface tension in a crystal is anisotropic.

2. It is generally found that the highest atom density planes have the lowest values of γ.

It is sometimes convenient to present a polar plot of the surface tension as is shown in Fig. 7.16 for a (100) plane of the simple cubic lattice. In this plot the radius vector equals the magnitude of the surface tension in the

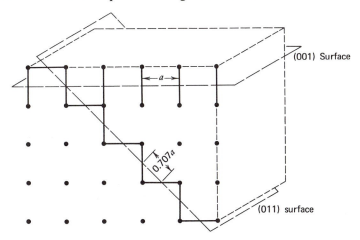

Figure 7.15 Diagram illustrating bonds across different (hkl) planes.

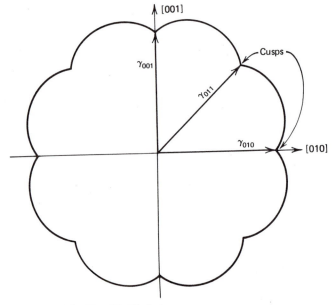

Figure 7.16 The Wulff plot.

plane of the crystal perpendicular to the radius vector. The low-index planes have much smaller surface tensions and the plots show a discontinuity at these planes, which are usually referred to as cusps; see Fig. 7.16. This type of polar plot is often called a Wulff plot and it may be used to determine the equilibrium shapes of surfaces.[4]

2. *Surface free energy: The change in the Helmholtz free energy of the system per unit area of interface generated, dA'/dA.* We use A' for Helmholtz free energy here to distinguish it from interface area, A.

It is shown in Refs. 4 that the relationship between surface tension, γ, and surface free energy is given for pure metals as

$$\gamma = \frac{dA'}{dA} \tag{7.10}$$

and for alloys as

$$\gamma = \frac{dA'}{dA} - \sum_i \mu_i \left(\frac{dn_i}{dA} \right) \qquad \text{at const } V, T, \text{strain} \tag{7.11}$$

where i refers to all components of the alloy. The quantity dn_i/dA refers to the change in the number of atoms of component i in the bulk grains due to a change of grain boundary area A. Equations 7.10 and 7.11 show that the surface tension is equal to the surface free energy in pure metals but not in alloys.

3. *Surface stress: Work required to deform (stretch) a surface.*

It was pointed out in connection with Fig. 4.25 that to describe the state of stress in a material, one must specify three normal stress components and three shear stress components. Similarly, to describe the state of stress on a surface in general, one must specify two normal stress components, f_{xx} and f_{yy}, and one shear stress component, f_{xy}. It is shown in Refs. 4 that the relationship between these surface stress components and the surface tension is given as $f_{xx} = \gamma + d\gamma/d\varepsilon_{xx}$, $f_{yy} = \gamma + d\gamma/d\varepsilon_{yy}$, and $f_{xy} = d\gamma/d\varepsilon_{xy}$, where ε_{xx} and ε_{yy} are normal strain components and ε_{xy} is the shear strain component. If the terms $d\gamma/d\varepsilon_{ij}$ are zero, then surface stress equals surface tension. Such a condition requires that γ not change with the stretching of the surface. As noted above, γ is related to the strength of the chemical bonds of the surface atoms. Therefore, if the surface atoms can maintain their same surface configuration via migration as the surface is stretched, $d\gamma/d\varepsilon_{ij}$ will be zero, and we will have $\gamma = f$. This equality of surface tension and surface stress generally holds in liquids, but it holds in solids only for processes occurring at high temperatures at relatively slow rates.

Although there is a distinction between surface stress and surface tension in solids, as we have pointed out here, the difference in magnitude is generally small, and we shall make little further mention of surface stress; it has been mentioned here mainly for the sake of completeness.

A. EQUILIBRIUM SHAPES OF SURFACES

From thermodynamics we know that at equilibrium the Helmholtz free energy for a system plus its surroundings will be a minimum for processes occurring at constant volume and temperature. Consider a bar of metal containing a single grain boundary separating two grains as shown in Fig. 7.17(a). We consider the region shown around the grain boundary as the system and take the remaining part of the bar as the surroundings. The grain boundary is now imagined to rotate to the position shown in Fig. 7.17(b). The change in the Helmholtz free energy of the system that accompanies this process may be written from Eq. 7.11 as

$$(dA')_{sys} = \gamma \, dA + \sum_i \mu_i (dn_i)_{sys} \qquad (7.12)$$

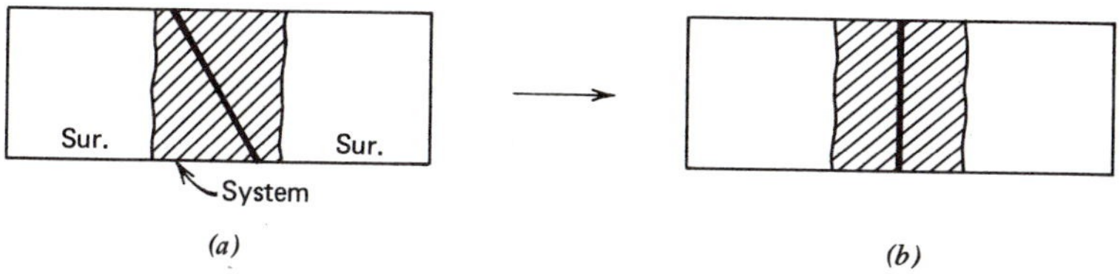

Figure 7.17 Rotation of a grain boundary between two crystals.

where dA is the change in area of the grain boundary, μ_i is the chemical potential of the components and $(dn_i)_{sys}$ is the number of atoms that cross the system boundary due to this process. As we will see later solute atoms frequently tend to absorb on grain boundaries so that when we decrease the boundary area we may cause solute atoms to move into the surroundings. The corresponding change in the Helmholtz free energy for the surroundings is

$$(dA')_{sur} = \sum_i \mu_i (dn_i)_{sur} \qquad (7.13)$$

The $\gamma\, dA$ term is zero because the surroundings contain no boundaries. Since all of the atoms that leave the system due to the area change must go into the surroundings we have $(dn_i)_{sys} = -(dn_i)_{sur}$. Using this relation we obtain by addition of Eqs. 7.12 and 7.13,

$$(dA')_{total} = (dA')_{sys} + (dA')_{sur} = \gamma\, dA \qquad (7.14)$$

Using the principle for thermodynamic equilibrium stated above we have the following condition for equilibrium,

Equilibrium = minimum $\gamma\, dA$ in the system

$$= \text{minimum} \left[\int_{sys\ area} \gamma\, dA \right] \qquad (7.15)$$

This is a very important result, which shows that the function $\gamma\, dA$ may be thought of as a thermodynamic potential that must be a minimum in the system at equilibrium. When γ is a constant the function is simply γA and for simplicity we will refer to this thermodynamic potential in general as γA.

   We are particularly interested in the shape of grain boundaries in metals and so we will now show what restrictions the above result places on grain-boundary junctions. Consider the three grains that meet at the common junction 0 shown in Fig. 7.18. The surface tensions of the grain boundaries between the three grains are all different and are labelled as γ_1, γ_2, and γ_3 on Fig. 7.18. The junction of the three grains at 0 is called a tripole junction, and we wish to determine its equilibrium configuration. To proceed we calculate the change in the thermodynamic potential, γA, of this system for a differential change in the geometrical configuration, and then equate this to zero to determine the minimum value of γA. Specifically we ask: What is the change in γA of this system when the tripole junction undergoes an infinitesimal displacement from 0 to P? With the junction at 0 and considering a unit length of the junction normal to the page, we have

$$\gamma A(\text{Jnct at 0}) = \gamma_1 \cdot o\alpha + \gamma_2 \cdot 0\beta + \gamma_3 \cdot 0\,\delta \qquad (7.16)$$

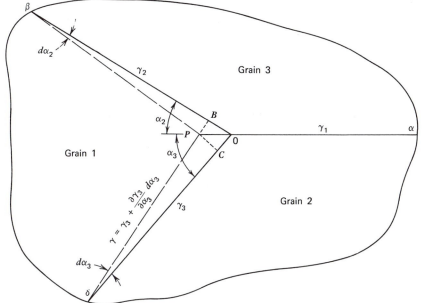

Figure 7.18 A differential displacement of a tripole junction from 0 to P.

When the junction moves from 0 to P the grain boundaries between grains 1 and 2 and between 1 and 3 will both rotate with respect to the grains and consequently both γ_2 and γ_3 may change. We may take this into account by writing the change of γ_3, for example, as $(\partial\gamma_3/\partial\alpha_3)\,d\alpha_3$. Consequently, for the junction at P we obtain

$$\gamma A(\text{Jnct at } P) = \gamma_1 \cdot P\alpha + \left[\gamma_2 + \frac{\partial\gamma_2}{\partial\alpha_2}\,d\alpha_2\right]P\beta + \left[\gamma_3 + \frac{\partial\gamma_3}{\partial\alpha_3}\,d\alpha_3\right]P\delta \quad (7.17)$$

Taking the difference between Eqs. 7.17 and 7.16 we have

$$\Delta(\gamma A) = \gamma_1(P\alpha - 0\alpha) + \gamma_2(P\beta - 0\beta) + P\beta \cdot \frac{\partial\gamma_2}{\partial\alpha_2}\,d\alpha_2$$

$$+ \gamma_3(P\delta - 0\delta) + P\delta\frac{\partial\gamma_3}{\partial\alpha_3}\,d\alpha_3 \quad (7.18)$$

In the limit of an infinitesimal displacement we may write

$$\begin{aligned}
&1. \quad P\beta - 0\beta = -0B = -0P\cos\alpha_2 \\
&2. \quad P\delta - 0\delta = -0C = -0P\cos\alpha_3 \\
&3. \quad PB = P\beta \cdot d\alpha_2 = 0P\sin\alpha_2 \\
&4. \quad PC = P\delta \cdot d\alpha_3 = 0P\sin\alpha_3
\end{aligned} \quad (7.19)$$

where relations 1 and 2 are obtained from simple geometry and relations 3 and 4 follow from the equation for arc lengths, $dS = R\, d\theta$. Substituting Eqs. 7.19 into Eq. 7.18 and equating to zero we obtain the following fundamental equation, which was first presented by Herring[5,6]:

$$\gamma_1 - \gamma_2 \cos \alpha_2 - \gamma_3 \cos \alpha_3 + \frac{\partial \gamma_2}{\partial \alpha_2} \sin \alpha_2 + \frac{\partial \gamma_3}{\partial \alpha_3} \sin \alpha_3 = 0 \qquad (7.20)$$

Thermodynamic equilibrium is obtained at a tripole junction only when Eq. 7.20 is satisfied. We now emphasize three points concerning this equation.

1. The terms $(\partial \gamma_i / \partial \alpha_i) \sin \alpha_i$ are called torque terms. They depend on the variation of γ with boundary orientation and, hence, will be zero when γ is isotropic as, for example, in liquid systems.

2. When the torque terms are zero Eq. 7.20 reduces to the form

$$\gamma_1 - \gamma_2 \cos \alpha_2 - \gamma_3 \cos \alpha_3 = 0 \qquad (7.21)$$

Figure 7.19 shows a "surface tension balance," which you may be familiar with from courses in physical chemistry. Here we are considering the surface tensions as a force per unit length normal to the junction boundary 0, and we see that a simple force balance in the γ_1 direction gives Eq. 7.21. Hence we conclude: If the torque terms are zero or sufficiently small, equilibrium is described by a force balance at the trijunction using the surface tension as the magnitude of force and the boundaries as the directions of the force vectors.

3. Many textbooks present the trijunction balance with the nomenclature of Fig. 7.20, which has some obvious advantages. Using this nomenclature Eq. 7.20 becomes

$$\gamma_{12} + \gamma_{23} \cos \theta_{13} + \gamma_{13} \cos \theta_{23} - \frac{\partial \gamma_{23}}{\partial \theta_{23}} \sin \theta_{23} - \frac{\partial \gamma_{13}}{\partial \theta_{13}} \sin \theta_{13} = 0 \quad (7.22)$$

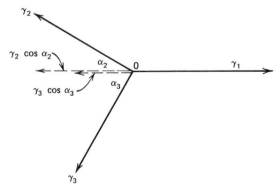

Figure 7.19 A surface tension force balance.

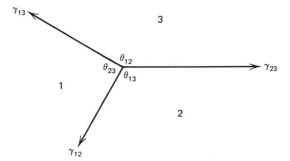

Figure 7.20 Alternative nomenclature used for tripole junction.

Also, if we neglect the torque terms it may be shown from a simple "surface tension force balance" that

$$\frac{\gamma_{12}}{\sin \theta_{12}} = \frac{\gamma_{13}}{\sin \theta_{13}} = \frac{\gamma_{23}}{\sin \theta_{23}} \qquad (7.23)$$

We may estimate the magnitude of the surface tension from the boundary energy. We have seen from Fig. 7.9 that the magnitude of the grain boundary energy is essentially constant for high-angle boundaries. Hence, for high-angle boundaries one expects little change in boundary energy as the boundary plane rotates, so that the torque terms will be essentially zero. Since, as pointed out earlier, over 90% of the boundaries will be high-angle boundaries in polycrystalline metals, the torque terms will frequently be negligible. In some cases, however, the torque terms may be quite important. Figure 7.13 shows a coherent and a partially coherent twin boundary. Table 7.2 gives some data for the energies of these boundaries. If one were to plot the energy of the boundary as a function of the angle ϕ shown in Fig. 7.13 a plot such as Fig. 7.21 would be obtained. It can be seen that the torque term for the coherent twin boundary, $\partial\gamma/\partial\phi$, is very high. When a coherent twin boundary rotates off the symmetry plane it will experience a force tending to rotate it back, and the torque terms of Eqs. 7.20 and 7.22 account for this force. It should be clear that any boundary orientation that produces a cusp in the γ plot will have

Table 7.2 Energy of Coherent and Partially
Coherent Twin Boundaries

Metal	Coherent	Partially Coherent
Cu	25 ergs/cm^2	440 ergs/cm^2
Fe	190 ergs/cm^2	705 ergs/cm^2

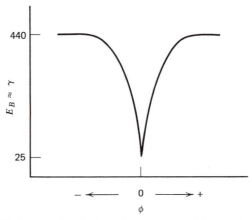

Figure 7.21   Variation of twin boundary energy with rotation of boundary.

strong torque terms. The coherent twin boundary and the tilt boundary are two examples.

Another important application of the minimization of the thermodynamic potential, γA, at equilibrium concerns the nature of the junction of grain boundaries at edges. Problem 7.4 shows four grains meeting at a common edge. By a slight motion of the boundaries this quadripole junction will break into two tripole junctions. In this problem you show that the thermodynamic potential is lowered by this process. Hence, we have the very interesting result that grain edges are three-rayed (tripole) at equilibrium. Examination of any polycrystalline microstructure reveals that the edges are virtually always three-rayed. Bubbles present an excellent model for grain boundaries. By observing soap bubbles or the bubbles in a baby bottle or the bubbles on a glass of beer one may learn a great deal about interfaces. C. S. Smith[7] describes a simple soap bubble model that may easily be constructed. Figure 7.22 presents a sequence of these bubble shapes as growth of the bubbles occurs. Notice that in going from Fig. 7.22(a) to 7.22(b) two tripole junctions came together along a horizontal direction and then separated along a vertical direction. When the two tripole junctions met they formed a quadripole junction. Observing this happen one sees that immediately upon forming this quadripole junction decomposes into two tripole junctions, thus illustrating that four-rayed junctions are unstable.

B.   PRESENCE OF A SECOND PHASE

Suppose you were to mix a small amount of lead into a bath of molten nickel and cool the melt to 350°C. The solubility of Pb in Ni is negligible

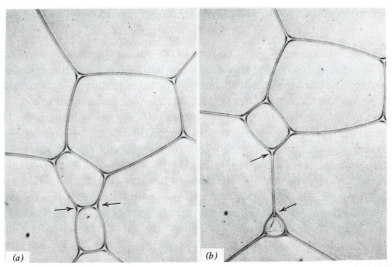

Figure 7.22 Boundary migration during growth of bubbles. [Reproduced by permission, from Ref. 10, American Society for Metals (1952).]

and at 350°C the Pb would be present as liquid within the polycrystalline Ni lattice. It is shown in Fig. 7.23 that there are essentially four types of locations within the polycrystalline Ni where the Pb might locate. In this figure we consider eight adjoining cubic grains of the Ni and note that the Pb could be located at (1) a corner of the grains, (2) an edge between four grains, (3) a face between two grains, or (4) within the bulk of a grain. Problem 7.4 illustrates that it is thermodynamically unstable for more than three grains to meet at a common edge. Hence, Fig. 7.23 is hypothetical and is used here only because of its simplicity in order to illustrate the four

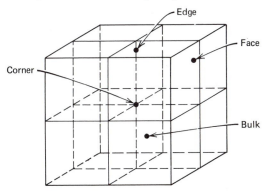

Figure 7.23 Location of second phase in a grain structure.

types of locations. We now consider the equilibrium shape of the second phase when it exists in each of these four locations.

1. Bulk. Consider a small volume of Pb within the bulk of a grain and ask: What is the expected shape of this Pb? If we neglect the variation of the surface tension between Pb and Ni (a solid–liquid surface tension) with crystal orientation we see that to minimize γA we must minimize A. Since the sphere has the minimum surface area per volume we expect the Pb to exist as spheres within the grains. If we consider the variation of γ with orientation then we expect the Pb to have the shape of a polyhedron whose faces are composed of the planes having low values of γ, that is, low index planes. This type of effect is observed in the stainless steel cladding of the fuel elements of nuclear reactors. The neutron flux in the reactor causes voids to form in the stainless and these voids have the shape of polyhedrons bounded mainly by (111) and (100) planes. Beautiful transmission electron microscope pictures of these voids are available.[8] When the second phase is a liquid the shape has also been observed to be polyhedral rather than spherical.[9] These examples illustrate the anisotropy in the solid–vapor γ and the solid–liquid γ, respectively.

2. Face. If the Pb goes into a grain boundary it might appear as shown in Fig. 7.24(a). If we construct a plane perpendicular to the edge of Fig. 7.24(a), we obtain a cross-sectional view as shown in Fig. 7.24(b). It should be clear that the point labeled "edge" in Fig. 7.24(b) is a tripole junction and we may apply Eq. 7.20 at this junction. We will neglect the torque terms and draw the surface tension force balance for the case of a second phase, β, in an α-α grain boundary as in Fig. 7.25. The angle that the second phase makes between the two α grains, labeled δ in Fig. 7.25, is frequently called the *dihedral angle*. Since the surface tension vectors

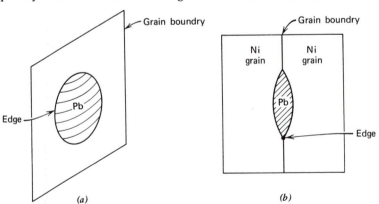

Figure 7.24 Two views of a second phase located at a grain face.

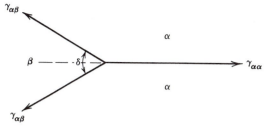

Figure 7.25 A surface tension force balance defining the dihedral angle δ.

acting to the left are identical, $\gamma_{\alpha\beta}$, we obtain from a force balance on Fig. 7.25

$$\gamma_{\alpha\alpha} = 2\gamma_{\alpha\beta} \cos \frac{\delta}{2} \qquad (7.24)$$

A plot of this equation is presented in Fig. 7.26. It is seen that when $\gamma_{\alpha\alpha} = \gamma_{\alpha\beta}$ the dihedral angle δ is 120°. This relation should hold in single-phase materials, which means that in single-phase alloys all three angles at a trijunction should be 120°. Using a soap bubble model as was described earlier you may quickly show yourself that this angular arrangement is obtained at an edge between three bubbles.

It may be seen from Fig. 7.26 that the following conditions must hold,

$$\gamma_{\alpha\beta} > \gamma_{\alpha\alpha} \qquad \delta > 120°$$
$$\gamma_{\alpha\beta} < \gamma_{\alpha\alpha} \qquad \delta < 120°$$

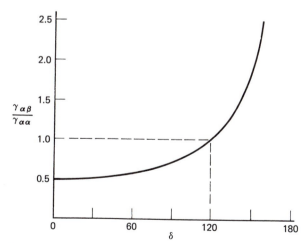

Figure 7.26 A plot of Eq. 7.24.

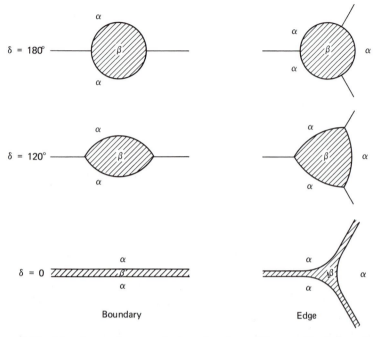

Figure 7.27 The shape of a second phase for three different dihedral angles at a grain boundary and a grain edge.

and for the two extreme cases we have

$$\gamma_{\alpha\beta} \gg \gamma_{\alpha\alpha} \qquad \delta \to 180° \qquad \textit{"No wetting"}$$
$$\gamma_{\alpha\beta} \le 0.5\gamma_{\alpha\alpha} \qquad \delta \to 0 \qquad \textit{"Complete wetting"}$$

The physical appearance of a second phase trapped at either a boundary or an edge for each of the above three cases is shown in Fig. 7.27. It can be seen from these sketches that the shape of the second phase is largely determined by the dihedral angle. When $\delta = 180°$ the second phase will remain spherical in either location. This condition is termed *no wetting*. It is similar to the case where a drop of water forms a nearly spherical bubble on a freshly waxed car. For this case of a liquid drop on a solid surface we have a surface tension balance, similar to Fig. 7.25, as shown in Fig. 7.28. The dihedral angle δ is defined as shown and when it is 180° the liquid drop becomes a sphere (neglecting distortion due to gravity). Notice that for a liquid drop on a solid the surface tension balance becomes

$$\gamma_{SV} = \gamma_{SL} + \gamma_{LV} \cos \delta \qquad (7.25)$$

and the no wetting condition becomes $\gamma_{SL} \ge \gamma_{SV} + \gamma_{LV}$.

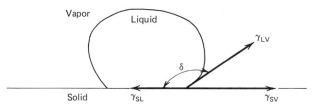

Figure 7.28 The surface tension force balance defining the dihedral angle δ for a liquid drop on a surface.

For the case where $\delta = 0$ the second phase "wets" the grain boundary, which means it is pulled along the grain boundary by capillary forces as shown in Fig. 7.27 and the entire grain-boundary surface becomes covered with a thin film of second phase. If this second phase is liquid the base metal will lose all its strength and simply fall apart. This phenomenon may be nicely illustrated by placing a small piece of gallium metal on a plate of aluminum. Heating to around 50–100°C will cause the Ga to melt and penetrate the grain boundaries if the Ga–Al interface is scratched to break the Al_2O_3 film on the Al. The Al plate is then easily broken and if the grain size is large one may remove individual grains of Al.

For the example of a liquid drop such as water on a surface, complete wetting occurs when the drop spreads freely over the solid. From Eq. 7.25 we see that this corresponds to the condition

$$\gamma_{SL} \leq \gamma_{SV} - \gamma_{LV} \tag{7.26}$$

3. Edge and Corner. We analyze this case by considering what happens when a second phase begins to penetrate along the edge of three grains as is shown in Fig. 7.29(a). Imagine that the β phase moves up along the edge and at the upper tip we have a corner between three α grains and one β grain. We imagine that there is an edge tension extending along each edge. Hence, we have four edge tension vectors extending out from the corner along each of the four edges as shown in Fig. 7.29(b) and these tensions must balance at the corner for equilibrium to obtain. Since the three edge tension vectors $\gamma_{\alpha\alpha\beta}$ are equal, the three angles labeled X are equal. The location of the dihedral angle δ is shown on Fig. 7.29(b). It should be clear that as X decreases the angle δ must also decrease so that one expects some functional relationship between X and δ. Within the above limitations one may show by simple geometry that the following equations must hold at the corner:

$$\cos \frac{X}{2} = \frac{1}{2 \sin (\delta/2)} \tag{7.27}$$

$$\cos (180 - Y) = \frac{1}{\sqrt{3} \tan (\delta/2)} \tag{7.28}$$

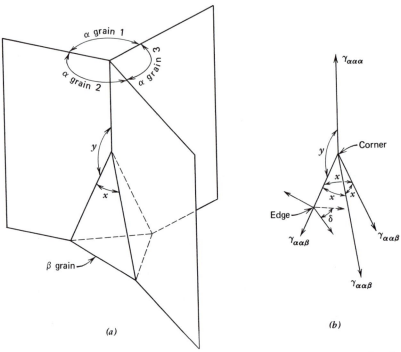

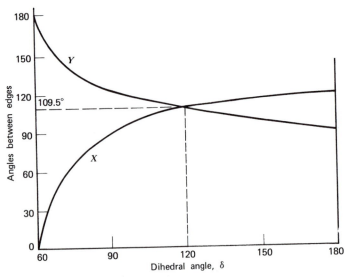

Figure 7.29 Three-dimensional view of a second phase moving up a grain edge.

Figure 7.30 Plots of Eqs. 7.27 and 7.28.

A plot of these two equations is presented in Fig. 7.30. To help illustrate the physical significance of these two equations, Figs. 7.31 and 7.32, from the work of C. S. Smith,[10] are presented. We will consider Fig. 7.30 for specific values of the dihedral angle.

1. $\delta = 180°$. Here $X = 120°$ and $Y = 90°$ so that the β phase becomes spherical, which corresponds to the result obtained above.

2. $\delta = 120°$. Here $X = Y = 109.5°$ so that the edges project from the corner at equal angles similar to the four bonds in the symmetrical tetrahedral bonding of SiO_4 groups. The resulting β-grain shape is illustrated at the corners of Fig. 7.31(a).

3. $\delta = 60°$. Here $Y \rightarrow 180°$ and $X \rightarrow 0$. It can be seen from Fig. 7.31(b) that in this case the β phase will penetrate along the edges of the grains and a "skeleton network" of β phase will be present throughout the structure. Figure 7.32(b) shows a microstructure under this condition. Notice that it is not at all obvious from the two-dimensional microstructure that the β phase is forming an interconnected, three-dimensional skeleton network. Such a network would be detrimental to the mechanical properties of the alloy.

4. $\delta = 0$. When $\delta \rightarrow 0$ the β phase spreads along the faces of the grain boundaries. Under these conditions a two-dimensional microstructure will appear as in Fig. 7.32(c).

We have now discussed the shape of the second phase at the four types of locations, bulk, face, edge, and corner, but we have not discussed at which of these locations the Pb phase in our example would prefer to exist. It is left to Problems 7.3 and 7.6 to answer this question. One finds that the thermodynamic potential γA is progressively lowered in going from the bulk to face to edge to corner location. Consequently one expects to find the Pb phase preferentially at the grain corners, assuming $\delta < 180°$.

C. APPLICATIONS

There are a number of cases where the above ideas have direct practical significance and three of them will be discussed here.

1. There are many applications in sintering. As an example consider carbide tool bits. In this case a powdered carbide such as tungsten carbide, WC, is bonded together with a metal such as Co. To produce good bonding one wants the bonding metal to wet the carbides and microstructures such as seen in Fig. 7.32(c) are desired. Hence, one must be able to produce a very low dihedral angle between the carbide and the bonding metal.

2. Hot shortness in steel. Sulfur combines with Fe to form FeS, which melts at 988°C. When the steel is hot rolled this compound melts. Since it

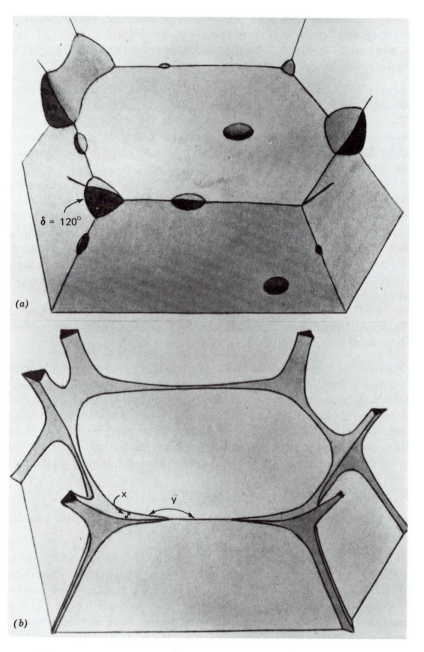

Figure 7.31 Appearance of a second phase at grain edges and corners. [From C. S. Smith, Trans. Met. Soc. AIME **175,** 15 (1948).]

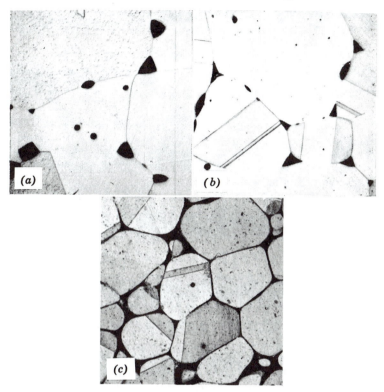

Figure 7.32   Microstructure of copper alloys containing a second phase liquid. Dihedral angle of liquid is (*a*) ~80°, (*b*) ~50°, and (*c*) ~0°. Magnification is (*a*) 500×, (*b*) 500×, and (*c*) 200×. [From C. S. Smith, Trans. Met. Soc. AIME **175**, 15 (1948).]

has a low value of δ in steel it wets the grain boundaries and the steel becomes brittle causing the failure known as hot shortness. It takes only a very small quantity of liquid to cover the grain boundaries and, consequently, only a small amount of sulfur impurity is required to cause this difficulty. Hot shortness is overcome by the addition of Mn to steel which causes MnS to form preferentially, and this compound melts at 1610°C.

3.   Soldering.   A successful solder must wet the surface of the metals to be soldered. Hence, the dihedral angle of a drop of liquid solder on the metal (see Fig. 7.28) must approach zero. As pointed out above this condition requires the surface tensions to satisfy Eq. 7.26. Consequently we require a low value for the surface tension between the liquid solder and the base metal. A successful flux will promote this condition.

7.3 THE SHAPES OF GRAINS IN TWO AND THREE DIMENSIONS

There are two known restrictions that place some limitations upon the shapes of grains:

1. Major requirement: The gains must fill space, that is, there can be no voids.

2. Secondary requirement: Within this restriction the thermodynamic potential γA must be a minimum.

We have seen above that the minimization of the thermodynamic potential requires that grain edges are three-rayed at 120° angles in single-phase structures; and, also, grain corners are four-rayed at 109.5° angles in single-phase structures. These conditions must describe a metastable equilibrium since complete equilibrium would require the elimination of all boundaries. These metastable equilibrium conditions are generally found to be closely approached in polycrystalline metal structures. C. S. Smith[7,10] has shown by combining restrictions 1 and 2 that the average two-dimensional section of a grain must be six-sided. By combining this result with the conclusion that in a single-phase material the dihedral angle must be 120° we find that the two-dimensional shapes of grains must be as shown in Fig. 7.33 where the sections are taken at a plane perpendicular to the edges. This result shows that grains with just a few sides (small grains) will have curved boundaries and the curvature will be concave inward; and grains with many sides (large grains) will have boundaries curved concave outward. Later we will see that this result has significance in grain growth.

One wonders whether the average three-dimensional shape of a grain may not approach some characteristic polyhedral shape. C. S. Smith has considered this question.[7,10] There is no regular polyhedron that will stack together to fill space and also form only four-rayed corners at 109.5°

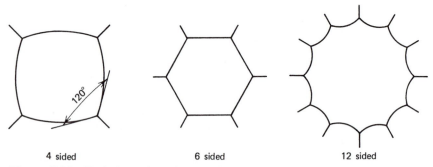

| 4 sided | 6 sided | 12 sided |

Figure 7.33 Variation of grain-boundary curvature with number of grain-boundary sides.

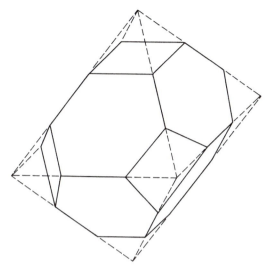

Figure 7.34 The tetrakaidecahedron (illustrating that it is a truncated octahedron having only square and hexagonal faces, thereby giving an average corner angle of 110°).

angles. There is one irregular polyhedron that comes very close to these requirements, called the tetrakaidecahedron; see Fig. 7.34. By introducing the proper curvature into the edges and faces one obtains 109.5° angles at the corners and this polyhedron, sometimes called the α-tetrakaidecahedron, stacks together on a bcc lattice to completely fill space. One might expect the average three-dimensional grain shape to approximate these α-tetrakaidecahedra. However, experiments have shown this not to be the case.[7] About all one can say is that the three-dimensional grain is an irregular polyhedron of no one kind with curved faces and four-rayed corners with *average* angles of 109.5°. Smith has shown from restrictions 1 and 2 that the average grain should have 5.143 edges per face and this condition appears to be verified by experiment.[10]

7.4 GRAIN-BOUNDARY SEGREGATION[11]

Figure 7.35(a) shows a sketch of two single crystals of a Cu–1% Sn alloy separated by a single grain boundary. This arrangement is usually called a bicrystal. If one were able to measure the composition across the boundary with a very fine probe the composition profile would probably appear as in Fig. 7.35(b). The Sn atoms will tend to segregate into the grain boundary. This phenomenon is called grain-boundary segregation and it generally

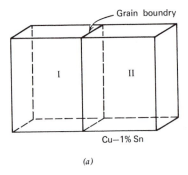

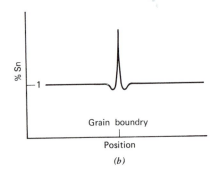

(a)   (b)

Figure 7.35 Variation of Sn solute concentration across a copper–copper grain boundary.

results in an excess solute concentration as shown in Fig. 7.35(b). However, a solute depletion at the boundary sometimes occurs.

To understand why grain-boundary segregation occurs we consider the effect of the Sn atom upon the Cu lattice when it dissolves. Figure 7.36(a) shows schematically how the larger Sn "bends" the planes of the Cu lattice. Consequently the Sn atom produces a considerable strain energy in its surroundings. Suppose we let the strain energy so produced by a dissolved Sn atom be E_s. We now ask: How does this strain energy vary as the Sn atom approaches a grain boundary? Since the grain boundary will have regions that are quite open [see Fig. 7.3(b)] we expect the Sn to locate there and to produce a very low strain energy. Hence, we expect the strain energy, E_s, due to the Sn atom to be significantly lowered at the grain boundary as is shown in Fig. 7.36(b). The Sn atom will experience a force due to variations of this energy which to a good approximation may be

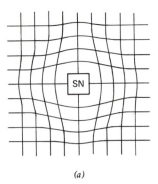

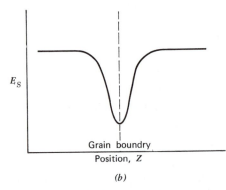

(a)   (b)

Figure 7.36 (a) Lattice strain produced by a Sn atom in Cu, and (b) the variation of this strain energy with position near a grain boundary.

written as $F_{Sn} = -dE_s/dZ$. Hence, in the region near the grain boundary the Sn atoms will experience a force pulling them toward the boundary. This force produces a segregation into the boundary, and we will see later that it also has a strong influence on the mobility of the grain boundary.

By consideration of the thermodynamics of surfaces[4] one may obtain the Gibbs absorption equation,

$$\left(\frac{dn}{dA}\right)_i = -\frac{1}{RT}\left[\frac{\partial \gamma}{\partial \ln N_i}\right] \tag{7.29}$$

where $(dn/dA)_i$ has been defined for Eq. 7.11 and is sometimes called surface excess, γ is interface tension, and N_i is the mole fraction i in solution in the neighboring grains. Equation 7.29 shows that if addition of atoms i lowers the surface tension, absorption of i occurs at the interface and vice-versa.

7.5 MOTION OF GRAIN BOUNDARIES

In order to discuss the motion of grain boundaries we will make use of the concept of mobility that was introduced in Chapter 6. Consider a bicrystal of two Cu grains as shown in Fig. 7.37. Suppose that somehow the chemical potential, μ, is higher in grain I than in grain II as shown in Fig. 7.37. We have shown in Chapter 6 that the force on an atom is given by the gradient of the Gibbs chemical potential, $-d\mu/dZ$. If we consider the interface to have a thickness of λ, then an atom of grain I on the interface experiences a force

$$F = -\frac{\mu_{II} - \mu_I}{\lambda} = -\frac{\Delta \mu}{\lambda} \tag{7.30}$$

These atoms will jump onto grain II causing the boundary to migrate to the left. Hence, the grain boundary will migrate to the left at a velocity equal to

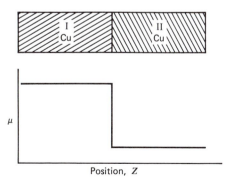

Figure 7.37 A chemical potential difference across a grain boundary.

the average velocity of the atoms to the right

$$v[\text{grain bdry}] = -v[\text{grain bdry atoms}] \qquad (7.31)$$

The average velocity of the grain boundary atoms may be taken as $v = BF$ where B is their mobility and F the force given in Eq. 7.30. Hence, from Eq. 7.31 we have

$$v_{gb} = B \cdot \frac{\Delta \mu}{\lambda} \qquad (7.32)$$

This equation serves very nicely as a basis for discussing the velocity of grain boundaries. It shows that the velocity depends upon the chemical potential difference across the boundary and the mobility of the atoms across the boundary. We will discuss each of these terms separately.

A. DRIVING FORCES

We consider three specific driving forces for grain-boundary movement, two of which are of practical interest and one mainly of theoretical interest.

1. Stored Energy. When a material is cold worked, a high density of defects is introduced into its lattice. These defects, mainly dislocations, increase the energy of the lattice. Consider a bicrystal as shown in Fig. 7.38 where grain I is a soft annealed grain and grain II is highly cold worked. Recognizing that the Gibbs chemical potential μ is identical to the partial molal free energy $\bar{G}$, we may write for the difference in chemical potential between the two grains $\Delta \mu = \Delta \bar{E} + P \Delta \bar{V} - T \Delta \bar{S}$. To a good approximation we may neglect the volume and entropy terms giving $\Delta \mu = \bar{E}_{II} - \bar{E}_{I}$. The stored energy of the soft annealed grain is essentially zero so that we may write

$$v_{gb} = B \frac{E_s / A}{\lambda} \qquad (7.33)$$

where A is Avogadro's number and E_s is stored energy per mole in grain II. This equation applies to the very practical problem of recrystallization and it indicates that the grain-boundary velocity goes linearly with the amount of stored energy.

Figure 7.38 A grain boundary between an annealed grain, I, and a cold worked grain, II.

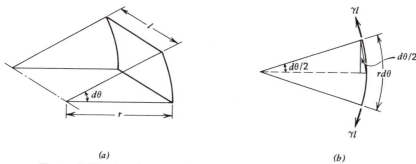

(b)

Figure 7.39   An element of a surface with cylindrical curvature.

2. Elastic Strain.   In Chapter 4 we showed that for the simple case of uniaxial tension the energy per unit volume in an elastically strained material is given as $\frac{1}{2}\sigma\varepsilon$ where σ is stress and ε is strain (Eq. 4.2). Substituting from Hooke's law this becomes $\sigma^2/(2E)$ where E is Young's modulus. Suppose now that two annealed grains of a bicrystal are subjected to a uniaxial stress σ, but that the two grains have different moduli, E_I and E_{II}. The stress would then produce a difference in the energy per atom in the two grains that would give rise to a boundary velocity as

$$v_{gb} = \frac{B}{\lambda} \cdot \frac{\bar{V}\sigma^2}{2}\left[\frac{1}{E_{II}} - \frac{1}{E_I}\right] \tag{7.34}$$

where $\bar{V}$ is the volume per atom. This example is of little practical significance because of its small magnitude. However, it does illustrate nicely that any phenomenon that will produce an energy difference between grains can give rise to a boundary migration.

3. Interface Curvature.   Because interfaces possess a surface stress, a curved surface must have a force acting normal to it to overcome the surface stress that tends to hold it planar. We now show that mechanical equilibrium across a curved surface requires a pressure difference across the surface. Consider an element of a cylindrical surface as shown in Fig. 7.39(a) having radius of curvature r. An edge view of the curved part of this surface is shown in Fig. 7.39(b). We assume now that the surface stress equals the surface tension γ. Then the force acting in the surface is $\gamma \cdot l$ as shown in Fig. 7.39(b). There are two such forces, and it is seen that each force has a horizontal component to the left, which for $d\theta$ small is $\gamma l \cdot \sin{(d\theta/2)}$. Mechanical equilibrium requires that the pressure on the concave side of the surface be slightly higher than on the convex side in order to balance these surface forces. If we let this pressure difference be

ΔP, then for mechanical equilibrium we have

$$2 \cdot \gamma l \sin \left(\frac{d\theta}{2}\right) = \Delta P \cdot l \cdot r \, d\theta \qquad (7.35)$$

For $d\theta$ small, $\sin (d\theta/2) \rightarrow d\theta/2$ and we have $\Delta P = \gamma/r$. In general, surfaces are not cylindrical and therefore have a more complex curvature. Consider a football, which may be thought of as an ellipsoid of revolution. Its radius of curvature is different along a plane through the minor axis and a plane through the major axis. We define two principle radii of curvature for any surface as follows. Construct two planes such that both planes pass through the same normal to the surface and are at right angles to each other. The radius of curvature of the surface intersections on these planes define two principle radii of curvature, r_1 and r_2. By arguments similar to those above we obtain

$$\Delta P = \gamma \left[\frac{1}{r_1} + \frac{1}{r_2}\right] \qquad (7.36)$$

It should be clear that for a sphere we would obtain $\Delta P = 2\gamma/r$, where r is the sphere radius. From thermodynamics we have at constant temperature

$$d\mu = \bar{V} \, dP \qquad (7.37)$$

where $\bar{V}$ is specific volume. If we take the specific volume as constant on either side of the interface we may integrate Eq. 7.37 across the interface to obtain

$$\mu^{\mathrm{I}} - \mu^{\mathrm{II}} = \bar{V}[P^{\mathrm{I}} - P^{\mathrm{II}}] = \bar{V} \, \Delta P \qquad (7.38)$$

where superscripts I and II refer to the two adjoining grains in Fig. 7.38. For a spherical grain boundary we obtain by combination of Eqs. 7.38 and 7.36

$$\mu^{\mathrm{I}} - \mu^{\mathrm{II}} = \frac{\bar{V}2\gamma}{r} \qquad (7.39)$$

This equation is frequently referred to as the Gibbs–Thomson equation. Since the pressure is always higher on the concave side of the boundary we see by Eq. 7.37 that the chemical potential is always higher on the concave side. This result is also apparent from Eq. 7.39. Consequently, the chemical potential difference will cause atoms to move toward the convex side. We conclude:

1.   Grain boundary curvature provides a driving force for growth.

2.   Since the atoms move toward the convex side under this force, curvature induced growth moves the boundary toward the concave side. This is illustrated in Fig. 7.40 where the boundaries have moved from positions 1 to 2 during a heat treatment.

Figure 7.40 Grain growth in aluminum. [From P. A. Beck and Ph. R. Sperry, J. Appl. Phys. **21**, 150 (1950).]

3. Small grains (less than six sides on a two-dimensional cut) become smaller and large grains (more than six sides on a two-dimensional cut) become larger, see Fig. 7.33, as a result of curvature-induced growth.

These results are nicely illustrated by the growth of an array of bubbles. The larger bubbles grow at the expense of the smaller bubbles, which shrink toward their center of curvature; see Fig. 7.22. In this case bubble growth results from gas atoms diffusing through the water membranes from the high-pressure to the low-pressure side. The pressure difference here is a factor of 2 larger than Eq. 7.36 because the bubble is a film with two surfaces.

It is frequently convenient to talk about curved surfaces relative to flat surfaces. In this case we may integrate Eq. 7.37 for the process of going from a flat surface to a curved surface. For a spherical surface we obtain

$$\mu(r) - \mu(\text{flat}) = \frac{2\bar{V}\gamma}{r} \tag{7.40}$$

From the above discussion we conclude that a grain boundary will have a driving force for motion any time the chemical potential of the atoms in two bordering grains is different. The two common causes of such a difference are

1. a difference in stored energy due to cold work,
2. curved grain boundaries.

B. MOBILITY

Four major factors that influence the mobility of grain boundaries will be discussed.

1. Impurity Atoms. In a classical set of experiments carried out in the late 1950s, Aust and Rutter[12] determined the mobility of grain boundaries as a function of a number of variables; some of their results are shown in Fig. 7.41. The important point to see in this result is that increasing the impurity content of tin in lead from less than 1 ppm to 60 ppm drops the grain-boundary velocity just over 4 orders of magnitude. Since the experiments were done with a constant driving force, this means that an increase of impurity content by a factor of 60 caused a decrease in grain boundary mobility by a factor of 10,000. It is amazing that as little as 60 ppm Sn in Pb can drop the grain boundary mobility by a factor of 10,000. Lead containing 60 ppm Sn would be listed as 99.994% pure.

Recent theories[13] have been fairly successful in explaining this large effect. Because of grain-boundary segregation a very low level of impurity within the bulk of the grains can still give rise to a significant impurity content within the grain boundaries. If the grain boundary starts to pull away from impurity atoms it will exert a force on the atoms as explained in Section 7.4 using Fig. 7.36(b). Consequently, as the boundary moves it will tend to drag along the impurity atoms. The drag force of this impurity atmosphere will greatly reduce the boundary mobility because the velocity of these atoms is limited by the rate at which they can diffuse through the matrix as they are dragged at the edge or just behind the boundary. The impurity drag theories (Ref. 13, p. 256) predict a velocity proportional to the first power of the driving force for high-purity metals with high driving forces (i.e., Eq. 7.32), where the boundary essentially breaks free of the impurity atoms; they also predict velocity linearly proportional to the driving force for impure metals with low driving forces. In the transition between these two regions the theory predicts a complex relation between

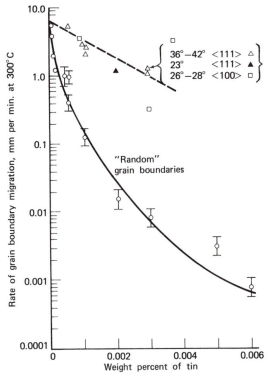

Figure 7.41 Influence of tin impurity level on grain-boundary migration rate in lead. The data for the upper curve occurred only for those boundaries having particular orientations, specified here by the degree of tilt about ⟨111⟩ and ⟨100⟩ directions. (From Ref. 12.)

velocity and driving force. Several experiments (Ref. 13, p. 405) have found the velocity to be related to driving force by the relation $v = B \cdot (\Delta\mu/\lambda)^m$, where m is larger than unity, ranging as high as 12. Agreement between theory and experiments is moderately good. It appears that most experimental work occurs in the transition region. Equation 7.32 only holds for ultrahigh-purity metals with high driving forces (i.e., stored energy rather than interface curvature).

2. Presence of Second-Phase Particles. When a moving boundary encounters a second-phase particle the particle will exert a restraining force upon the boundary that will cause the boundary to be pulled back at the particle location as shown in Fig. 7.42. The restraining force is calculated from the component of the surface tension acting in the vertical direction. From Fig. 7.42 and a little geometry we find that the restraining force is

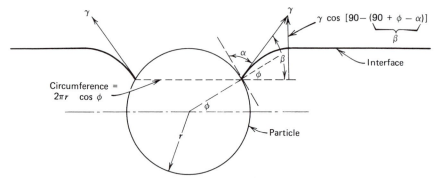

Figure 7.42 Interaction of a second-phase spherical particle with a grain boundary. The grain boundary interface is moving upward and being restrained by the particle.

given by

$$F_{res} = 2\pi r \cos \phi \cdot \gamma \cos (\alpha - \phi) \qquad (7.41)$$

Maximizing this force with respect to ϕ we obtain[14]

$$F_{res}(max) = \pi r \gamma (1 + \cos \alpha) \qquad (7.42)$$

Hence, depending on the value of α, the maximum pinning force varies from $\pi r \gamma$ to $2\pi r \gamma$. (Note: You should be able to show yourself from a surface tension balance that unless the particle/matrix surface tension is changed as the interface passes the value of α will be 90°.) It is apparent that the pinning force depends primarily on the size of the particles and the number of particles and to a lesser extent on the particle material, which only affects the value of α. The magnitude of the pinning force in some metal systems has been found to agree with Eq. 7.42.[14]

3. Temperature. We can relate the mobility of an atom jumping across the grain boundary to the grain-boundary diffusion coefficient, $B \approx D_{gb}/kT$. The grain-boundary diffusion coefficient increases exponentially with temperature; see Fig. 6.12(b). Consequently the temperature dependence of B would go as $e^{-Q/kT}$. The exponential term essentially "wipes out" the $1/T$ term and we find that the grain-boundary mobility is very temperature dependent.

4. Orientation of Grains across Boundary. As the orientation mismatch drops to zero the diffusion coefficient in the boundary drops from the value of the grain-boundary diffusion coefficient to the matrix diffusion coefficient. Hence, we expect low-angle boundaries to have a smaller mobility than high-angle boundaries.

As shown in Fig. 7.41 it has been found that certain high-angle boundaries have very high mobilities in the presence of solute. This result shows that there is a definite variation in the structure of high-angle boundaries. These special boundaries have been explained as coincident boundaries that have a good atomic fit and therefore do not allow much solute segregation.[15] The high mobility results from the lowered impurity drag.

To summarize, the velocity of a grain boundary may be given as $v = B \cdot (\Delta\mu/\lambda)^m$, where except for the case of very high purity metals m is greater than 1. The chemical potential difference, $\Delta\mu$, will generally be due to either stored energy or boundary curvature. The mobility is a function of (1) solute content, (2) presence of second-phase particles, (3) temperature, and (4) relative crystal orientation across the boundary.

C. NORMAL GRAIN GROWTH

The growth of grains under the driving force of grain boundary curvature is referred to here as *normal grain growth*. In the absence of stored energy due to cold work, this is the driving force causing grain growth. We will present an approximate derivation for the rate of normal grain growth.

From above we may write the velocity of growth of a spherical grain as

$$v = B\left(\frac{\Delta\mu}{\lambda}\right)^m = \frac{D_B}{kT} \cdot \left(\frac{\bar{V}\,2\gamma/r}{\lambda}\right)^m \qquad (7.43)$$

At constant temperature the terms in this equation are constant so that we obtain the velocity proportional to radius of curvature as $v = c/r^m$, where c is the proportionality constant. If we assume that the radius of curvature of a grain is proportional to the diameter of the grain, D, we may write

$$v = (\text{const})\left(\frac{1}{D}\right)^m = \frac{dD}{dt} \qquad (7.44)$$

Integrating this equation we obtain

$$D^{m+1} - D_0^{m+1} = kt \qquad (7.45)$$

where k is a constant and D_0 is the grain size at time zero. For the case of D_0 negligibly small we have

$$D = kt^n \quad \text{and} \quad n = \frac{1}{m+1} \qquad (7.46)$$

Equation 7.46 was first presented by Hu and Rath[16] and they have compared it to several experimental studies which show that n is generally less than $\frac{1}{2}$. It is only equal to $\frac{1}{2}$ for very high purity metals at high

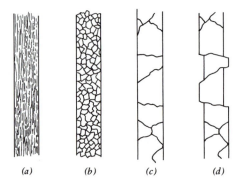

(a)   (b)   (c)   (d)

Figure 7.43   Change in microstructure of a pure tungsten filament heated by alternating current (from Ref. 17).

temperatures. This agrees with the impurity drag theories because one only expects $m = 1$ for such conditions.

An interesting practical example where the above ideas are quite important involves tungsten filaments in light bulbs. Since the filaments operate at an extremely high temperature, considerable grain growth may be expected to occur. When large grains extending across the filaments develop as shown in Fig. 7.43 the filaments in some cases (see Ref. 17, p. 138) become very brittle and break apart under the stresses produced by the thermal expansion on heating and cooling. Consequently, second-phase particles of thoria are doped into the tungsten to limit this grain growth. For a random distribution of second-phase particles of volume fraction f and radius r the number of particles intercepted by 1 cm^2 of interface is $3f/2\pi r^2$.[1] Consequently these particles will produce a force per unit area restraining the grain boundary motion of $(3f/2\pi r^2) \cdot \pi r \gamma [1 + \cos \alpha]$, from Eq. 7.42. When this force balances the pressure force due to curvature, grain growth will stop and we will obtain a limiting grain size. If we assume spherical interfaces of radius R, we obtain using Eq. 7.36 *the limiting radius of curvature*,

$$R = \frac{4r}{3f(1+\cos \alpha)} \tag{7.47}$$

Assuming the grain diameter to be proportional to the radius of curvature we see from Eq. 7.47 that the limiting grain size produced by second-phase particles depends upon their volume fraction, their radius, and the contact angle α. This result was first presented by Zener.[1] The effectiveness of a small amount of ThO$_2$ in limiting grain growth in tungsten is illustrated in Fig. 7.44.

Figure 7.44 Tungsten rods annealed at 2700°C for 2 min. (*a*) Pure, (*b*) 0.75% ThO₂ added (100×) (from Ref. 17).

REFERENCES

1. D. McLean, *Grain Boundaries in Metals,* Oxford Univ. Pr., Oxford, 1957.

2. J. Weertman and J. R. Weertman, *Elementary Dislocation Theory,* Macmillan, New York, 1964.

3. W. T. Read, Jr., *Dislocations in Crystals,* McGraw-Hill, New York, 1953.

4. (a) W. W. Mullins, Solid Surface Morphologies Governed by Capillarity, in *Metal Surfaces,* W. D. Robertson and N. A. Gjostein, Eds., American Society of Metals, New York, 1963, Chapter 2. (b) R. K. Trivedi, Theory of Capillarity, in *Key Topics in the Theory of Phase Transformations,* American Institute of Metallurgical Engineers, New York, 1975, Chapter 2.

5. C. Herring, in *The Physics of Powder Metallurgy,* W. E. Kingston, Ed., McGraw-Hill, New York, 1951, Chapter 8.

6. C. Herring, in *Structure and Properties of Solid Surfaces,* R. Gomer and C. S. Smith, Eds., Univ. Chicago Press, Chicago, 1953, Chapter 1.

7. C. S. Smith, Grain Shapes and Other Metallurgical Applications of Topology, in *Metal Interfaces,* American Society of Metals, New York, 1952, pp. 65–113.

8. C. W. Chen, Phys. Stat. Sol. (a) **16,** 197 (1973).

9. J. Basterfield, W. A. Miller, and G. C. Weatherly, Can. Met. Quart. **8,** 131 (1969).

10. C. S. Smith, Some elementary principles of polycrystalline microstructure, Metallurgical Reviews **9,** No. 33, 1 (1964).

11. J. H. Westbrook, Segregation at grain boundaries, Metallurgical Reviews **9,** No. 36, 415 (1964).

12. K. T. Aust and J. W. Rutter, Trans. Met. Soc. AIME **215,** 119 (1959).

13. *The Nature and Behavior of Grain Boundaries,* H. Hu, Ed., Plenum, New York, 1972.

14. M. F. Ashby, J. Harper, and J. Lewis, Trans. Met. Soc. AIME **245,** 413 (1969).

15. K. T. Aust and B. Chalmers, Met. Trans. **1,** 1095 (1970).

16. H. Hu and B. B. Rath, Met. Trans. **1,** 3181 (1970).

17. C. J. Smithells, *Tungsten,* Chapman & Hall, London, 1952.

ADDITIONAL READING

1. C. S. Smith, The shapes of things, Scientific American **190**, 58 (1954).

2. C. V. Boys, *Soap Bubbles—Their Colours and the Forces Which Mold Them*, Dover, New York, 1959.

PROBLEMS

7.1 Two blocks of cold-worked copper (fcc, lattice parameter a) are given a recovery anneal and all of the dislocations form subgrain boundaries.

(a) The average separation of the dislocations in the boundaries is 4×10^{-4} cm in block A and 4×10^{-3} cm in block B. What is the angle of rotation between the lattices of the subgrains in the two blocks?

(b) Which block would have the higher subgrain boundary energy? Give an approximate expression for the ratio of the subgrain boundary energies in the two blocks.

7.2 (a) The atoms in the interior of an fcc crystal have 12 nearest neighbors. As an approximation the chemical energy per bond may be taken as $\frac{1}{12}$ of the latent heat of vaporization, that is, $\frac{1}{12}$ times L. Using this as a measure of the bond energy, calculate the surface tension of the (1 1 1), (1 0 0), and (0 1 1) surfaces in an fcc crystal as a function of L and the lattice parameter a. State any assumptions.

(b) Using your results from these calculations, which faces would be expected to form the surfaces of a small polyhedron of Cu if it were heated just below its melting point for a long period of time until its equilibrium surface was obtained?

7.3 Suppose a small amount of a second phase (B) is contained within the volume of the primary phase (A) and has a spherical shape. Let the total volume of phase B be designated as V. When this volume of phase B moves to the corner of four adjacent A grains, it assumes the shape of a regular tetrahedron having a side length of L.

(a) What is the dihedral angle of B in A?

(b) Let γ_{AA} be the surface tension between A grains and γ_{AB} be the surface tension between A and B grains. Show analytically whether B will be stable as a sphere within the bulk of A or as a tetrahedron at a corner of four A grains. State clearly all assumptions you have to make to solve this problem. Hint: Be sure to account for the AA boundary area that is wiped out when B moves into the corner.

$$\text{Volume tetrahedron} = 0.1178 L^3$$

$$\text{Surface tetrahedron} = 1.732 L^2$$

Answer: $(\gamma A)_{\text{BULK}} / (\gamma A)_{\text{CORNER}} = 1.671$.

7.4

The four grains shown below, numbered 1, 2, 3, and 4, meet along a common edge O. This type of junction is called a quadripole junction. If the boundary CO shifts to CP and the boundary DO shifts to DP the quadripole junction will have

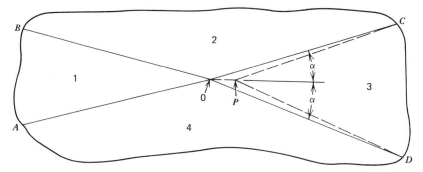

dissociated into two tripole junctions, DPC and AOB, and a new boundary, OP, will have been generated.

(a) The problem is to determine which is more stable, the two tripole junctions or the single quadripole junction. Take α to be 30° and assume that all of the grain boundaries are high-angle boundaries. This latter assumption allows you to neglect torque terms and take γ constant for all five boundaries—do you know why?

(b) Suppose α were 70°. Would the quadripole junction now be stable? Explain your answer.

7.5   From a surface tension balance on the tripold junction of Fig. 7.20, derive Eq. 7.23.

7.6   A second phase, B, exists in metal A. A small volume of B has the shape of a sphere when it lies entirely within a grain. However, when it lies in a face separating two A grains it has the shape of a double spherical cap.

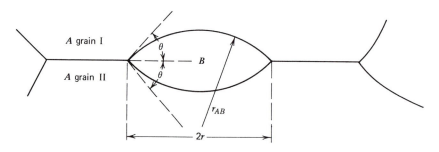

$$\text{Volume double spherical cap} = 2\left[\pi r_{AB}^3\left(\frac{2-3s+s^3}{3}\right)\right]$$

$$\text{Area double spherical cap} = 2[2\pi r_{AB}^2(1-s)]$$

$$\text{where}\quad s = \cos\theta, \quad r = r_{AB}\sin\theta$$

A spherical cap is simply a section of a sphere, that is, a cap of a sphere. Hence, r_{AB} is the radius of the sphere from which the cap is taken, or alternatively, it is the radius of curvature of the $A-B$ interface.

Question: (a) If the dihedral angle of B in A is exactly 120° determine analytically whether B would prefer to lie in the grain face or the grain interior.

Answer: $(\gamma A)_{BULK}/(\gamma A)_{FACE} = 1.296$.

(b) Repeat part (a) for the case where the dihedral angle is zero.

7.7   Suppose that the normal to a high-angle grain boundary in a copper wire of 0.1 mm diameter is inclined to the wire axis at an angle of 20°. Upon annealing, what would you expect to happen to the grain boundary?

7.8   In Problem 7.7, the grain boundary in the initial state is a simple 4° tilt boundary. Will the process predicted under the conditions of Problem 7.7 occur? Explain.

7.9   Derive the relation

$$\cos\left(\frac{X}{2}\right) = \frac{1}{2\sin(\delta/2)}$$

that was given as Eq. 7.27.

CHAPTER 8

NUCLEATION

We will find in the succeeding chapters that the principal means at our disposal for the control of the properties of alloys are phase transformations. Consequently, before proceeding further some brief generalizations concerning phase transformations will be presented. Table 8.1 lists the different possible types of phase transformations along with some familiar examples. Notice that each of the transformations could also be considered when proceeding in the opposite directions. In these cases examples 1, 2, and 3 would correspond to boiling, sublimation, and melting, respectively. We will be mainly concerned with transformations 3 and 4, which are liquid → solid (solidification) and solid → solid transformations, respectively.

All of these phase transformations are accompanied by a change or a rearrangement in the atomic structure. In addition to a structure change, a phase transformation may produce a composition change and/or a strain formation. In Table 8.2 transformations are listed in an order of increasing complexity going from (a) to (d).

Table 8.1 Major Types of Phase Transformations

Type of Transformation	Example
1. Vapor → liquid	Condensation of moisture
2. Vapor → solid	Formation of frost on a window
3. Liquid → crystal	Formation of ice on a lake
4. Crystal 1 → crystal 2	
(a) Precipitation	Formation of Fe_3C on cooling austenite
(b) Allotropic	α-Fe → γ-Fe at 910°C
(c) Recrystallization	Cold-worked Cu → new grains at high temperatures

Table 8.2   Degree of Complexity Involved in Phase Transformations

(a) Structure change
(b) Structure change+composition change
(c) Structure change+strain formation
(d) Structure change+strain formation+composition change

In order to illustrate these types of transformations a sketch of the iron–carbon phase diagram is presented in Fig. 8.1. The simplest type of transformation, a, involves only a structure change and it is represented on Fig. 8.1 by the solidification or melting of pure iron. Solidification or melting of any alloy represents a b-type transformation since a composition change is always involved in the alloys. (There are a few rare exceptions to this rule such as the aziotrope composition in alloys such as Au–Ni and some compositions in In–Tl alloys.) In most solid–solid phase transformations the two phases have a different specific volume so that the new phase occupies a different volume than the parent phase and this results in strain formation. The c-type transformation involving both structure and strain changes is represented by the $\alpha \rightarrow \gamma$ transformation in pure iron and the formation of martensite upon quenching austenite. Most solid–solid transformations are type d (the most complex) and three examples on Fig. 8.1 are $\alpha \rightarrow \alpha + Fe_3C$, $\gamma \rightarrow$ pearlite $(\alpha + Fe_3C)$, $\gamma \rightarrow Fe_3C + \gamma$.

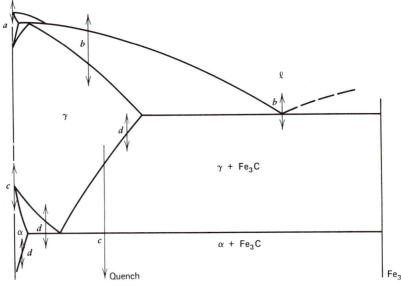

Figure 8.1   Various transformations of Table 8.2 illustrated on the Fe–C system.

We now discuss a specific phase transformation as an example. Consider a bar of pure iron, which will have a bcc structure at room temperature. If we heat this metal the atoms will vibrate more rapidly and thereby cause the average separation distance between atoms to increase. Consequently, the lattice parameter will increase as the bcc lattice expands. However, when a temperature of 910°C is reached the bcc lattice transforms into a fcc lattice because, for reasons not completely understood, above 910°C the fcc structure has a lower free energy. This transformation from bcc to fcc does not just happen spontaneously throughout the α(bcc) matrix when a temperature of 910°C is reached. What does happen is that small regions of γ(fcc) iron form (nucleate) at different locations within the α iron as shown in Fig. 8.2. These new grains then grow and the remaining α matrix is transformed into a γ matrix by growth of these new grains plus nucleation and growth of additional γ grains. Essentially, all phase transformations occur by a nucleation and growth of nuclei. The rate at which a transformation occurs is controlled by these two processes, nucleation and growth. The growth of the new phase is described by an equation such as was developed in Chapter 7 for the rate of interface motion. In this chapter we are interested in describing the rate of nucleation. We define this rate of nucleation I for the above example as

$$I = \frac{\text{No. } \gamma \text{ regions formed/unit time}}{\text{volume of } \alpha \text{ phase}}$$

and our purpose is to consider what is involved in developing an expression for the rate I.

When the new phase nucleates it generally forms at some discontinuity in the parent matrix such as a grain boundary or a dislocation, but it also may form within a uniform region of the parent matrix. The distinction

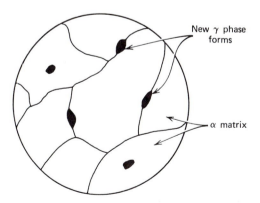

New γ phase forms

α matrix

Figure 8.2   Nucleation of the fcc γ phase in the bcc α phase of iron.

between these two types of nucleation is quite important and is defined as follows:

Homogeneous nucleation: New phase forms uniformly throughout bulk of parent phase.

Heterogeneous nucleation: New phase forms preferentially at in-homogeneities in parent phase.

8.1 HOMOGENEOUS NUCLEATION

We will now present a rather simplified version of what has come to be called the classical theory of nucleation. This theory originated from the ideas of Volmer and Weber in 1925.[1] As an example transformation, we will consider freezing and take the free-energy change upon freezing as $\Delta G_B = G_s - G_l$ where G_s and G_l are the free energy of solid and liquid, respectively. If the liquid is cooled below the melting point the free energy of the solid will become smaller so that ΔG_B becomes a negative number below the melting temperature. This is shown graphically in Fig. 8.3. Here the *bulk* free energy is plotted as a function of temperature for both the solid and the liquid phases. At the freezing temperature both phases must have the same free energy because at this temperature they are in

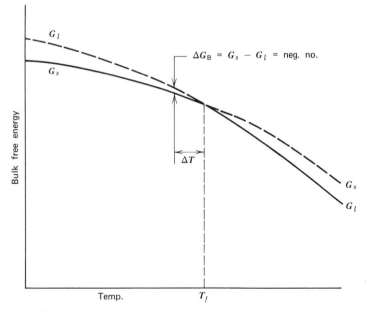

Figure 8.3 The temperature dependence of the bulk free energy of the solid and liquid phases.

thermodynamic equilibrium. Suppose that somehow the liquid is cooled below the freezing point by an amount ΔT without freezing. We call this supercooling. The liquid will then have a free energy that is higher than the solid by the amount $|\Delta G_B|$ as shown on Fig. 8.3. We would therefore expect the liquid to freeze since the free energy would thereby be lowered. However, it is quite easy to supercool a number of liquids for indefinite periods, and this shows that there is some barrier to freezing. For example, if you try to freeze high-purity tin you find that it invariably will supercool 5–20°C before freezing starts. To understand the nature of the barrier one must consider what happens physically when a liquid freezes. We will first consider homogeneous nucleation. In this case the solid begins to form as very small regions throughout the bulk of the liquid. These small regions will be called nuclei and they will be considered to be spherical in shape. When these nuclei first form they are extremely small, probably on the order of 10 Å or so in diameter. Consequently their surface-to-volume ratio is very high. Since there will be a positive free energy associated with the formation of the surface area of the nuclei, this surface energy will act as a barrier to the formation of small nuclei. To account for this surface effect we write the free-energy change to form a nucleus as

$$\Delta G = \tfrac{4}{3}\pi r^3 \, \Delta G_B + 4\pi r^2 \gamma \tag{8.1}$$

where the surface free energy has been simply taken as the surface tension γ. This is correct in pure materials but introduces a small error for alloys, which we will neglect. ΔG_B is the bulk free-energy change, which is defined on Fig. 8.3, for the case where the ordinate is free energy per unit volume. The two terms of Eq. 8.1 are plotted on Fig. 8.4 showing the negative bulk free-energy term that causes the process to occur and the positive surface free-energy term inhibiting the process, and it is seen that the total free-energy change contains a maximum at r^*. Figure 8.4 gives us the total free-energy change for a process where some n number of liquid atoms combine together to give a solid of size r. The actual nucleation process is not thought to occur by this large-scale type of fluctuation where n liquid atoms suddenly become solid. It is thought that a size distribution of small clusters of atoms exists in the liquid at any time and these clusters are considered potential nuclei. Due to thermal fluctuations these clusters continually gain and lose atoms. A nucleation event occurs when one of these clusters continues to gain more atoms than it loses. Consequently, we are interested in the following reaction:

$$\left\{\begin{array}{c} \text{Nucleus of } n \text{ atoms} \\ \text{and radius } r \end{array}\right\} + 1 \text{ atom} \Rightarrow \left\{\begin{array}{c} \text{Nucleus of } n+1 \text{ atoms} \\ \text{and radius } r+\Delta r \end{array}\right\}$$

If the free-energy change for this reaction is negative then a nucleation

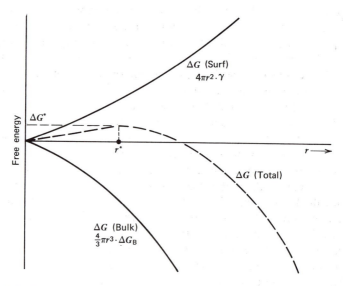

Figure 8.4   Free energy of formation of a nucleus as a function of its radius.

event is favored. Examination of Fig. 8.4 shows that for all values of $r > r^*$ this reaction has a negative free-energy change; that is, above r^* growth of a nucleus lowers the free energy. Therefore r^* is a critical radius and its value is determined by differentiation of Eq. 8.1 to be

$$r^* = \frac{-2\gamma}{\Delta G_B} \tag{8.2}$$

The free-energy change to form a critical sized nucleus, ΔG^*, is found by substituting into Eq. 8.1,

$$\Delta G^* = \frac{16\pi\gamma^3}{3(\Delta G_B)^2} \tag{8.3}$$

Nucleation results from the growth of clusters; whenever a cluster happens to reach the critical size radius it becomes the nucleus for the new phase.

We are now in a position to determine an expression for nucleation rate I. Let

$$C_n = \frac{\text{No. of nuclei of critical size } r^*}{\text{volume}}$$

$$\frac{dn}{dt} = \frac{\text{No. of atoms joining a nucleus}}{\text{sec}}$$

If a nucleation event occurs every time an atom adds to a cluster of critical size, we may write

$$I = C_n \frac{dn}{dt} = \frac{\text{nucleations}}{\text{sec.-vol.}} \tag{8.4}$$

First, consider the determination of an expression for dn/dt. This term involves the rate at which atoms add to a nucleus as shown in Fig. 8.5(a). In principle we should consider two reactions,

 (a) $S(n^* \text{ atoms}) + 1 \text{ atom} \rightarrow S(n^* + 1 \text{ atoms})$
 (b) $S(n^* + 1 \text{ atoms}) - 1 \text{ atom} \rightarrow S(n^* \text{ atoms})$

For a first approximation we will neglect the reverse reaction (b), which is also consistent with the form of Eq. 8.4. This will be called *assumption A*. Figure 8.5(b) shows a hypothetical free-energy curve for an atom as it moves from the liquid to the solid nucleus at a temperature below the freezing temperature. We will let ΔG_A be the activation energy for an atom jumping from the liquid onto the solid nucleus. We now list,

 1. $e^{-\Delta G_A/kT}$ = fraction of liquid atoms with free energy greater than ΔG_A; see Chapter 6, p. 161
 2. ν = vibration frequency of a liquid atom
 3. s = number of liquid atoms facing a solid nucleus across the interface
 4. $p = f \cdot A$
  f = probability the liquid atom is vibrating toward the nucleus
  A = probability that it does not bounce back by an elastic collision
with a solid atom

If every time a liquid atom at the interface vibrated, it jumped onto the solid we would have $dn/dt = \nu s$. However, we must reduce this number by

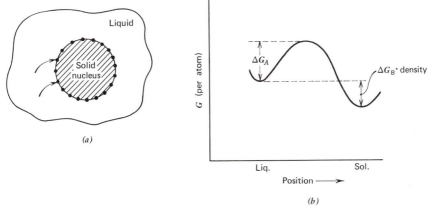

      (a)   (b)

Figure 8.5 (a) Schematic view of atoms joining a solid nucleus from the liquid. (b) Variation of free energy per atom as it moves from liquid onto solid nucleus.

the probability factor p of item 4 and the probability that an atom will have sufficient energy to jump, item 1. Hence we have

$$\frac{dn}{dt} = vs \cdot p \cdot e^{-\Delta G_A/kT} \tag{8.5}$$

In the second part of the analysis we determine an expression for the number of critical sized nuclei, C_n. In Chapter 5 we determined an expression for the equilibrium number of vacancies, Eq. 5.9,

$$\frac{n_e}{N} = e^{-\Delta G_v/kT} \tag{8.6}$$

Dividing the left-hand side of Eq. 8.6 by the volume v, we obtain

$$\frac{n_e/v}{N/v} = \frac{C_v}{d} = e^{-\Delta G_v/kT} \tag{8.7}$$

where C_v is a volume concentration of vacancies and d is a density. We may calculate the equilibrium number of critical sized nuclei by an analysis similar to that presented for vacancies in Chapter 5. We obtain

$$\frac{C_n}{d} = e^{-\Delta G^*/kT} \tag{8.8}$$

where ΔG^* is the free energy of formation of critical sized nuclei and d is a site density.

We imagine that an equilibrium distribution of clusters exists in the liquid,

$$Q_9 \rightleftarrows Q_{10} \rightleftarrows Q_{11} \ldots Q_n \rightleftarrows Q_{n+1} \ldots Q_{max-1} \rightleftarrows Q_{max} \tag{8.9}$$

where Q_n is a cluster of n atoms and Q_{max} is the largest size cluster. As the temperature is lowered to produce nucleation, this distribution shifts to the right. Once nucleation begins one expects the volume concentration of clusters to be slightly shifted from their equilibrium values. However, we neglect this effect here and call this *assumption B*: The volume concentration of critical sized nuclei equals their equilibrium concentration.

With the aid of this assumption we combine Eqs. 8.4, 8.5, and 8.8 to obtain

$$I = K_v \exp\left(-\frac{\Delta G_A + \Delta G^*}{kT}\right) \tag{8.10}$$

where $K_v = vs^*pd$.

As the temperature is lowered the equilibrium distribution of Eq. 8.9 shifts to larger size clusters. If one assumes that this shift occurs at a steady state (time independent), then it is possible to remove both assumptions A and B above. This shift at a constant rate would necessarily be a quasi-steady-state since it could not last too long. This quasi-steady-state analysis was originally proposed by Becker and Doring[2] in 1935 and it was extended to solids by Turnbull and Fisher.[3] A detailed review is given by Christian[4] that shows

$$K_v = \frac{vs^*pd}{n^*}\left[\frac{\Delta G^*}{3\pi kT}\right]^{1/2} \tag{8.11}$$

We are mainly interested in the temperature dependence of Eq. 8.10 because it is by control of temperature that we regulate and manipulate phase transformations. It has been estimated that the K_v/v term is within one or two powers of ten for all problems of interest.[3,4] Hence, we will take the K_v term constant and consider the temperature dependence of the two exponential terms in Eq. 8.10. The term involving ΔG_A is plotted in Fig. 8.6 and it can be seen that the temperature dependence of this term is similar to that of diffusion coefficients. We may take ΔG_A as temperature independent and, hence, the function rises sharply with temperature. The magnitude of this term gives us a measure of the mobility of the atoms.

The term involving ΔG^* is more difficult to treat because ΔG^* is temperature dependent. Equation 8.3 shows that ΔG^* depends on γ and ΔG_B. The surface tension γ is only a weak temperature function so that we have

$$[e^{-\Delta G^*/kT}] \propto [e^{-1/T\,\Delta G_B^2}] \tag{8.12}$$

To illustrate the temperature dependence of ΔG_B we examine the bulk

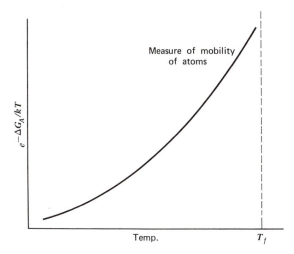

Figure 8.6 Temperature dependence of the ΔG_A term in Eq. 8.10.

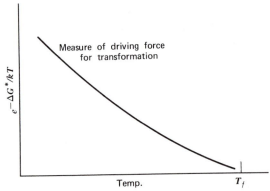

Figure 8.7   Temperature dependence of the ΔG^* term in Eq. 8.10.

free-energy curves shown in Fig. 8.3. It can be seen that ΔG_B is zero at the freezing temperature and it increases as the metal is supercooled below the freezing temperature. Therefore, Eq. 8.12 must be zero at the phase transformation temperature, T_f, and increase as the temperature drops, as shown in Fig. 8.7. Notice that this term is proportional to ΔG_B, which is really the source of the driving force causing the transformation to occur. Hence this term $e^{-\Delta G^*/kT}$ is a measure of the driving force. Taking the product of the terms in Fig. 8.6 and 8.7 we expect the nucleation rate of Eq. 8.10 to contain a maximum as shown in Fig. 8.8(a). As shown on this figure the nucleation rate is suppressed at low temperatures because of the low atomic mobility, whereas at high temperatures the driving force becomes small suppressing the rate. Consequently, we expect nucleation rates to peak at intermediate temperatures between the transformation

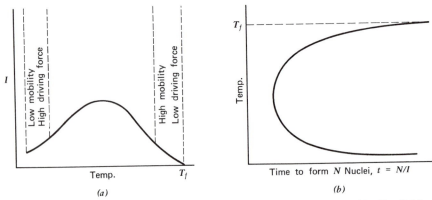

Figure 8.8   (a) Temperature dependence of nucleation rate predicted by Eq. 8.10. (b) Alternative plot of Eq. 8.10.

temperature and absolute zero. An alternative way of presenting this temperature dependence is shown in Fig. 8.8(b), where instead of plotting nucleation rate we plot the time to achieve a certain number of nuclei. This curve has the same form as the TTT curves used to describe the kinetics of phase transformations, which will be discussed later.

8.2 HETEROGENEOUS NUCLEATION

It should be clear from the above discussion that the reason that nucleation does not occur immediately upon reaching the transformation temperature is the barrier presented by the surface energy requirements of the nuclei. Consequently, physical systems undergoing phase transformations attempt to reduce this surface energy barrier by having nucleation occur upon a preexisting interface. In this way the preexisting interface is "wiped out" and consequently the net surface energy change may be reduced somewhat. We illustrate this process by considering the formation of a new phase, β, out of a parent phase, α, at the container wall. Figure 8.9(a) shows the new β phase forming as a spherical cap on the wall, that is, the β phase is a portion of the sphere having radius $r_{\alpha\beta}$. A top view of Fig. 8.9(a) would show the β phase as a circle with projected radius R. We now ask: What is the change in surface free energy produced by formation of this β phase? As above, we take the surface free energy equal to the surface tension and we have

$$\Delta G(\text{surf}) = [A_{\alpha\beta}\gamma_{\alpha\beta} + A_{\beta w}\gamma_{\beta w}] - A_{\beta w}\gamma_{\alpha w} \qquad (8.13)$$

where $A_{\alpha\beta}$ is the area of the $\alpha-\beta$ interface and $A_{\beta w}$ is the area of the $\beta-w$ interface. Notice that we must subtract the surface energy of the $\alpha-w$

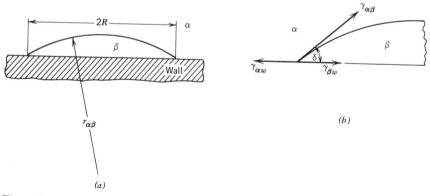

Figure 8.9 (a) Geometry of the spherical cap. (b) Surface tension diagram at the edge of the cap.

interface that was wiped out when the β phase formed, and the area of this wiped-out α-w interface is exactly $A_{\beta w}$. To proceed we construct the "surface tension force balance" at the edge of the three phases as shown in Fig. 8.9(b),

$$\gamma_{\alpha w} = \gamma_{\beta w} + \gamma_{\alpha \beta} \cos \delta \tag{8.14}$$

Letting $S = \cos \delta$ and recognizing that $A_{\beta w} = \pi R^2$ we obtain by combining Eqs. 8.13 and 8.14,

$$\Delta G(\text{surf}) = A_{\alpha \beta} \gamma_{\alpha \beta} - \pi R^2 (\gamma_{\alpha \beta} \cdot S) \tag{8.15}$$

The expression for the total free-energy change upon forming the spherical cap nucleus is written as

$$\Delta G = \Delta G(\text{bulk}) + \Delta G(\text{surf})$$
$$= V_\beta \, \Delta G_B + (A_{\alpha \beta} - \pi R^2 S) \gamma_{\alpha \beta} \tag{8.16}$$

To expand this equation we need the following expressions for the volume and surface area of a spherical cap:

$$V_\beta = \pi r_{\alpha \beta}^3 \left[\frac{2 - 3S + S^3}{3} \right]$$
$$A_{\alpha \beta} = 2 \pi r_{\alpha \beta}^2 [1 - S] \tag{8.17}$$
$$R = r_{\alpha \beta} \sin \delta$$

Substituting into Eq. 8.16 we obtain

$$\Delta G = \pi r_{\alpha \beta}^3 \left[\frac{2 - 3S + S^3}{3} \right] \cdot \Delta G_B + [2 \pi r_{\alpha \beta}^2 (1 - S) - \pi r_{\alpha \beta}^2 (\sin^2 \delta) S] \gamma_{\alpha \beta} \tag{8.18}$$

Recognizing that $\sin^2 \delta$ is $1 - S^2$ we obtain by algebra

$$\Delta G = [\tfrac{4}{3} \pi r_{\alpha \beta}^3 \Delta G_B + 4 \pi r_{\alpha \beta}^2 \gamma_{\alpha \beta}] \left[\frac{2 - 3S + S^3}{4} \right] \tag{8.19}$$

Comparing this result to Eq. 8.1 it is seen that the only difference is the term in the right-hand bracket. To obtain the critical value for $r_{\alpha \beta}^*$ we take the derivative, and similar to the results on Eq. 8.1 we obtain

$$r_{\alpha \beta}^* = -\frac{2 \gamma_{\alpha \beta}}{\Delta G_B} \tag{8.20}$$

Comparing this result to Eq. 8.2 it is seen that the radius of curvature of the spherical cap is identical to the radius of the sphere that would be obtained in homogeneous nucleation. Consequently, comparing Eqs. 8.1 and 8.19 it is apparent that the ΔG^* for heterogeneous nucleation will be identical to homogeneous nucleation except for the bracket term involving S in

Eq. 8.19,

$$\Delta G^*(\text{het}) = \Delta G^*(\text{hom})\left[\frac{2-3S+S^3}{4}\right] \qquad (8.21)$$

The term in brackets varies from 0 to 1 as the dihedral angle δ of Fig. 8.9(b) varies from 0° to 180°. Therefore, $\Delta G^*(\text{het}) < \Delta G^*(\text{hom})$, which shows that less energy is required for heterogeneous nucleation and consequently it should occur more easily. This result may also be seen by noting that

$$R^* = -\frac{2\gamma_{\alpha\beta}}{\Delta G_B}\sin \delta = r^*_{\alpha\beta}\sin \delta \qquad (8.22)$$

As δ decreases the value of R^* decreases, which indicates that the volume of the heterogeneous nucleus will become smaller and, hence, require fewer atoms for its formation. At $\delta = 0$ the volume becomes zero, so that one expects nucleation without any supercooling required.

There are many examples of the applicability of nucleation. Perhaps the most famous example involves rainmaking. Clouds consist of small drops of water plus water vapor. When the cloud temperature is below 0°C the water droplets become supercooled and ice crystals nucleate in the cloud and grow at the expense of the droplets. When the ice crystals are sufficiently large they fall as snow or are converted to rain or sleet depending on the temperature of the atmosphere.[5] Causing a supercooled cloud to release its water content as rain is essentially a nucleation problem. One desires to seed the cloud with particles that will cause nucleation to occur with a minimum of supercooling. Consequently one desires a seed material that has a very small dihedral angle δ with ice; see Fig. 8.9(b). From Eq. 7.26 we see that this requires a small value of $\gamma_{\text{ice-seed}}$. From our discussion of the previous chapter it should be clear that interfaces of small γ are characterized by a fairly good crystal matching at the interface. Hence, we need a seed crystal having a crystal structure that contains crystal planes similar in structure and lattice spacing to the lattice planes of ice. Crystals of silver iodide have proven effective in seeding clouds, and these crystals satisfy this crystal-matching requirement.[5] However, experiments have shown[6,7] that generally more is involved in successful seeding than just good crystal matching, but the exact nature of the additional factors involved remains obscure.

A pleasant example illustrating heterogeneous nucleation involves the formation of bubbles of CO_2 in a glass of beer or champagne. Observation reveals that the bubbles originate from certain select points, frequently on the glass wall. These points are probably small invisible cracks or other inhomogeneities on the wall that act as heterogeneous nucleation sites. We will be concerned with different types of solid–solid phase transformations and in these cases it is almost always true that nucleation occurs heterogeneously at grain boundaries or dislocation clusters in the metal.

There has been criticism of the classical theory of nucleation because the nuclei are so small as to be in the range of the interatomic force distance. Consequently, one suspects that applying macroscopic thermodynamic concepts, as in Eq. 8.1, may be an invalid extrapolation. This question[8] goes beyond the scope of this book but at the present time the classical theory of nucleation offers us the best framework for understanding the important phenomenon of nucleation.

REFERENCES

1. M. Volmer and A. Weber, Z. Physik. Chem. **119**, 277 (1925).

2. R. Becker and W. Doring, Ann. Phys. **24**, 719 (1935).

3. D. Turnbull and J. C. Fisher, J. Chem. Phys. **17**, 71 (1949).

4. J. W. Christian, *The Theory of Transformations in Metals and Alloys*, Pergamon, New York, 1965, Chapter 10.

5. L. J. Battan, *Cloud Physics and Cloud Seeding*, Anchor Books, Doubleday and Co., New York, 1962, pp. 60–64, 91–95.

6. A. Walton, Int. Sci. Tech. **2**, 28 (December 1966).

7. K. A. Jackson, Ind. Eng. Chem. **57**, 29 (1965).

8. D. R. Uhlmann and B. Chalmers. Ind. Eng. Chem. **57**, 19 (1965).

ADDITIONAL READING

1. D. Turnbull, The undercooling of Liquids, Scientific American Volume 212, 38 (January 1965).

2. A. Walton, Nucleation, Int. Sci. Tech. Volume 2, 28 (December 1966).

PROBLEMS

8.1 It was shown that for a spherical nucleus

$$\Delta G^* = \frac{16\pi\gamma^3}{3\Delta G_R^2} \quad \text{and} \quad r^* = \frac{-2\gamma}{\Delta G_R}$$

Show that ΔG^* is related to the volume of the critical sized nucleus, V^*, by the equation

$$\Delta G^* = -\frac{V^*}{2}\Delta G_B$$

8.2   Expressions for ΔG^* and r^* were also derived for the case where the nucleus was in the shape of a spherical cap (heterogeneous case). Show that for this case ΔG^* is related to the volume of the critical sized nucleus, V^*, by the same equation as found in Problem 8.1.

8.3   Suppose that a second-phase β nucleates from a primary phase in a polycrystalline metal. Consider two possible sites for nucleation of the second phase.

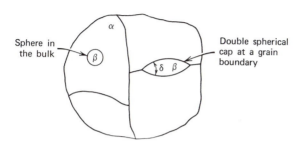

(a) It nucleates as a sphere within the bulk of a grain.

(b) It nucleates as a double spherical cap at a grain boundary. From Problems 8.1 and 8.2 above we have the following relations that apply to both case (a) and (b):

$$\Delta G^* = -\frac{V^*}{2}\Delta G_{\scriptscriptstyle B}, \qquad r^* = \frac{-2\gamma}{\Delta G_{\scriptscriptstyle B}}$$

The second equation shows that the critical radius of curvature r^* is the same, whether the nucleus is a sphere or a double spherical cap. To form a critical nucleus, a small cluster grows by fluctuations until it attains a critical radius of curvature that is independent of its shape. The critical volume will not be independent of shape.

Problem: (a) Determine an expression for the ratio of ΔG^*(sphere)$/\Delta G^*$(double spherical cap).

(b) If the dihedral angle δ were $120°$ would you expect the β nucleus to form first at the grain boundaries or in the bulk? Explain. When would you expect the nucleus to form first within the bulk?

8.4   (a) Show that when the surface tension of the α–α grain boundary equals twice the surface tension of the α–β grain boundary ($\gamma_{\alpha\alpha} = 2\gamma_{\alpha\beta}$) the critical volume V^* for the double spherical cap is zero. (b) Under these conditions would you expect supercooling? If not, explain why not in terms of the physical quantities involved here and also in terms of the mathematical equations.

8.5   Equation 8.2 is an expression for the radius of a critical sized nucleus. In the derivation the volume and the area of the nucleus were expressed as a function of the radius.

Derive an expression for the number of atoms in a critical sized nucleus for the cases where the nucleus is (a) a sphere and (b) a cube. In your derivation let

n = No. atoms in the nucleus

ΔG_B = bulk free-energy change per atom

γ = surface free-energy per area

V = volume per atom

Your answer should give the number of atoms in a critically sized nucleus n^* as a function of ΔG_B, γ, and V.

CHAPTER 9
SOLIDIFICATION

The solid–liquid phase transformation is perhaps the most important phase transformation that we study because almost all metals must undergo this transformation before becoming useful objects. We will examine solidification of alloys in a fair amount of detail because of its commercial importance and also because of its close similarity to other transformations to be discussed later.

9.1 NUCLEATION

Numerous experiments have been conducted on the nucleation of freezing in liquid metals within the past 20 years and the major findings of these studies are listed below.

1. Bulk samples (1 lb or so) can only be supercooled a few degrees centigrade. (However, metals having more complex crystal structures such as Sn, Bi, Ga, and the like will generally supercool 5–15°C.)

2. If a metal is subdivided into small particles (for example, by filing) then a very large supercooling is obtained in some of the particles. For example, Bi, 90°C; Ag, 227°C; Ni, 319°C.[1]

3. Bulk samples contained in molten glass can also be supercooled very much; for example, Bi, 102°C[2]; Ag, 250°C.[3] Also, bulk samples solidified in levitation in a hydrogen atmosphere have achieved large supercoolings, for example, Ni, 480°C.[4]

From Chapter 8 we know that metals supercool because the formation of the surface of the nuclei acts as a barrier to the nucleation. The more we supercool, the larger the free energy, ΔG_B, available to force the transformation, as may be seen graphically in Fig. 8.3. At the freezing temperature, ΔG_B is exactly zero so that we have

$$(\Delta G_B)_f = 0 = \Delta H_f - T_f \, \Delta S_f \tag{9.1}$$

where the subscript f refers to the freezing temperature. If the heat capacities of the liquid and solid metal are the same ($\Delta C_p = 0$) then from thermodynamics we know that the enthalpy and entropy changes upon freezing are independent of temperature. As examples, the average value of ΔC_p for freezing of Cu, Fe, Sn, Pb, Ni, and Al is only 3.6%. Hence, to a good approximation we may take ΔH and ΔS constant with temperature, as ΔH_f and ΔS_f,

$$\Delta G_B(T) = \Delta H_f - T \Delta S_f \tag{9.2}$$

Combining Eqs. 9.1 and 2 we obtain

$$\Delta G_B = \Delta H_f \cdot \frac{\Delta T}{T_f} = \Delta S_f \cdot \Delta T \tag{9.3}$$

where $\Delta T = T_f - T$. This is a very important equation because it provides us with a good first-order approximation of the amount of free energy available for any phase transformation as a function of the amount of supercooling. Combining this equation with Eq. 8.2 we obtain the temperature dependence of the critical size nucleus as

$$r^* = -\frac{2\gamma_{sl}}{\Delta S_f \cdot \Delta T} \tag{9.4}$$

which is plotted in Fig. 9.1. This result shows that as supercooling increases, the size of a critical nucleus decreases. Consequently, nucleation becomes easier at lower temperatures because fewer atoms are required for a critical nucleus to form.

In homogeneous nucleation the nucleus forms from a cluster in the liquid. X-ray studies of liquid metals have shown that there is no long-range order in the liquid but that a short-range order does persist in the liquid. On a statistical average there is a definite grouping of atoms around any given atom. When a fcc metal melts, the number of nearest-neighbor atoms drops from 12 to around 8. Consequently, a statistical

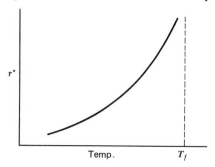

Figure 9.1 Variation of critical sized radius for nucleation with temperature.

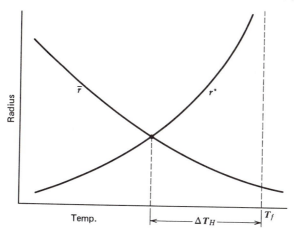

Figure 9.2 Illustration of homogeneous nucleation temperature at point where critical radius equals maximum cluster size radius.

distribution of clusters exists in the liquid metal, and we will take the largest cluster size at any temperature to have radius $\bar{r}$. As the temperature is lowered, the value of $\bar{r}$ will increase because of the reduction in thermal energy. The shape of this curve will depend on the structure of the liquid, but it must appear qualitatively as shown in Fig. 9.2. Figure 9.1 has been superimposed upon this $\bar{r}$ plot, and it can be seen that at temperatures below $T_f - \Delta T_H$ the maximum cluster size, $\bar{r}$, is greater than the critical size nucleus, r^*. Consequently, ΔT_H is the supercooling required for homogeneous nucleation. It would be impossible to supercool below ΔT_H because the structure of the liquid provides nuclei larger than r^* below this amount of supercooling. It should be clear from this presentation that the homogeneous nucleation temperature will depend on the liquid structure.

From Chapter 8 we know that heterogeneous nucleation occurs more easily and, hence, at smaller supercooling. It only takes one single nucleus to cause a sample to solidify, because once nucleated in a supercooled liquid, the solid quickly grows to transform the entire liquid. Consequently, it is thought that the reason that bulk liquids can only be supercooled a few degrees is that it is very improbable that no impurity particles will exist in a bulk sample. This would require particle purities on the order of 1 part in 10^{15} by volume. Since it is difficult to achieve purities of 1 part in a million, it seems reasonable to expect that a sufficient number of impurity particles will always be present in bulk samples to give heterogeneous nucleation.

If a high-purity bulk sample is subdivided into small liquid drops, then it is reasonable to expect that some of the liquid drops will be free of impurity particles. Consequently, these liquid drops should display homogeneous nucleation if they are not nucleated at the surface or by some other effect. This reasoning accounts for the large supercoolings described under point 2 above, and it was felt for some time that these small liquid drops displayed homogeneous nucleation.

The fact that bulk samples have recently demonstrated large supercoolings when contained in molten glass or levitated in hydrogen is probably due to the effective removal of the nucleating particles under these conditions. Since the supercoolings under these conditions are actually more than in the fine-particle experiments (compare 2 and 3 above) there is presently some doubt whether true homogeneous nucleation has been achieved in any of these experiments. It is clear, however, that in commercial operations nucleation within liquid metals virtually always occurs heterogeneously.

It is generally desirable in casting metals to achieve as small a grain size as possible because small grain size significantly improves the mechanical properties of metal castings. A significant grain refinement can be achieved by the addition of innoculents to the liquid metal. These innoculents are added in a suitable form to be distributed uniformly throughout the liquid and they act as nucleating agents to increase the nucleation rate throughout the casting. In aluminum alloys, for example, Ti and B are added in the form of potassium fluoride salts, which decompose to form elemental Ti and B, which react with each other and the Al. Recent research[5] indicates that the grain refinement is produced by nucleation upon the TiAl₃ compound that forms by a peritectic reaction. The exact mechanism by which innoculents enhance nucleation is generally unknown and the most effective innoculents are determined by trial-and-error methods.

In view of the fact that liquid metals are easily supercooled, one might wonder if solid metals can be superheated, that is, heated above their melting points without melting. Experiments show that the answer to this question is generally no. Figure 9.3(a) shows a solid metal in contact with a vapor phase. If a small quantity of liquid metal forms on the solid surface as in Fig. 9.2(b) it will quickly cover the entire surface as shown in Fig. 9.3(c) because it has been observed that liquid metals always wet their own solids.

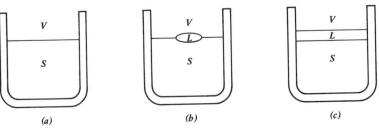

Figure 9.3   Melting at a free surface.

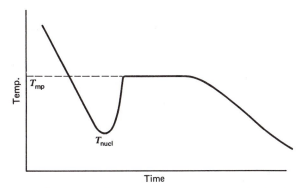

Figure 9.4 A cooling curve illustrating supercooling and lack of superheating.

Consequently, if we make a "surface tension balance" on the drop as in Fig. 7.28 we find that wetting requires $\gamma_{SV} \geq \gamma_{LV} + \gamma_{SL}$ as discussed on p. 193. Suppose now we consider the surface free-energy change for the melting reaction illustrated in going from Fig. 9.3(a) to 9.3(c):

$$\Delta G(\text{surf}) = G(\text{final}) - G(\text{initial})$$
$$= A[\gamma_{LV} + \gamma_{SL} - \gamma_{SV}] \qquad (9.5)$$

where the area of the interfaces, A, is the same for all three interfaces. From the above discussion it should be clear that the term in brackets will be negative since liquid metals wet their solids. Consequently, for melting at the surface of solids in contact with vapor, the surface free-energy change is negative, which means that the free energy drops on melting. Therefore, there will be no surface energy barrier to melting and no superheating. If one can melt a solid so the liquid formed does not contact a vapor phase then it is possible to superheat the solid. However, this is difficult to achieve even in laboratory experiments.

An interesting application of these ideas involves the simple cooling curve as shown in Fig. 9.4 that is obtained upon freezing pure metals. The liquid metal will invariably supercool below the freezing temperature before nucleation occurs. After nucleation occurs at $T = T_{nucl}$ the temperature quickly rises due to the latent heat evolved upon freezing. However, this rise in temperature abruptly stops when the melting point temperature, T_{mp}, is reached. Since superheating may not occur, the temperature can only rise above T_{mp} if the solid just formed is melted. The latent heat by itself is insufficient to cause this melting and so the temperature rise stops at T_{mp}.

There are three main points in this section:

1. Liquid metals supercool due to a surface energy barrier of the solid nuclei.

2. Solid metals do not superheat because there is no surface energy barrier when the melting occurs at a surface.

3. Nucleation for solidification is virtually always heterogeneous. Consequently, supercooling is increased by removal of particles by such means as purification or subdivision. Nucleation is increased and fine-grained structures promoted by addition of suitable heterogeneous nucleation sites.

9.2 SOLIDIFICATION OF PURE METALS

Once a solid nucleus forms in a cooling metal the transformation is often completed by growth of this solid nucleus rather than by further nucleation. We will now consider various aspects of the growth of the solidifying metal.

A. KINETICS OF THE ATOMIC PROCESSES AT THE SOLID–LIQUID INTERFACE

Consider a moving solid–liquid interface as shown in Fig. 9.5. We may imagine that two atomic processes occur at this interface,

Sol. atom → Liq. atom: Melting reaction

Liq. atom → Sol. atom: Freezing reaction

By comparison with the presentation on p. 223 we may write the rate per unit area of these two processes as

$$\left(\frac{dn}{dt}\right)_M = \text{Rate melting}(S \to L) = p_M s_s \nu_s e^{-\Delta G_M/kT} \tag{9.6}$$

$$\left(\frac{dn}{dt}\right)_F = \text{Rate freezing}(L \to S) = p_F s_l \nu_l e^{-\Delta G_F/kT} \tag{9.7}$$

where s_s, s_l = no. atoms/area on solid and liquid interface, respectively;

ν_s, ν_l = vibration frequency of solid and liquid atoms, respectively;

$p_{M,F} = f_{M,F} \cdot A_{M,F}$

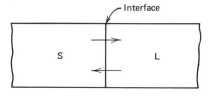

***Figure* 9.5** The atomic reactions at the solid–liquid interface.

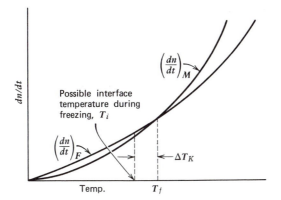

Figure 9.6 The temperature dependence of the melting and freezing rates.

where $f_{M,F}$ is the probability that an atom of sufficient energy is moving toward the interface and $A_{M,F}$ is the probability that an atom is not kicked back by an elastic collision upon arrival.

ΔG_M = activation energy for an atom jumping $S \rightarrow L$.

ΔG_F = activation energy for an atom jumping $L \rightarrow S$.

The product $s\nu$ is the jump frequency per area if the atoms jumped everytime they vibrated. The term p accounts for the fact that the atoms will only jump if they are moving in the correct direction and if they do not bounce back. The term $e^{-\Delta G/kT}$ accounts for the fact that only those atoms having sufficient energy will be able to jump.

At equilibrium the flux of atoms jumping into the liquid must be exactly matched by the atom flux onto the solid, that is, $(dn/dt)_M = (dn/dt)_F$, at T_f. Consequently, a plot of Eqs. 9.6 and 9.7 must have the form shown in Fig. 9.6. It should be clear that in order to have freezing, more atoms must jump from the liquid to the solid than vice versa, that is, $(dn/dt)_F > (dn/dt)_M$. Figure 9.6 illustrates that it is not possible to freeze a metal if the solid–liquid interface temperature is exactly at the freezing temperature. For the interface to move it must be at some temperature below T_f in order to satisfy the condition $(dn/dt)_F > (dn/dt)_M$. Consequently a solidifying interface must always be undercooled by some finite amount, which we term ΔT_K. The following points are listed for emphasis:

1. The temperature of a freezing interface must be less than the equilibrium freezing temperature, T_f.

2. This undercooling is called the *kinetic* undercooling, ΔT_K.

3. $\Delta T_K = T_f - T_i$ is the undercooling at the interface required to produce a *net* flux of atoms from liquid to solid.

It is generally true that there is a kinetic undercooling or a kinetic superheating associated with any moving phase boundary, whether it be a solid–liquid boundary, a solid–solid boundary, a solid–vapor boundary, and so on.

B. TEMPERATURE DISTRIBUTION AT THE SOLID–LIQUID INTERFACE

We will see below that the temperature profile at the solid–liquid interface is an important factor in the control of the shape of the interface. Therefore, it is helpful to make a clear distinction between two types of profiles.

1. Positive Gradient. For simplicity consider a bar of metal solidifying under conditions of unidirectional heat flow. Generally, the temperature of the liquid is higher than the solid and the temperature profile is as shown in Fig. 9.7. Notice that the interface is undercooled by ΔT_K in order to drive the interface reactions, and the temperature gradient in the liquid is positive.

2. Negative Gradient. Suppose an ingot of pure liquid tin is slowly cooled so that the entire ingot is at a constant temperature about 15°C below T_f as shown by curve 1 of Fig. 9.8(a). Solid tin now nucleates at the walls and begins to grow inward. The temperature of the solid-liquid interface will be $T_f - \Delta T_K$, and since ΔT_K is generally less than 1°C, the temperature profile after nucleation must be as shown by curve 2 of Fig. 9.8(a). A negative temperature gradient is formed in the liquid for an interface moving to the right, and this case is illustrated for unidirectional heat flow in Fig. 9.8(b). Hence, it is possible to achieve both negative and positive temperature gradients in the liquid ahead of an advancing solid–liquid interface.

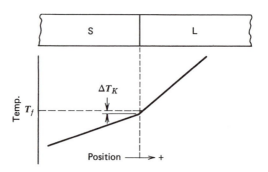

Figure 9.7 The positive temperature gradient at a solid–liquid interface.

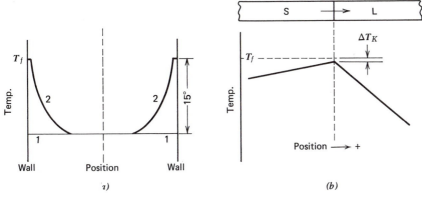

Figure 9.8 The negative temperature gradient at a solid–liquid interface. (*a*) Origin due to supercooling; (*b*) unidirectional heat flow case.

C. INTERFACE MORPHOLOGY (SHAPE)

The discussion of interface shape will be divided into two sections depending on the type of temperature gradient in the liquid at the interface.

1. Positive Gradient. Two distinct types of interfaces are observed; they are shown in Fig. 9.9. The faceted interface of Fig. 9.9(*a*) displays a jagged interface consisting of well-defined planes (facets) that clearly show the crystalline character of the solid. The general solid–liquid interface lies parallel to the freezing temperature isotherm, but the faceted planes are generally at angles to the T_f isotherm. The nonfaceted interface of Fig. 9.9(*b*) is simply a planar interface that lies parallel to the T_f isotherm.

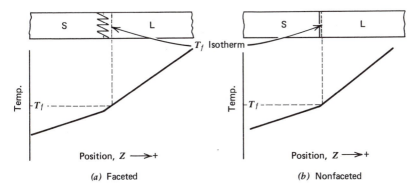

Figure 9.9 The two interface morphologies observed in pure metals with a positive gradient. (*a*) The faceted interface; (*b*) the nonfaceted or planar interface.

Table 9.1 Correlation of Interface Morphology with the Entropy of Melting, ΔS_m ($R =$ gas constant)

Materials	$\Delta S_m/R$	Morphology
All regular metals and some organics	<2	No facets
Semimetals and semi-conductors, Bi, Sb, Ga, Ge, Si	2.2–3.2	Facets observed
Most inorganics	>3.5	Facets observed

a. Occurrence. Due largely to the work of K. A. Jackson[6] we now know that an extremely good correlation exists between the entropy of melting, ΔS_m, of a material and whether it freezes with a faceted or nonfaceted interface. As shown in Table 9.1 all regular metals freeze with nonfaceted interfaces. Although the semimetals and semiconductors usually freeze with a faceted interface, they will freeze with a nonfaceted interface under certain conditions of freezing rate and temperature gradient. In a book devoted to metals one might be tempted to disregard faceted growth since all regular metals solidify with a nonfaceted interface. However, the two most important casting alloys from a commercial point of view are cast iron and Al–Si alloys, and the chemical elements responsible for their useful properties, C and Si, both freeze with a faceted interface.

b. Kinetic Undercooling. The results here are nicely distinct, and fall into the following ranges.

Faceted growth: $\Delta T_K \approx 1\text{–}2°C$

Nonfaceted growth: $\Delta T_K \approx 0.01\text{–}0.05°C$

The kinetic undercooling in nonfaceted growth is so small that to the present time there has been no accurate measurement of it. Hence, in pure metals the solid–liquid interface will lie essential upon the T_f isotherm in the system.

c. Mechanism. The difference in the morphologies is due to a distinctly different atomic mechanism by which the liquid atoms attach themselves to the solid interface. The faceted interface is thought to grow by the motion of small ledges as shown in Fig. 9.10(a). The atoms add only at the ledges so that the interface advances only when a ledge passes along it. The nonfaceted interface is thought to advance by addition of atoms at all points of its surface as indicated in Fig. 9.10(b). The nonfaceted interface is sometimes called a diffuse interface; we know very little about the structure of such interfaces.

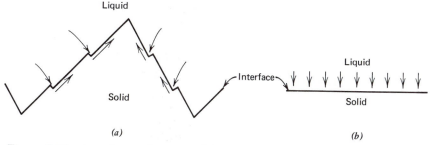

Figure 9.10 Atomic attachment at (*a*) the faceted interface and (*b*) the non-faceted interface.

d. Preferred Orientation. The faceting planes of the faceted interface are certain preferred crystallographic planes, that is, certain {*hkl*} planes. Consequently, faceted interfaces display a definite preferred orientation of the solid crystal structure relative to the interface plane.

The nonfaceted interface may lie on any crystal plane. However, if grown over a distance of a few centimeters the solid crystal will generally rotate to cause one preferred set of {*hkl*} planes to lie in the S/L interface plane. Consequently, nonfaceted growth displays a weak preferred orientation that has been found to be sensitive to very small impurity additions.[7]

2. Negative Gradient. We first consider the case of nonfaceted materials. It can be seen from Fig. 9.11 that the liquid in front of the advancing interface is supercooled since the temperature gradient is negative. Let the amount of supercooling be defined as $S = T_f - T$. It has been determined experimentally that when $dS/dZ > 0$ the planar interface becomes unstable and degenerates into a dendritic interface. Dendrite is the Greek word for tree, and it describes the treelike projections that form on the solid surface after it becomes unstable. For the case where the supercooling is quite

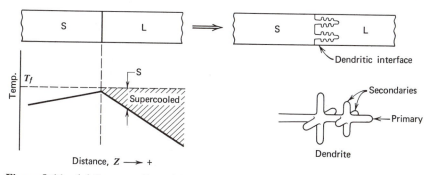

Figure 9.11 (*a*) Supercooling with a negative temperature gradient. (*b*) Resulting dendritic morphology.

Table 9.2 Preferred Crystallographic Growth
Directions of Dendrites

Crystal Structure	Crystallographic Direction along Axis of Dendrites
fcc	$\langle 100 \rangle$
bcc	$\langle 100 \rangle$
bct (Sn)	near $\langle 110 \rangle$
hcp	$\langle 10\bar{1}0 \rangle$

small, interface morphologies intermediate between planar and dendritic
are obtained. These morphologies are called cellular and they will be
discussed briefly later.

Now consider the case of faceting materials growing in a negative
gradient. Whether or not dendrites form depends on the value of $\Delta S_m/R$.
Faceting materials in the lower range of $\Delta S_m/R$ form dendrites in a
negative gradient, whereas faceting materials in the upper range of $\Delta S_m/R$
continue to freeze with the faceted morphology rather than the dendritic
morphology.

Figure 9.11 shows a sketch of a dendrite illustrating a main branch and
several secondary side branches. Dendrites exhibit a preferred orientation.
Each dendrite is a single crystal, and it has been found that in each crystal
system the crystallographic direction parallel to the dendrite axis is the
same. This is illustrated by the data presented in Table 9.2.

The kinetic undercooling at the tips of the advancing dendrite interface
is of the same order of magnitude as for the positive-gradient case. It will
be somewhat larger, however, because dendrites generally grow at a higher
rate than planar interfaces.

D. INTERFACE RATE CONTROLLED BY HEAT FLOW

Since the kinetic undercooling, ΔT_K, is quite small in metals, the interface
temperature will always lie very closely along the freezing temperature
isotherm in the system. The location of the freezing temperature isotherm
is controlled by the heat transfer conditions in the system. Consequently,
the rate of solidification is controlled by the heat flow. This may be
illustrated by a simple heat balance at the interface of a solidifying rod such
as shown in Fig. 9.9. The heat balance involves three terms:

1. Diffusive heat flux from the liquid into the interface, $-k_l(dT/dZ)_l$.

2. Diffusive heat flux from the interface into the solid, $-k_s(dT/dZ)_s$.

3. Heat flux generated by the latent heat at the interface, $-L\rho \cdot R$

where L is the latent heat per gram, ρ is the density, R is the interface rate, and k_s and k_l are the thermal conductivities of the solid and liquid, respectively:

$$-k_s\left(\frac{dT}{dZ}\right)_s = -k_l\left(\frac{dT}{dZ}\right)_l - L\rho R \qquad (9.8)$$

All three terms are written with a minus sign because the fluxes are all in the negative direction and each of the individual terms are positive quantities. Solving for the interface rate we have

$$R = \frac{k_s(dT/dZ)_s - k_l(dT/dZ)_l}{L\rho} \qquad (9.9)$$

This equation gives us the rate of advance of a solidifying interface, and it applies in general if we replace the derivatives with normal derivatives at the interface in question. We will find that the rate of solid–solid phase transformations is generally *not* controlled by heat flow as is this solid–liquid transformation.

9.3 SOLIDIFICATION OF ALLOYS

When an alloy solidifies, the solid that forms generally has a different composition than the liquid from which it is freezing. Therefore, the distribution of a solute in the solid will generally be different than it was in the liquid prior to freezing. This redistribution of the solute produced by solidification is frequently termed *segregation*. In this section we are interested in quantitatively describing segregation. The discussion will be limited to binary alloys in order to simplify the presentation.

A. EQUILIBRIUM FREEZING

Figure 9.12 presents the solid–liquid region of a eutectic-type phase diagram. At any temperature such as T_1 this diagram tells us the fractional compositions of the solid and of the liquid that will be in equilibrium with each other. The equilibrium distribution coefficient, k_0, is defined as the ratio of the equilibrium solid and liquid compositions:

$$k_0 = \frac{X[\text{solidus at } T]}{X[\text{liquidus at } T]} = \frac{X_s}{X_l} \qquad (9.10)$$

Consequently, k_0 is the ratio of the solidus to liquidus composition at any temperature of interest. You may show yourself that k_0 is a constant independent of composition if the solidus and liquidus are straight lines.

If one cools a volume of liquid of uniform composition X_0 to temperature T_1, then solid composition $k_0 X_0$ will begin to form if no nucleation barrier interferes. If the temperature is lowered to T_2, and if equilibrium

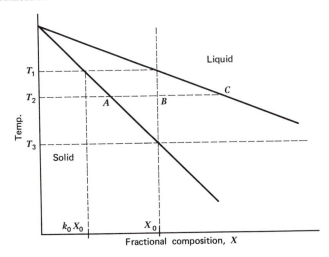

Figure 9.12 A phase diagram.

freezing occurs, then from the well-known lever law we have

$$\text{Wt fraction liquid} = \frac{\text{Length } A\text{-}B}{\text{Length } A\text{-}C} \qquad (9.11)$$

This result is obtained from mass balances on the liquid and solid for (1) the solute component, and (2) both components together. One obtains two equations that combine to give Eq. 9.11, the lever law. In deriving this result one assumes that *all* of the liquid is at composition C and *all* of the solid is at composition A. Consequently, the lever law requires *all* of the liquid and *all* of the solid to have uniform composition. This latter requirement is infrequently obtained in solidification, so that the lever law has only very limited usefulness in describing the solidification of real systems. For instance, the lever law indicates that the last liquid freezes at temperature T_3 and at this point *all* of the solid has composition X_0. This condition is infrequently obtained because the first solid formed with a composition $k_0 X_0$, and it is difficult for this solid to increase its composition to X_0 by the time the last liquid freezes.

B. NONEQUILIBRIUM FREEZING

In this discussion we make two major assumptions that will later be examined in detail:

I. Uniform liquid composition
II. A flat solid–liquid interface

Consider now a horizontal cylinder of liquid alloy having uniform composition X_0. This initial uniform composition is plotted below the alloy in Fig. 9.13(a). We now cool the left end of the rod allowing a small amount of solid to form. The solid–liquid interface is then located at position 1. The area under the curve of Fig. 9.13(a) between any two points Z_1 and Z_2 is proportional to the mass of solute in the rod between points Z_1 and Z_2. After the initial small volume of liquid freezes, the composition in this volume has dropped from X_0 to k_0X_0. Therefore, a mass of solute proportional to the cross-hatched area to the left of position 1 on Fig. 9.13(a) has been removed from the solid and rejected into the liquid. This must cause the liquid composition to rise above X_0 a small amount as shown. We know from the previous section that at the solid–liquid interface the temperature will be within 0.01–0.05°C of the equilibrium temperature. Hence, we consider that a *local equilibrium* exists at the solid–liquid interface. This means that Eq. 9.10 applies at the interface so that we may write $X_s = k_0X_l$. The composition of the solid and liquid are "tied together" at the interface by this equilibrium relation. Consequently, as the liquid composition rises due to the solute rejection at the interface, the solid composition must also rise. After the interface has advanced to position 2, the solid composition will have gradually risen as shown in Fig. 9.13(a) due to the increasing liquid composition, and the interface temperature will have dropped from 1 to 2 as shown in Fig. 9.13(b). Notice that this is not equilibrium freezing because the solid has a nonuniform composition. A uniform solid composition could only be obtained by solid-state diffusion, which is frequently so small that it has an insignificant effect on the solid composition profile. After the entire rod has solidified

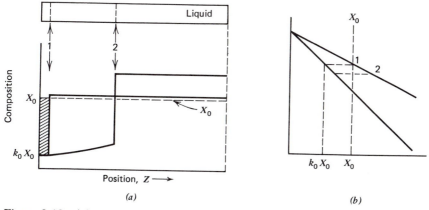

(a)   (b)

Figure 9.13 (a) Composition profile along a rod solidified from the left end. (b) Corresponding phase diagram for alloy of (a).

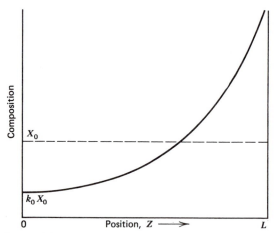

Figure 9.14 Composition profile after entire rod of Fig. 9.13(a) has solidified, illustrating normal solidification.

the composition profile will appear as shown in Fig. 9.14. Notice that we obtain a purification in the left end of the bar and consequent segregation of solute to the right end. This type of segregation is called *macrosegregation* because it extends over long distances, on the order of the length of the bar. This type of behavior is also frequently called *normal solidification*, because one normally expects the first part of the solidified rod to contain solid of composition near to $k_0 X_0$. Later when we relax assumption II we will see that one may actually reverse Fig. 9.14.

1. Derivation of the "Normal Freeze Equation." It is fairly simple to derive the equation for the curve of Fig. 9.14. In this derivation we will use volume concentration, C, instead of weight fraction composition X. The relation between the two is simply $C = X \cdot$ (density). A mass balance will be made on the solute within a differential volume of liquid, $A\,dZ$, which freezes. This amount of solute is given by Eq. 9.12 and it is proportional to the cross-hatched area shown on Fig. 9.15(a):

$$dM[\text{before freezing}] = C_l A\,dZ \tag{9.12}$$

After freezing the solute is redistributed between the liquid and solid as shown in Fig. 9.15(b), so that we may write

$$dM[\text{after freezing}] = C_s A\,dZ + dC_l \cdot A[L - Z - dZ] \tag{9.13}$$

Notice that we have not allowed any solute to diffuse into the solid out of the volume element. Hence, our derivation neglects all solid-state diffusion. We will also assume equal liquid and solid density so that

$$k_0 = \frac{X_s}{X_l} \approx \frac{C_s}{C_l} \tag{9.14}$$

Equating 9.12 and 9.13 and employing Eq. 9.14 we obtain after a little algebra

$$\int_0^Z \frac{(1-k_0)\,dZ}{L-Z} = \int_{C_0}^{C_1} \frac{dC_l}{C_l} \qquad (9.15)$$

To integrate Eq. 9.15 we recognize that at $Z=0$ the liquid composition must be the original composition, C_0, because no solid has yet formed. Assuming k_0 constant we integrate this equation to obtain

$$C_l(Z) = C_0 \left[\frac{L-Z}{L}\right]^{k_0-1} \qquad (9.16)$$

Since the solid composition at any position Z is simply $k_0 C_l(Z)$ we obtain the *normal freezing equation*,

$$C_s(Z) = k_0 C_0 \left[1-\frac{Z}{L}\right]^{k_0-1} \qquad (9.17)$$

In addition to assumptions I and II this equation is restricted to

III. Negligible solid-state diffusion
IV. k_0 = constant
V. Equal solid and liquid densities

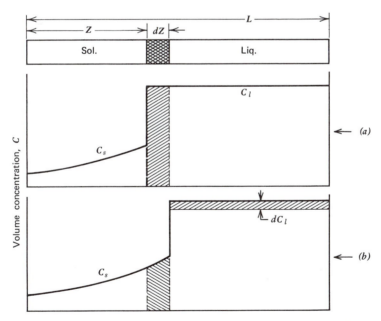

Figure 9.15 Solute redistribution during solidification of a volume element dZ. (*a*) Solute profile before volume element freezes. (*b*) Solute profile after volume element freezes.

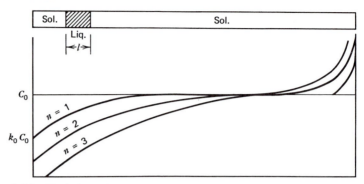

Figure 9.16 Variation of composition along a rod produced by zone melting; n equals the number of zone passes.

If we replace volume concentration C by weight fraction concentration X, and Z/L by the weight fraction of the alloy solidified then it may be shown that Eq. 9.17 is obtained in terms of these variables without needing assumption V. Restrictions III and IV are frequently satisfied in practice. Equation 9.17 presents us with a first-order approximation to nonequilibrium freezing, which for some cases is quite adequate.

C. ZONE MELTING

It was pointed out above that normal freezing produces a purification in the first portion of the rod to solidify, as shown in Fig. 9.14. In 1952 W. G. Pfann[8,9] published a famous paper showing that a much more effective way of purification could be obtained by the technique of zone melting. In this technique a small zone of length l is melted in the rod and it is passed down the rod as illustrated in Fig. 9.16. After the zone has passed down the rod the first time ($n = 1$) a purification is obtained as shown. By carrying out a mass balance similar to above for the solute in the zone, you may show yourself that after one pass of the zone the concentration in the rod would be

$$C_s = C_0[1 - (1 - k_0)e^{-k_0 Z/l}] \tag{9.18}$$

The purification after one pass is less than that obtained in normal freezing. However, with zone melting one may repeatedly pass the zone down the rod and an increased purification is obtained after each pass, until after many passes an ultimate distribution is reached. The equations for concentration versus length after more than one pass of the zone[8] are more complex than Eq. 9.18 and will not be discussed here.

Zone melting is an extremely effective method of removing impurity elements having k_0 values of less than 0.5. For example, with an impurity of $k_0 = 0.1$ the average concentration of the impurity in the first half of the rod is reduced by a factor of around 1000 after only five passes (Ref. 8, p. 287). It is important to remember, however, that this result as well as Eq. 9.18 are subject to the same five restrictions that apply to the normal freeze equation as listed above. Again, the last three assumptions generally have very little influence in practice. However, the first two restrictions, I and II, are quite important and are discussed separately in the next two sections.

D. MIXING IN THE LIQUID

In the derivations of both the normal freeze equation (9.17) and the zone melting equation (9.18) the assumption of uniform liquid composition was employed (assumption I, p. 246). This turns out to be a very severe restriction, as we shall now see.

Convection currents in the liquid fraction of a solidifying alloy will tend to produce a uniform liquid composition. Natural convection is difficult to eliminate from liquid metal alloys because of their low viscosity and high density, so that one might expect a uniform liquid composition to generally be obtained. However, there is a fundamental characteristic of fluid flow that prevents this. When a fluid flows past a solid surface the velocity of the fluid right at the surface is always found to be zero. This is called the no-slip condition. It is not at all obvious that it should obtain. If one passes a fluid down a pipe at a low velocity as in Fig. 9.17(a) the fluid flows parallel to the pipe wall at all points and is termed laminar flow. The flow velocity is a maximum at the center and drops parabolically to zero as the wall is approached. At very high velocities the flow becomes turbulent, involving many eddies and swirls in the center of the pipe where the mean velocity is almost constant across the diameter. However, the velocity must always drop to zero at the wall. Hence, a thin boundary layer of laminar flowing fluid is always present at the wall. Such a boundary layer is always present in the liquid at the solid–liquid interface, and it inhibits a uniform liquid composition.

As explained above, solute is continually being rejected from the solid into the liquid at the solid–liquid interface during solidification. In order to obtain a uniform liquid composition this solute must be transported very quickly throughout the liquid. Two transport mechanisms are available to

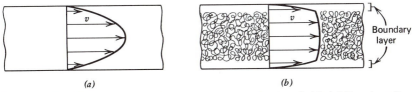

(a)   (b)

Figure 9.17 The velocity profile across a tube of flowing fluid. (a) Laminar flow. (b) Turbulent flow.

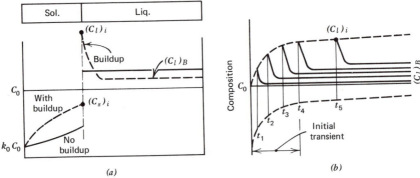

Figure 9.18 (a) Effect of a solute buildup in the liquid upon composition of solidified rod. (b) Establishment of the solute buildup during the initial transient.

mix this solute into the liquid, diffusion and fluid flow (convection). Of these two, diffusion is a much slower mechanism. In the boundary layer at the interface there can be no convective transport normal to the interface because of the laminar flow parallel to the interface. Solute can only be transported through the boundary layer into the convecting liquid by the slow mechanism of diffusion. Consequently, a buildup of solute is obtained in the boundary layer region as is shown by the dashed curve of Fig. 9.18(a). Beyond the boundary layer the bulk liquid composition is uniform at the value $(C_l)_B$ due to mixing. Since local equilibrium is closely approached at the interface we have $(C_s)_i = k_0(C_l)_i$ (where we use volume concentrations and assume constant density). The solute buildup causes $(C_l)_i$ to rise rapidly and, hence, $(C_s)_i$ must also rise rapidly. Therefore, the solid composition rises more rapidly than for the case of no buildup, as is illustrated in Fig. 9.18(a). It should be apparent that this buildup in the boundary layer region has a very strong effect upon the solid composition profile

Consider the boundary layer region at the start of freezing. Solute is being rejected into the boundary layer region from the moving interface, which causes the solute to build up in this region as shown in Fig. 9.18(b). As the solute builds up, however, its concentration gradient across the boundary layer becomes steeper, and the rate of transport through the boundary layer by diffusion increases until eventually a balance is obtained between the input and output to the boundary layer region. At this point the buildup stops rising such that the ratio $(C_l)_i/(C_l)_B$ becomes constant. The region over which the buildup occurs is called the initial transient; see Fig. 9.18(b).

It is useful to define a term called the *effective distribution coefficient, k_e*:

$$k_e = \frac{(X_s)_i}{(X_l)_B} \approx \frac{(C_s)_i}{(C_l)_B} \qquad (9.19)$$

After completion of the initial transient the effective distribution coefficient is constant. The value of k_e tells a great deal about the effect of liquid mixing upon the solute profile that is frozen into the solid. The two limiting cases are shown in Figs. 9.19(a) and 9.19(b). In the first case $k_e = 1$ so that $(C_s)_i = (C_l)_B$. Here the buildup carries the solid composition all the way up to the original composition, C_0, and the mixing is not sufficient to raise $(C_l)_B$ above C_0. This result is obtained for the case of very little or no mixing. In case (b), $k_e = k_0$, we must have $(C_l)_i = (C_l)_B$ (equate Eqs. 9.10 and 9.19), which means there is no buildup. Hence, there is uniform liquid composition and we have the result of Eq. 9.17 discussed above. For the intermediate case, Fig. 9.19(c), the initial transient is followed by a gradually rising solid composition similar to case (b) but at a higher level of composition. Since k_e is so important we now derive an equation for its dependence upon measurable parameters.

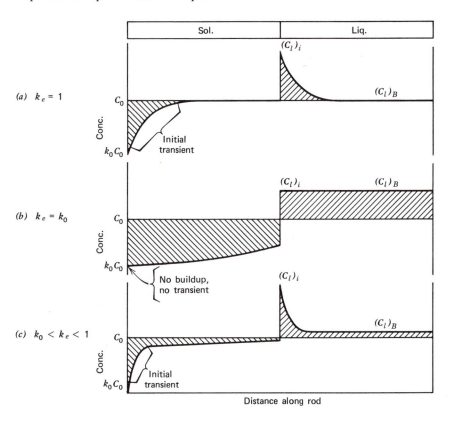

Figure 9.19 Solute profiles for various values of the effective distribution coefficient k_e.

1. Derivation of the Equation for k_e. To proceed it is first necessary to derive the differential equation that governs the liquid concentration, C_l, in the boundary layer (Fig. 9.20). We take the interface as our reference point. The liquid flows toward an observer on the interface so that the flux of solute due to fluid flow at any point in the liquid is $-RC_l$, where C_l is the local liquid volume concentration, and R is the rate at which the liquid moves toward the observer (interface rate). The minus sign is required because this flow flux is directed in the minus Z direction and we are taking R itself as a positive number. The total flux due to diffusion and fluid flow is

$$J = -RC_l - D\frac{dC_l}{dZ}$$ (9.20)

We now apply the continuity equation (Eq. 6.7) just as we did on p. 140 to obtain Fick's second law; here we obtain

$$D\frac{\partial^2 C_l}{\partial Z^2} + R\frac{\partial C_l}{\partial Z} = \frac{\partial C_l}{\partial t}$$ (9.21)

After the initial transient has passed, the amount of solute in the boundary layer remains relatively constant and so we assume $\partial C_l/\partial t = 0$ after the transient. This is an assumption because the value of $(C_l)_B$ continues to rise after the transient. However, the *solute profile* in the boundary layer becomes nearly constant after the transient, and experiments have shown this assumption to be quite good. Therefore the differential equation describing the solute level, C_l, in the boundary layer after the transient becomes

$$\frac{d^2 C_l}{dZ^2} + \frac{R}{D}\frac{dC_l}{dZ} = 0$$ (9.22)

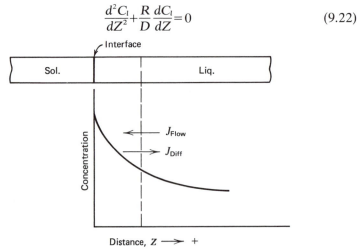

Figure 9.20 The fluxes of solute in the liquid relative to the solid–liquid interface.

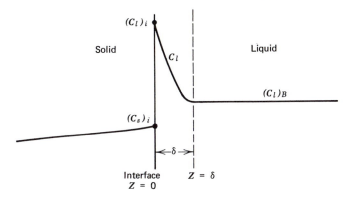

Figure 9.21   The boundary layer model for solute transport into the stirred liquid.

We must now solve this equation for the conditions that prevail at the two boundaries of the boundary layer region. The left-hand boundary is the solid–liquid interface and we obtain this boundary condition from a flux balance of solute across the interface,

$$|\text{Flux directed into interface}| = |\text{Flux directed out from interface}| \tag{9.23}$$

Again, we take the interface as our reference point, and from Fig. 9.21 we see that the only mechanism by which solute enters the interface is fluid flow from the liquid, so we have

$$|\text{Flux directed into interface}| = R(C_l)_i \tag{9.24}$$

Notice, however, that a flux is directed away from the interface by three mechanisms, flow into the solid, diffusion into the liquid, and also diffusion into the solid. As before, we neglect diffusion into the solid and we obtain

$$|\text{Flux directed out from interface}| = R(C_s)_i - D\left(\frac{dC_l}{dZ}\right)_i \tag{9.25}$$

We need the minus sign on the diffusion term because $(dC_l/dZ)_i$ is negative. Combining Eqs. 9.23, 9.24, and 9.25 with the relation $(C_s)_i = k_0(C_l)_i$ we obtain

Boundary condition 1: $D\left(\dfrac{dC_l}{dZ}\right)_i + R(C_l)_i(1-k_0) = 0$ at $Z=0$

$$\tag{9.26}$$

(You will note that we have implicitly assumed equal solid and liquid densities.) As shown on Fig. 9.21 the boundary layer thickness is defined as δ, and we assume that beyond a distance δ the liquid concentration is

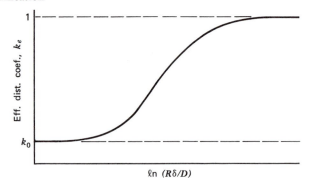

Figure 9.22   A plot of the Burton–Prim–Slichter equation, Eq. 9.28.

uniform at $(C_l)_B$. Hence, we have

Boundary condition 2: $C_i = (C_l)_B$ at $Z = \delta$ (9.27)

It is now a simple matter to solve Eq. 9.22 for boundary conditions 9.26 and 9.27 and obtain $C = \text{function}[R, D, k_0, \delta, Z, (C_l)_B]$. From the solution at $Z = 0$ it is simply a matter of algebra to show

$$k_e = \frac{k_0}{k_0 + (1 - k_0)e^{-R\delta/D}}$$ (9.28)

This is a famous equation first derived in 1953 by Burton, Prim, and Slichter.[10] It shows that the effective distribution coefficient is a function of k_0 and the dimensionless group $R\delta/D$. A plot of Eq. 9.28 is shown in Fig. 9.22 for a specific value of k_0. It is illustrated that k_e extends from a minimum value of k_0 to a maximum of 1 as $R\delta/D$ increases.

Consider the normal freezing of a rod under two conditions:

1. $R\delta/D$ sufficiently small to give $k_e = k_0$.
2. $R\delta/D$ such that $k_e = 2 \cdot k_0$ (but $k_e < 1$ if $k_0 < 1$).

It was shown in discussing Fig. 9.19(b) that the condition $k_e = k_0$ corresponds to no buildup, and the solid composition profile frozen out follows the normal freeze equation (Eq. 9.17) as shown in Fig. 9.23. We would also like to have an equation for the composition profile when $k_e > k_0$ such as case 2 here. The above analysis has shown that after the initial transient is over, the value of k_e becomes a constant as given by Eq. 9.28. Therefore, for the case of solute buildup we can perform a mass balance similar to that done on Fig. 9.15, and if we assume the total solute content in the boundary region is constant, we obtain

$$C_s = k_e C_0 \left[1 - \frac{Z}{L} \right]^{k_e - 1}$$ (9.29)

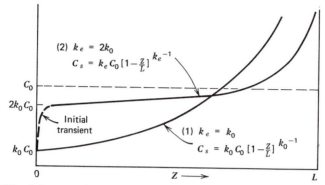

Figure 9.23 Solute profiles with, (1) complete mixing, $k_e = k_0$, and (2) partial mixing, $k_e = 2k_0$.

This equation gives the curve for case (2) shown on Fig. 9.23 for all points beyond the initial transient. Since the initial transient is generally only about 1 mm long when mixing is present, Eq. 9.29 is usually quite adequate to give the solute profile for cases where $k_e > k_0$.

By similar reasoning one may show that for this case of $k_e > k_0$ the equation for zone melting (Eq. 9.18) is modified to become

$$C_s = C_0[1 - (1 - k_e)e^{-k_e Z/l}] \tag{9.30}$$

We now consider two cases for the application of Eqs. 9.28, 9.29 and 9.30. But it is emphasized that these equations are still subjected to the restriction of a flat interface, that is, assumption II, p. 246. First consider the case where we desire to produce maximum purification, that is, zone melting. It should be clear from the foregoing, particularly Fig. 9.23, that we desire that k_e approach k_0 as closely as possible. From Fig. 9.22, then, we see that we desire $R\delta/D$ as small as possible. Hence, we require a low rate of interface motion and a high degree of mixing to minimize the boundary layer thickness, δ. Second, consider the problem where we desire a rod of constant composition, which would require $k_e = 1$. Again referring to Fig. 9.22 it is seen that we require a high interface rate and no mixing in order to produce maximum δ. A constant composition is then obtained after the initial transient, which for this case of no mixing can be on the order of a centimeter or two.[11] It turns out that unless the alloy contains less than around 1% of solute it is generally not possible to maintain a flat interface (assumption II violated), and neither of the above examples then apply for reasons that we shall see in the next section.

E. INTERFACE SHAPE

You will recall that for pure metals freezing under a negative temperature gradient the flat interface became unstable and formed dendrites because

dS/dZ was positive as shown in Fig. 9.11. To understand why the interface is unstable, consider what happens when a small bump forms on the surface. The velocity of the tip of the bump will be given by Eq. 9.9 written as

$$V_{\text{tip bump}} = \frac{-k_l(dT/dZ)_l}{L\rho} \qquad (9.31)$$

where we have neglected the temperature gradient in the solid because it is near zero and $(dT/dZ)_l$ is the liquid temperature gradient normal to the tip of the bump. Because of the small kinetic undercooling, ΔT_K, the temperature at the tip of the bump will be quite near to the freezing temperature, T_f. Therefore when the bump protrudes ahead of the flat interface the value of $(dT/dZ)_l$ will increase as shown by the dashed curve of Fig. 9.24. This will cause the velocity of the bump to increase (see Eq. 9.31), and as long as dS/dZ is positive we expect the bump to propagate ahead of the general interface.

In pure metals we may only obtain supercooling ahead of the interface if the real temperature T_R has a negative gradient so that it falls below the constant freezing temperature T_f. However, in alloys the freezing temperature is not a constant, but rather it is a function of composition as given by the liquidus line on the phase diagram. Hence, in alloys we may obtain supercooling with a positive temperature gradient. This is illustrated in Fig. 9.25. The upper figure shows the solute buildup profile in the boundary layer at the interface. We consider a eutectic system, $k_0 < 1$, as shown on the right. Therefore, the solute buildup must decrease the freezing

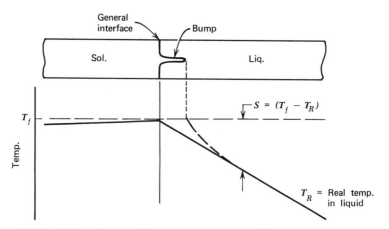

Figure 9.24 Effect of bump formation on the interface upon the temperature profile for the case of solidification in a negative gradient in a pure metal.

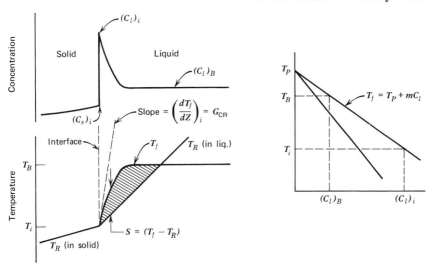

Figure 9.25 The supercooled region (shaded area) produced by the solute buildup in alloy solidification in a positive gradient.

temperature because of the inverse relationship between T_f and C for systems with $k_0 < 1$. The lower figure shows how the freezing temperature, T_f, must vary due to the solute buildup. We now superimpose a plot of the real temperature, T_R, assuming a positive gradient. Because ΔT_K is quite small the real temperature of the system must be at the freezing temperature where freezing occurs, that is, at the interface. Therefore the two curves must match at the interface, and for the conditions of Fig. 9.25 it is seen that a region of the liquid is supercooled. This has been called constitutional supercooling. Suppose we let the temperature gradient in the liquid be called G_l. It can be seen from Fig. 9.25 that if $G_l \geq (dT_f/dZ)_i$, where $(dT_f/dZ)_i$ is the slope of the freezing temperature curve at the interface, then no supercooling occurs. Consequently, $(dT_f/dZ)_i$ is a critical value of the temperature gradient and we define

$$G_{CR} = \left(\frac{dT_f}{dZ}\right)_i \qquad (9.32)$$

Suppose now that we take the slope of the liquidus curve as m, as shown for a straight-line liquidus on Fig. 9.25. Then we have $G_{CR} = m(dC_l/dZ)_i$, where both terms on the right are negative. The derivative of the liquid concentration at the interface has already been determined from the interface flux balance, Eq. 9.26. Consequently, we obtain

$$G_{CR} = \frac{-mR}{D}(C_s)_i \left[\frac{1-k_0}{k_0}\right] \qquad (9.33)$$

where we have taken $(C_l)_i$ in Eq. 9.26 as $(C_s)_i/k_0$. This is a famous equation first essentially derived by Tiller, Jackson, Rutter, and Chalmers in 1953.[12] Numerous experiments have shown that it predicts the stability of solidifying planar interfaces remarkably well.

In the derivation given here the volume concentration C was used and it was implicitly assumed that the liquid and solid densities were equal, that is, $\rho_l = \rho_s$. The derivation may be obtained without any assumption about constant density[13] and one then obtains the result in terms of fractional concentration X as

$$G_{CR} = \frac{-mR}{D}(X_s)_i \left[\frac{1-k_0}{k_0}\right] \cdot \frac{\rho_s}{\rho_l} \qquad (9.34)$$

There is a fundamental problem with this constitutional supercooling theory that may be demonstrated by an analogy. Suppose that you know that the free-energy change for some chemical reaction, ΔG, is a large negative number. This information alone is not sufficient to tell you whether or not the reaction will in fact occur. It may be that some kinetic barrier will inhibit the reaction from proceeding to the lower free-energy state. For example, the activation energy for some part of the atomic processes involved in the reaction mechanism may be too high at the temperature of interest. Similarly, in the present example the fact that dS/dZ is positive into the liquid only tells us that the interface will be unstable provided no kinetic barrier inhibits the formation of a bump on the planar interface. Consequently, one must perform some kind of an analysis of the kinetic processes that occur when a perturbation (a bump) forms on a flat interface.

Suppose a small bump forms on a planar interface as shown in Fig. 9.26(a). The temperature profile along the center line of the bump is also shown. We know that the interface temperature at the tip, T_i(tip), must be less than the freezing temperature of the liquid at the tip, T_f(tip), in order to cause the bump to grow (see discussion on p. 239). Furthermore, the velocity at which the bump grows must be proportional to this kinetic undercooling, $V \propto [T_f(\text{tip}) - T_i(\text{tip})]$. There are two important effects that will influence the freezing temperature at the tip of the bump, T_f(tip), and hence the velocity of the bump.

1. Curvature effect. If the bump becomes more pointed its radius of curvature decreases, and this changes the free energy in the region of the

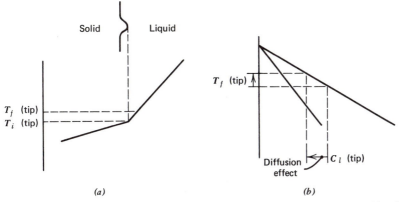

Figure 9.26 (a) Relation of a bump on interface to temperature profile. (b) Influence of diffusion effect upon freezing temperature of bump tip.

tip according to Eq. 7.40:

$$\Delta G \approx \frac{2\gamma_{s-l}\bar{V}}{r} \qquad (7.40)$$

derived in Chapter 7. Here γ_{s-l} is the solid–liquid surface tension, $\bar{V}$ the specific volume of the liquid, and r is the radius of curvature of the tip. This change in free energy reduces the freezing temperature of the tip, $T_f(\text{tip})$, by the amount $\Delta G/\Delta S_f$ (see Eq. 9.3 and discussion on pp. 370–373. Consequently, an increased curvature (smaller r) tends to reduce $T_f(\text{tip})$, which in turn reduces the velocity of a perturbation on a flat interface. Therefore, curvature tends to *enhance planer stability*.

2. Diffusion effect. If the bump becomes more pointed then more of the solute rejected from the tip of the bump can diffuse off to its sides rather than being piled up ahead of the advancing tip. This sideways diffusion reduces the solute buildup ahead of the tip and consequently reduces the concentration in the liquid at the tip, $C_l(\text{tip})$, which in turn raises $T_f(\text{tip})$, as is shown on Fig. 9.26(b). An increase in $T_f(\text{tip})$ will tend to increase the velocity of the perturbation. Therefore, the diffusion effect tends to *reduce planar stability*.

It can be seen that the above two effects act in opposite directions. The curvature effect reduces the tip temperature by $2\gamma_{s-l}\bar{V}/r\,\Delta S_f$, and the diffusion effect increases the tip temperature by $m \cdot \Delta C$, where m is the slope of the liquidus and ΔC is the composition decrease at the tip produced by the bump shape. A correct theoretical analysis must analyze both effects and determine the physical conditions under which the velocity

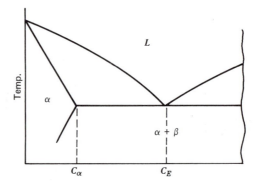

Figure 9.27   A eutectic phase diagram defining compositions C_α and C_E.

of the perturbation is less than the velocity of the advancing planar interface, and these conditions then define the range of stability of the planar interface. The stability analysis for this problem was first presented in 1963 and the values of the critical temperature gradient above which a planar interface is stable is given as[14]

$$G_{CR} = \frac{-mR(C_s)_i}{D} \left[\frac{1-k_0}{k_0} \right] \cdot S \left[\frac{1+k_s/k_l}{2} \right] \qquad (9.35)$$

where S is a constant dependent on the value of γ_{s-l}, and k_s and k_l are the thermal conductivities of the solid and liquid, respectively. For compositions above a few weight percent the value of S is essentially unity. The ratio k_s/k_l ranges from around 2 for metals to $\frac{1}{2}$ for semiconductors. Hence, the stability criterion of Eq. 9.35 is nearly the same as the constitutional supercooling criterion of Eq. 9.33. In fact, the predicted values of G_{CR} from these two equations are sufficiently close and experiments sufficiently difficult that, to date, it has not been shown in metal alloys that Eq. 9.35 gives a better fit to the data, although its validity has been demonstrated in nonmetallic systems.[15]

Many experiments have been carried out in which G_l, R, and C_s have been controlled and the shape of the solidifying interface determined by a variety of techniques. These experiments show that the mode of transition from a flat to dendritic interface depends upon the composition of the alloy. Consider a simple eutectic system represented by the phase diagram of Fig. 9.27.

I.   Alloys of Compositions $C_0 < C_\alpha$.   It is found for alloys in this range that as G_l drops below G_{CR} the flat interface undergoes a gradual transition from planar to dendritic. The first instability occurs at the grain boundaries

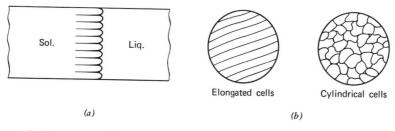

Elongated cells   Cylindrical cells

(a)   (b)

Figure 9.28 Schematic illustrations of cellular interfaces. (a) Longitudinal view and (b) transverse view.

and spreads across the surface, forming what is called a cellular interface.[16,17] Longitudinal sections on the cells reveal them to be smooth sided and parabolic-like in shape as shown in Fig. 9.28(a). Transverse sections reveal two common morphologies that differ according to whether the cells are elongated or cylindrical as shown in Fig. 9.28(b). Upon further decreasing G_l, the cells become irregular and eventually develop side branches, characteristic of the dendrite morphology. The cells do not necessarily grow with a preferred crystallographic orientation along their axes as do the dendrites. The transition for these alloys may be characterized as,

1. $G \geq G_{CR}$ Flat interface
2. G slightly less G_{CR} Cellular interface
3. $G \ll G_{CR}$ Dendritic interface.

II. Alloys of Compositions $C_\alpha < C_0 < C_E$. Alloys in this range of compositions will produce a two-phase solid consisting of α and β phases. Consequently, when the interface is planar these alloys will have a eutectic-like microstructure. Recent studies have found[18,19] that the cellular morphology does not always occur, depending on composition and values of G_l/R. The interface breaks down from a planar morphology directly to a dendritic morphology. The critical gradient for planar interface stability of these alloys is not given by Eq. 9.32. This case presents an example where the constitutional supercooling criterion does not apply and one must use a perturbation analysis as was discussed above in regard to Eq. 9.35. The stability of the planar morphology in these alloys is considered further on p. 277.

It is quite instructive to numerically evaluate Eq. 9.34 for conditions that generally prevail in castings and ingots. We consider the following values:

$D \approx 10^{-5}$ cm²/sec at liquidus temperature for most metals
$|m| > 1°C/wt \%$
$R > 2.5 \times 10^{-3}$ cm/sec (2.2 in./hr) for ingots and castings

Table 9.3 Values of G_{CR} for Various k_0 and $(X_s)_i$ Values Calculated from Eq. 9.34 Assuming $\rho_s = \rho_l$.

k_0 \ $(X_s)_i$	10 wt %	1 wt %	0.01 wt % (100 ppm)
0.4	3,750°C/cm	375°C/cm	3.75°C/cm
0.1	18,500°C/cm	1850°C/cm	18.50°C/cm

Evaluation for three compositions and two k_0 values are given in Table 9.3. Since minimum values for both R and m were considered it may be seen that extremely high gradients are required to suppress dendrites. Two conclusions may be drawn.

1. The temperature gradients necessary for flat interfaces are generally too high to be obtained unless (a) very low rates are used, and (b) solute content is less than 1%.

2. The gradients in ingots and castings are usually less than 3–5°C/cm. Therefore, in commercial practice, alloys virtually always freeze with dendritic interfaces. Experiments confirm this conclusion.

9.4 SOLIDIFICATION OF EUTECTIC ALLOYS

The basic difference in solidification of eutectic alloys and those discussed above results from the fact that in eutectic growth two solid phases form simultaneously from the liquid. Perhaps the two most important casting alloys from a commercial point of view are near eutectic composition alloys, cast iron and Al–Si. Cast eutectic alloys are also of interest for their possible use as composite materials.

A. MICROSTRUCTURE

One of the interesting characteristics of eutectic alloys is their great variety of very interesting and beautiful microstructures. Over the years a number of different schemes have been proposed for classifying eutectic microstructures. A scheme that has gained favor in recent years will be presented here. The microstructures are classified into three categories.

I. Regular. There are basically two types of regular microstructures, lamellar or fibrous. The lamellar structure consists of parallel plates of the two phases as shown in Fig. 9.29(a). The fibrous microstructure consists of rods (sometimes blades) of one phase contained in a matrix of the second phase as shown in Fig. 9.29(b). This latter type of structure has many potential applications as a composite material. If the rod phase is very

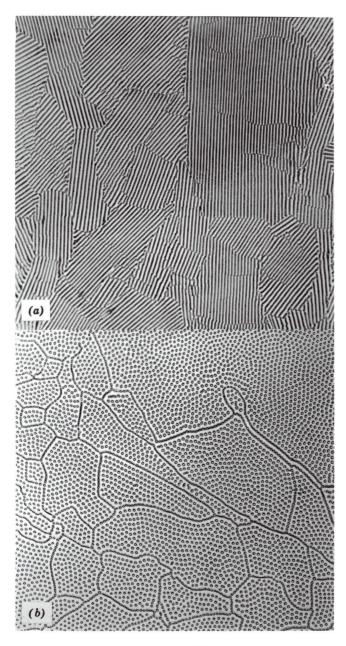

Figure 9.29 Regular eutectic structures. (*a*) Lamellar eutectic composite, Pb–23 wt % Cd (256×). (*b*) Rod eutectic composite, Sn–18 wt % Pb (150×).

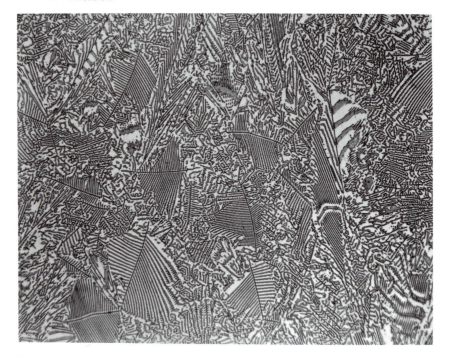

Figure **9.30**   Complex-regular eutectic structure. Aluminum–germanium eutectic, Al–35 wt % Ge (500×). Dark phase is germanium.

strong and the matrix phase quite ductile, one may obtain a composite material having both high strength and ductility.[20] There are also many electromagnetic applications.[21] Such materials have to be carefully grown, however, because the rods or lamellae have to grow in parallel arrays.

II. Complex Regular.   In the complex-regular microstructure one observes two types of regions, (1) regions of a repeating regular pattern and (2) regions of random orientation. These two regions may be seen in Fig. 9.30. (Typical photos may also be found in Ref. 22, Figs. 41, 50, and 54, and Ref. 23, Figs. 15 and 37.)

III. Irregular.   In the irregular type of microstructure the structure consists essentially of a random orientation of the two phases (see, for example, Figs. 8 and 39 of Ref. 23).

One may obtain a modestly good correlation between the type of microstructure and whether the individual phases form faceted or nonfaceted interfaces when grown as single phases. This correlation is illustrated in Table 9.4. For example, it was illustrated in Table 9.1 that all regular

Table 9.4   Correlation between Eutectic Morphology and Tendency of Phases to Facet

Microstructure	Growth Morphology of Eutectic Phases	Examples
(I) Regular	Nonfaceted/Nonfaceted	Sn–Pb, Al–Zn
(II) Complex-Regular	Nonfaceted/Faceted	Al–Si, Sn–Bi
(III) (a) Irregular	Nonfaceted/Faceted	Al–Si, Fe–C
(b) Irregular	Faceted/Faceted	—

metals form nonfaceted interfaces and most semiconductors form faceted interfaces. Eutectic alloys composed of a metallic phase and a semiconductor phase generally have a type II or III structure, for example, Al–Si or Al–Ge. The reason for the correlation of Table 9.4 is probably the different kinetic undercooling for faceted versus nonfaceted materials. When both phases are nonfaceted the tips of each phase are within about 0.02°C of the eutectic temperature and the general solid–liquid interface is essentially isothermal and therefore planar as shown in Fig. 9.31(*a*). For the faceted–nonfaceting case, however, the tips of the faceted phase must undercool around 1 or 2°C compared to only around 0.02°C for the nonfaceted phase. Consequently, the tips of the faceted phase must grow at temperatures around 1–2°C cooler than the tips of the nonfaceting phase. With a positive temperature gradient this means the faceting phase will lag slightly behind the nonfaceting phase as shown in Fig. 9.31(*b*). Apparently, this configuration is unstable and the leading nonfaceting phase overgrows the faceting phase in a random manner. This causes the faceting phase to branch out where it becomes overgrown and this branching leads to the irregular

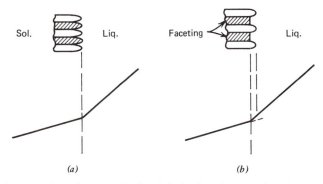

(a) *(b)*

Figure 9.31   The interface profile for (*a*) nonfaceting–nonfaceting eutectics and (*b*) nonfaceting–faceting eutectics.

microstructures. Although not apparent in Fig. 9.30, all of the faceted Ge particles (dark phase) are interconnected in the third dimension.

There are a number of important exceptions to the correlation given in the above table. It has been found that in a number of faceting–nonfaceting systems a regular microstructure is obtained, for example, Ag–Bi, Al–Al₃Ni, superalloys–TaC. The reasons that these systems deviate from the general rule are not well understood at this time. The occurrence of a regular microstructure in the latter system is quite fortuitous for practical reasons. Directional solidification produces aligned TaC fibers in the superalloy matrix, and the resulting composite alloy has significantly improved tensile and stress rupture properties.[24]

B. KINETICS OF LAMELLAR GROWTH

There are three common types of phase transformations that produce a regular, paired two-phase microstructure. These are eutectic, eutectoid (i.e., pearlite formation), and discontinuous precipitation reactions. In all of these transformations a very interesting diffusion pattern is obtained at the transforming interface that controls the spacing of the coupled phases formed during transformation.

The eutectic phase diagram, such as Fig. 9.32(b), shows that equilibrium solidification of eutectic liquid transforms liquid of composition, C_E, into two solid phases, α and β, of compositions C_α and C_β. Also, the proportions of the two phases must be such that their average composition is the same as the liquid from which they formed, C_E. Consider now the

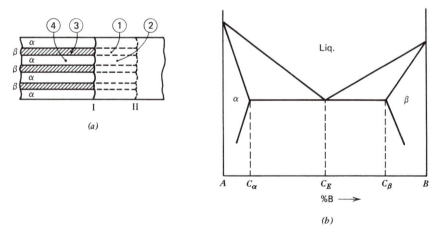

Figure 9.32 (a) The phase changes produced as a lamellar eutectic solidifies. (b) Corresponding phase diagram.

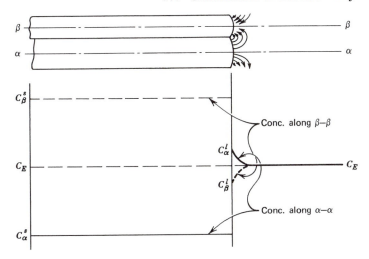

Figure 9.33 The diffusion path required to produce the solute redistribution, and the concentration profiles along the centerlines of the α and β lamellae.

motion of the solid–liquid interface during regular eutectic growth as shown in Fig. 9.32(a). At points 1 and 2 the liquid composition is at the eutectic composition, C_E. At point 3 we expect a solid composition of C_β and at point 4 a solid composition of C_α. After the interface advances to position II the composition at region 1 must increase from C_E to C_β and the composition at region 2 must decrease from C_E to C_α. This means that the composition at 2 must suddenly drop to around 15% B as the interface passes while at 1 it must increase to around 80% B. This drastic change in composition between regions 1 and 2 must occur by diffusion of B from in front of the α plate to in front of the β plate as shown at the top of Fig. 9.33. As the α plates freeze they spew out B atoms, which diffuse around to in front of the β plates, while the β plates suck up B atoms as they freeze. This diffusion requires the concentration of the liquid in front of the α plates to be higher than in front of the β plates, since a concentration gradient is necessary for diffusion. Consequently, the liquid composition in front of α and β plates cannot be the same, C_E, as indicated by the equilibrium phase diagram, Fig. 9.32(b). The interface actually undercools a small amount, ΔT_E, as shown on Fig. 9.34. By simply extrapolating the boundaries of the phase diagram we see that the composition of the supercooled liquid in front of the α plate, C_α^l, will be higher than in front of the β plate, C_β^l, as required for the diffusion. Notice also that the solid phases will have compositions just slightly different from the equilibrium compositions (that is, $C_\alpha^s > C_\alpha$ and $C_\beta^s < C_\beta$; compare Figs. 9.32(b) and 9.34).

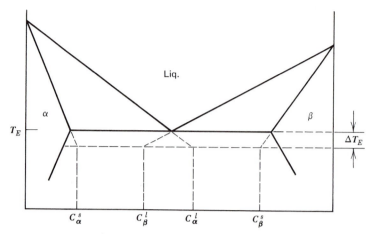

Figure 9.34 Extrapolation of the phase boundaries to the supercooled interface temperature. (Often called the Hultgren extrapolation after, A. Hultgren, *Hardenablilty of Alloy Steels*, American Society for Metals, Cleveland, 1938, p. 55.

In Fig. 9.33 the solute profile is shown along the centerline of the α and β plates. Notice that the liquid concentration gradients cause B atoms to move away from the α plate and toward the β plate. The liquid compositions C_α^l and C_β^l are actually very close to C_E because the plate spacing is so small that large concentration differences are not required to produce sufficient diffusion. It should be clear that the faster the rate of interface motion the smaller the plate spacing will be in order to accomplish the required diffusion. The flux of B atoms that is ejected from the α plate may be written as

$$J(\text{reject}) = \frac{1}{A_\alpha} \frac{dm}{dt} = R(C_\alpha^l - C_\alpha^s) \approx R(C_E - C_\alpha) \qquad (9.36)$$

where A_α is the area of the α–liquid surface, R is the rate of interface motion, and C_E and C_α are defined on Fig. 9.32(b). To calculate the lateral diffusion of this ejected solute in a rigorous manner it is now required to determine the three-dimensional solute profile in the liquid. To do this it is necessary to write Eq. 9.22 for three-dimensional diffusion and solve it for the boundary conditions at the interface region. Such analyses have been carried out[25] but we will only present a first-order analysis here, after the ideas of Zener.[26] The lateral diffusion between α and β plates will occur predominantly in the Y direction on Fig. 9.35, and it is approximated as

$$J(\text{diff}) = D \frac{\Delta C}{S_0/2} \qquad (9.37)$$

where ΔC is some mean composition difference in the liquid between the α and β fronts, and the diffusion distance has been taken as $S_0/2$. At steady state the flux of B atoms rejected from an α plate will equal the lateral diffusion flux that transports these atoms into the β plates, so that equating Eqs. 9.36 and 9.37 one obtains

$$R = \frac{2D\Delta C}{S_0(C_E - C_\alpha)} \tag{9.38}$$

It was shown in Section 9.2A that in order for a pure metal to freeze the $l \rightarrow s$ reaction must exceed the $s \rightarrow l$ reaction at the interface, and this is achieved by a small supercooling, ΔT_K, that we termed the kinetic undercooling. When the metal freezes at a supercooling of ΔT_K the free energy $\Delta G = \Delta S_f \cdot \Delta T_K$ is released upon freezing. This free energy is required to cause a net atomic reaction rate from liquid to solid, an irreversible process. When a eutectic freezes two additional irreversible processes occur, so that for eutectic growth we must have three irreversible processes:

a.   A net atomic reaction rate from liquid to solid.
b.   Diffusion in the liquid ahead of the α and β phases.
c.   Generation of α–β interface area between the two solid phases.

Just as process (a) requires energy, so will processes (b) and (c). This additional energy for eutectic growth is obtained by additional supercooling, so that when a eutectic freezes the free energy released is given by

$$\Delta G_B = \Delta S_f \cdot \Delta T_E \tag{9.39}$$

where ΔS_f is the freezing entropy per volume of the eutectic liquid and ΔT_E is the undercooling of the liquid at the freezing interface for this case of eutectic solidification.

Consider the interfacial region shown in Fig. 9.35, which has a height of S_0 and a depth into the page of 1 cm. Notice that S_0 is the spacing between

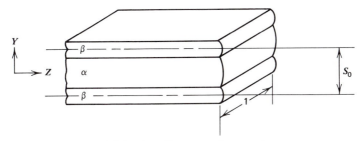

Figure 9.35   Model for lamellar eutectic growth.

β plates, and the area $S_0 \cdot 1$ is a "unit cell" area. The free energy released, ΔG_B, is used to drive the three irreversible processes described above and labeled (a), (b), and (c). For the case of regular eutectic growth (nonfaceted–nonfaceted) process (a) is negligible compared to processes (b) and (c) so that it will be neglected here. The interface of Fig. 9.35 is advanced a distance dZ and an energy balance is carried out on the volume $[S_0 \cdot 1 \cdot dZ]$:

   I.   Free energy released is given by $\Delta G_B \cdot S_0 \cdot 1 \cdot dZ$.
  II.   Free energy to produce two α–β interfaces is given by $2\gamma_{\alpha\beta} \cdot 1 \cdot dZ$.
 III.   Free energy to drive diffusion is given by $\Delta G_d \cdot S_0 \cdot 1 \cdot dZ$.

Here we have taken ΔG_d as the free energy per volume used to drive the diffusion process. Making the free energy balance we obtain

$$\Delta G_B = \frac{2\gamma_{\alpha\beta}}{S_0} + \Delta G_d \qquad (9.40)$$

Suppose now that all of the free energy released were used to generate α–β interface. In this hypothetical case ΔG_d would be zero and the spacing would have to be a minimum, $S_0 = S_{min}$, because we have the maximum interface area. Making these two substitutions into Eq. 9.40 we obtain

$$S_{min} = \frac{2\gamma_{\alpha\beta}}{\Delta G_B} \qquad (9.41)$$

This equation tells us the minimum possible spacing of the eutectic plates. Substituting Eqs. 9.39 and 9.41 into 9.40 we obtain for the free energy to drive diffusion

$$\Delta G_d = \Delta S_f \cdot \Delta T_E \left[1 - \frac{S_{min}}{S_0} \right] \qquad (9.42)$$

We now interpret what this means in terms of supercooling below the eutectic temperature. The total supercooling, ΔT_E, determines the amount of free energy available for the eutectic reaction; see Eq. 9.39. We may divide the total supercooling into a part that provides the free energy to generate surface, ΔT_s, and a part that provides the free energy to drive diffusion, ΔT_d; see Eq. 9.43b:

$$\Delta G_B = \Delta G_d + \Delta G_s \qquad (9.43a)$$

$$\Delta S_f \Delta T_E = \Delta S_f \Delta T_d + \Delta S_f \Delta T_s \qquad (9.43b)$$

This result may be displayed geometrically on a phase diagram as shown in Fig. 9.36. The concentration difference available for diffusion, ΔC, is shown on Fig. 9.36 and it may be calculated from the liquidus slopes m_α

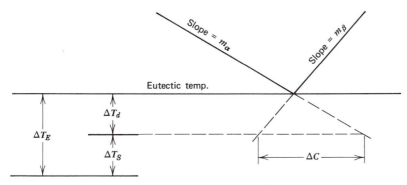

Figure 9.36   Relation of total undercooling, ΔT_E, to the phase diagram.

and m_β, as

$$\Delta C = \Delta T_d \left[\frac{1}{|m_\alpha|} + \frac{1}{m_\beta} \right] \tag{9.44}$$

Equation 9.42 may be rewritten in terms of undercooling as

$$\Delta T_d = \Delta T_E \left[1 - \frac{S_{min}}{S_0} \right]$$

and substituting this expression and Eq. 9.44 into Eq. 9.38 we obtain

$$R = \left[\frac{1}{|m_\alpha|} + \frac{1}{m_\beta} \right] \frac{2D \, \Delta T_E}{(C_E - C_\alpha)S_0} \left[1 - \frac{S_{min}}{S_0} \right] \tag{9.45}$$

We rewrite this equation in the following form:

$$\Delta T_E = (\text{const}) \cdot R \cdot \frac{S_0}{1 - S_{min}/S_0} \tag{9.46}$$

In the solidification of eutectic alloys the rate of solidification is controlled by the heat flow as explained on p. 244. Hence, we may easily grow a eutectic at constant rate R. Under such conditions Eq. 9.46 shows that the undercooling, ΔT_E, is only a function of the spacing, S_0, as is shown graphically in Fig. 9.37. Following the ideas of Zener,[26] it was first suggested by Tiller[27] that the spacing would probably adjust so as to minimize the undercooling at the interface. There has been much discussion of the validity of this optimization principle over the years, but it still is used even though there is no clear theoretical principle on which it is based. The optimum spacing corresponding to the minimum ΔT_E on Fig. 9.37 is found by taking the derivative of Eq. 9.46 with respect to S_0 and equating to zero. One obtains

$$S_{opt} = 2S_{min} \tag{9.47}$$

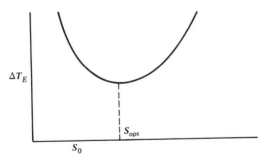

Figure 9.37 Temperature dependence of Eq. 9.46.

We may now obtain a rate equation by substituting into Eq. 9.45 $S_{min} = S_{opt}/2$, $S_0 = S_{opt}$, and $\Delta S_f \, \Delta T_E = \Delta G_B = 2\gamma_{\alpha\beta}/S_{min} = 4\gamma_{\alpha\beta}/S_{opt}$:

$$R = \left[\frac{1}{|m_\alpha|} + \frac{1}{m_\beta}\right] \frac{4\gamma_{\alpha\beta}D}{\Delta S_f(C_E - C_\alpha)} \cdot \frac{1}{S_{opt}^2} \tag{9.48}$$

This final result may be written in the form

$$S_{opt} = \frac{const}{\sqrt{R}} \tag{9.49}$$

This result predicts that the observed spacing, S_{opt}, should vary inversely with R. The spacing decreases with increasing rate in order to allow the diffusion to redistribute the solute between the two phases. More complete analyses of this problem[25] give the same result as Eq. 9.49 but with a different constant.

In recent years there have been numerous experiments investigating eutectic solidification, and the results generally agree quite well with Eq. 9.49. The results for experiments on Pb–Sn are shown in Fig. 9.38. Notice

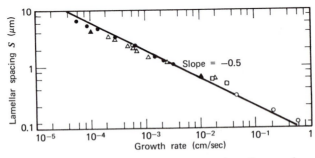

Figure 9.38 Summary of experimental data on lamellar spacing versus rate of solidification in Pb–Sn alloys [From H. E. Cline and J. D. Livingston, Trans. Met. Soc. AIME **245**, 1990 (1969).]

that the equation holds over growth rates of four orders of magnitude. However, there appears to be a limited range of spacing over which it is possible to achieve regular eutectic structures. It is generally found that above $S \sim 10$ μm the lamellar plates become very wavy and begin to break up. Below $S \sim 0.5$ μm it is generally very difficult to keep the regular structure from breaking up. Consequently, the usual eutectic spacing is around 1–3 μm. This is a very small spacing requiring magnifications up to 500× for ease of viewing in an optical microscope.

C. STABILITY OF EUTECTIC INTERFACES

The stability of the planar solid–liquid interface during growth of single phase alloys was discussed in Section 9.3E. It was shown that both the constitutional supercooling analysis and the perturbation analysis give very nearly the same numerical values for the critical gradients required for planar interface stability. However, this is not the case for the stability of the planar solid–liquid interface during growth of two-phase alloys. The constitutional supercooling approach does not work very well, depending on composition, and the perturbation analysis is much more complex for this case. We will consider three specific conditions.

1. Pure Binary Eutectic.   When a pure binary liquid of eutectic composition freezes, the average composition of the solid is identical to the composition of the liquid from which it forms. Consequently, there is no solute buildup such as is shown in Fig. 9.25 and no constitutional supercooling from such a buildup.* However, there is a small solute buildup in front of the α plates and a solute depletion in front of the β plates as shown in Fig. 9.33, and these solute profiles can produce constitutional supercooling. However, this small constitutional supercooling is not generally a sufficient condition for planar instability, and the following results are found:

I.   For regular microstructures [nonfaceted–nonfaceted], the planar interface is stable for all $G_l \geq 0$. Hence, one does not usually find dendrites in casting these alloys.
II.   Complex-regular microstructures [nonfaceted–faceted] sometimes become unstable and form dendrites at low G_l.[28]

2. Impure Binary Eutectic.   If one adds an impurity element to a pure eutectic alloy, the impurity element will have some average distribution coefficient between the two solid phases and the liquid, k_{avg}. If $k_{avg} < 1$, the impurity element will build up in front of the eutectic liquid interface and

---

* Theoretical analysis[25] shows that a very small solute boundary layer may be present in some cases even though the entire liquid is of eutectic composition.

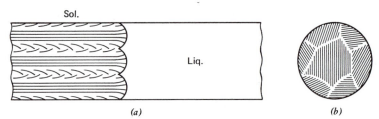

(a)   (b)

Figure 9.39 Schematic representation of cellular growth, called colony growth. (a) Longitudinal view of solid–liquid interface. (b) Transverse view of colony structure.

may give rise to constitutional supercooling. It was originally shown by Weart and Mack[29] that this condition causes the planar interface to become cellular as shown in Fig. 9.39(a) in a manner similar to that for single-phase growth, Fig. 9.28. The lamellae tend to grow perpendicular to the solid–liquid interface as shown in Fig. 9.39(a) and, consequently, each cell is easily distinguishable in a transverse section, as illustrated in Fig. 9.39(b). It has become customary to call these cells *colonies* and to refer to this type of growth as colony growth. (See Fig. 14 of Ref. 23 for actual micrographs.)

If sufficient impurity is added to form dendrites it is usually found that the dendrites are composed of either pure α phase, pure β phase, or a phase of the impurity. However, experiments on ternary systems[30] have shown that dendrites sometimes form with a two-phase, or eutectic-like substructure, such as the cells of Fig. 9.39. (See Figs. 10–15 of Ref. 30 for actual micrographs.)

3. Eutectic Structures at Off-Eutectic Compositions. Consider an alloy of composition C_0 on the phase diagram of Fig. 9.40(a). One often assumes that this alloy would consist of α phase plus eutectic, where the eutectic

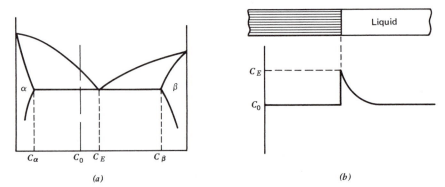

(a)   (b)

Figure 9.40 (a) An off-eutectic composition C_0. (b) Solute buildup required for growth of a regular eutectic structure at off-eutectic composition.

fraction is given by the lever law as $(C_0 - C_\alpha)/(C_E - C_\alpha)$. In a cast alloy the α phase would be present as dendrites with the eutectic between the dendrites (see Fig. 2.1). However, it has recently been demonstrated[18] that it is often possible to solidify this alloy of composition C_0 so that no dendrites are present and the entire microstructure is exactly similar to the eutectic microstructure. We will refer to such a structure as a *eutectic composite*. The composition of a eutectic composite is necessarily an average composition, $\bar{C}$, over the two phases of the composite, $\bar{C} = f_\alpha C_\alpha + (1 - f_\alpha)C_\beta$, where f_α is the volume fraction of the α phase in the eutectic composite, and C_α and C_β are the volume concentrations of the α and β phases in the eutectic composite as determined from the phase diagram, Fig. 9.40(a). For a solid of eutectic composition, C_E, the value of f_α is determined from the phase diagram as $(C_\beta - C_E)/(C_\beta - C_\alpha)$. However, if one can make the α plates of the eutectic alloy thicken, then the volume fraction of α phase, f_α, will increase and the average eutectic composite composition $\bar{C}$ drops below C_E. Hence, a eutectic composite can, in principle, have any average composition between C_α and C_β depending only on the relative volume fractions of the α and β plates in the composite microstructure. Figure 9.29 presents two examples of eutectic composites. The Sn–Pb alloy is 20 wt % off the eutectic composition and the Pb–Cd alloy is 6 wt % off eutectic.

The next question then is how can one produce such a eutectic composite at off-eutectic compositions. Consider the rod of composition C_0 solidifying to the right in Fig. 9.40(b). In order for the solid to freeze out with a regular composite microstructure it is necessary that,

1. the liquid at the solid–liquid interface be near to the eutectic composition, C_E, and
2. the planar interface must be stable (no dendrites).

The first criterion is met by the presence of a solute buildup at the interface as shown on Fig. 9.40(b). This solute buildup in the liquid can produce constitutional supercooling depending on the value of G/R. Proceeding as in Section 9.3E it is a simple matter to show that the critical condition for constitutional supercooling is $G_{CR} = -mR(C_E - C_0)/D$. However, it has been found experimentally that the presence of constitutional supercooling is not always a sufficient condition to cause the planar interface to degenerate into dendrites. Recent experimental results[19] show that the constitutional supercooling criterion works quite well if C_0 is close to C_α. However, for C_0 close to C_E it is often possible to maintain a planar interface at much lower G values than predicted by the constitutional supercooling criterion. At the present time a correct stability criterion based on kinetic perturbation analysis is the subject of considerable debate.

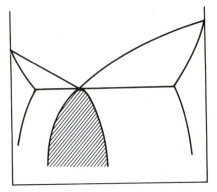

Figure **9.41**   The coupled growth region (cross hatched) on the phase diagram.

Consequently, there is no a priori way of knowing the conditions that will produce a eutectic composite microstructure at off-eutectic compositions. Often, however, experimental results are available in the form such as is shown in Fig. 9.41. If one can supercool the alloy into the shaded region on the phase diagram it will be found to solidify with a eutectic composite microstructure; outside of this region it will have a eutectic composite plus dendrite microstructure. The shaded region is called the "coupled-growth" region. The diagram of Fig. 9.41 is an operational diagram that really does not tell us the temperature at which the actual growth occurs. However, under the conditions at which these diagrams are determined the temperature gradient is very low (probably less than 1–5°C/cm). Therefore, the diagrams show the compositions where eutectic composites will grow at very low gradients and fairly high rates.

D. CRYSTALLOGRAPHY

It is generally found that a preferred crystallographic orientation exists between the plates of the two phases of a regular eutectic structure. Figure 9.42 shows the relationships which have been found for the lead–tin system.[23] These results may be written as,

$$\text{Interfacial plane} \parallel (1\bar{1}\bar{1}) \text{ Pb} \parallel (0\bar{1}1) \text{ Sn}$$
$$[211] \text{ Pb} \parallel [211] \text{ Sn}$$

Notice that all five degrees of freedom of the boundary are specified by these two statements. There is fairly good evidence that preferred orientations occur in order to reduce the surface energy of interfaces.

Eutectics exhibit a weak preferred growth direction. Over a distance of a few centimeters the grains having the preferred orientation will tend to

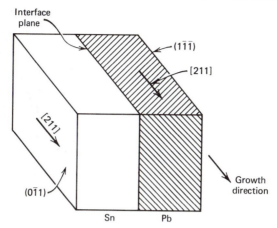

Figure 9.42   The preferred crystallographic relationship between the Pb and Sn lamellae in the lead–tin eutectic.

crowd out their neighbors. For example, in the Pb–Sn alloy the [211] direction is the preferred growth direction for both phases.[23]

9.5   CAST METALS

Most of the metal objects produced each day throughout the world are shaped by some type of mechanical forming process, and only about 20% are used directly in their as-cast form. These shaped products are formed from metal slabs that in turn were produced by solidification of ingots or continuously cast cylinders of the parent metal. The properties of the final metal product frequently depend strongly upon the quality of the original cast metal. Consequently, the quality of cast metals is important not just in foundry products but in almost all metal products.

The solidification process exerts a very strong influence upon the following three properties of the cast metal:

1.   The segregation of alloying elements.
2.   The microstructure (i.e., grain size and phases present).
3.   The soundness (i.e., porosity in the metal).

A thorough analysis of metal casting is a study in itself. Consequently, our aim here is simply to indicate somewhat qualitatively the major physical factors that control the above three properties.

A.   DENDRITIC SOLIDIFICATION

It was emphasized in Section 9.3 that metal alloys virtually always freeze with a dendritic interface. Figure 9.43 is a longitudinal section showing an

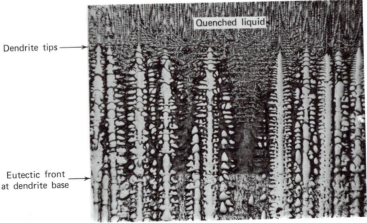

Figure 9.43 An array of tin dendrites growing vertically. Dendrites were preserved by rapid quench of the liquid at the solid–liquid interface. Sn–20 wt % Pb alloy, rate $= 12 \mu$m/sec, gradient $= 310°$C/cm, $(38\times)$.

array of tin dendrites in a Sn–Pb alloy advancing vertically into the liquid. Rapid quenching has preserved the dendrite shapes as well as the eutectic front that advances as a planar front at the dendrite base. A section transverse to Fig. 9.43 below the eutectic front is given in Fig. 2.1. It shows a cross section of the dendrites in a eutectic matrix.

Cast metals normally solidify with temperature gradients much lower than for this Sn–Pb alloy and the dendrites are therefore much longer. A practical case is illustrated in Fig. 9.44(a), which presents data on the

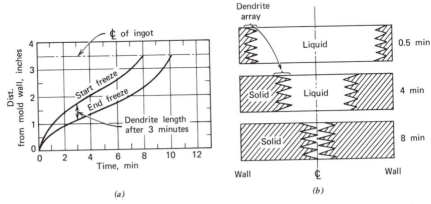

Figure 9.44 (a) Advance of solidification front in an 0.6% carbon steel cast in a 7-inch-square cast-iron mold (from Ref. 31). (b) Schematic view of the advance of the dendrite array from wall to centerline of mold.

solidification of an 0.6% carbon steel cast into a 7-inch square gray iron mold.[31] As illustrated in Fig. 9.44(b) the metal begins to freeze at the mold wall and the solid–liquid interface consists of an array of dendrites that advances toward the center of the ingot as freezing proceeds. The location of the tip and base of the dendrite array was found with thermocouples and plotted in Fig. 9.44(a) as the start-freeze and end-freeze curves, respectively. These curves allow one to determine the length of the dendrite array at any time or location in the ingot. The length of the dendrite array is probably the most important parameter in controlling the casting characteristics of an alloy and it will be discussed in some detail later.

A quantitative description of dendritic solidification is quite difficult because of the complex morphology of the dendrites. If we assume a very simplified shape for dendrites we obtain a picture of dendritic solidification that is qualitatively correct. We take the dendrites to be simple plates that taper to points with no side branches as shown in Fig. 9.45(a). The volume element shown in Fig. 9.45(a) may be thought of as a "unit cell" because when stacked together after solidification it generates the entire solid structure. As the array advances to the right the volume element freezes from bottom to top as shown in Fig. 9.46(a). Notice that we may analyze

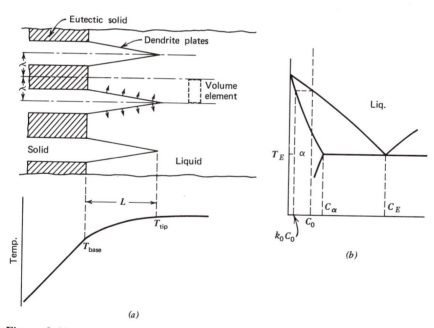

Figure 9.45 (a) The plate model for the dendrite array. (b) Phase diagram showing initial composition C_0.

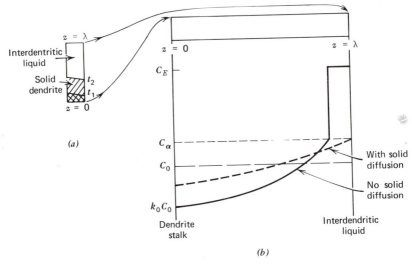

(a)

(b)

Figure 9.46 (a) Position of solid–liquid interface in the volume element at times t_1 and t_2. (b) Composition profiles produced by solidification of volume element.

the segregation within this element just as we did for the unidirectionally solidified rod of Fig. 9.14. First, we make two assumptions that have been shown[32] to be reasonably good. We assume that (a) all of the solute rejected from the solid in each volume element stays within the volume element, and (b) the composition of the interdendritic liquid remains uniform. Assumption (b) holds because the dendrite spacing (2λ) is so small that diffusion in the liquid is fast enough to maintain uniform composition. Consider an alloy of composition C_0 as shown on Fig. 9.45(b). As the volume element freezes the composition in the element rises according to the normal freeze equation until the solid composition reaches C_α. At this point the entire interdendritic liquid has risen to composition C_E so that the remaining liquid in the volume element freezes as eutectic solid at temperature T_E with composition C_E, as shown by the solid curve on Fig. 9.46(b). (It may be helpful to work Problem 9.4 before this discussion.) Since the final liquid must all freeze at one temperature, T_E, the interdendritic base must be flat as found in Fig. 9.43 and also as shown on Fig. 9.45(a). The final cast structure in this case consists of dendrites surrounded by interdendritic eutectic. Notice on Fig. 9.45(b) that the equilibrium alloy for composition C_0 should be a single-phase α alloy. In our analysis we used the normal freezing equation to analyze the segregation in the volume element. Hence, we neglected solid-state diffusion, which is only a moderately good assumption for this case because λ is so small.

Allowing solid-state diffusion to occur, the segregation in the volume element is decreased as shown by the dashed curve of Fig. 9.46(b). Notice that with sufficient solid-state diffusion for alloys of composition C_0 on Fig. 9.45(b) no interdendritic eutectic will form. It should be clear from your understanding of normal segregation that if $C_0 \geq C_\alpha$ the interdendritic regions will always contain some eutectic. For alloys with $C_0 < C_\alpha$ whether or not interdendritic eutectic forms depends on the solid-state diffusion coefficient, the value of k_0, and other factors. It is generally not possible to predict mathematically whether eutectic will form when $C_0 < C_\alpha$. Often there is some composition $C_0 < C_\alpha$ below which interdendritic eutectic will no longer form. The interdendritic eutectic introduces a second phase into the solid alloy, and if not removed by annealing, the second phase may be detrimental to the mechanical properties of the metal. This is particularly true for precipitation hardening alloys where one desires to strengthen a phase by controlled formation of the second phase, and the uncontrolled form of the second phase introduced between dendrites serves as a stress concentrator.

B. DENDRITE LENGTH

The length of the dendrite array will be termed L as shown on Fig. 9.45(a). If we define the temperature at the tip and base of the array as shown, then we may write

$$L = \frac{T_{\text{tip}} - T_{\text{base}}}{\bar{G}_A} \tag{9.50}$$

where $\bar{G}_A$ is the average temperature gradient in the two-phase dendrite array.

Consider a dendrite advancing into a liquid of composition C_0. Virtually all of the solute is rejected laterally into the interdendritic liquid as shown by the small arrows on Fig. 9.45(a). Consequently, there is only a very small solute buildup at the tip of the dendrite as shown in Fig. 9.47(a). This means that the liquidus temperature at the dendrite tips is quite close to the liquidus temperature of the alloy of composition C_0, and so the temperature at the tip of the dendrite array should be close to the C_0 liquidus temperature. Experiments have shown that for normal dendrite growth conditions the tip temperature is within 1–2°C of the C_0 liquidus temperature so that we take, $T_{\text{tip}} = T_{\text{liquidus}}$. (In addition to the suppression of the tip temperature below the liquidus temperature due to the solute buildup, we expect a depression due to kinetic undercooling, ΔT_K, and also due to curvature, as discussed on p. 260. Both these effects are quite small and will be neglected; they are mainly of theoretical interest.) We now define

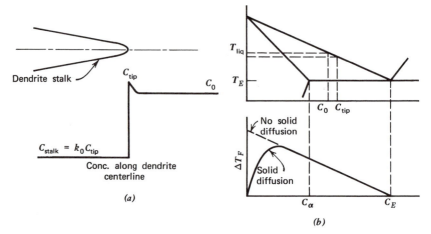

Figure 9.47   (a) Concentration profile along centerline of a dendrite. (b) The freezing range ΔT_F as a function of composition relative to the phase diagram.

the freezing range of an alloy as ΔT_f, where

$$\Delta T_f = T_{\text{liquidus}} - T_{\text{base}} \qquad (9.51)$$

and we now have $L = \Delta T_f / \bar{G}_A$. The temperature at the base of the dendrites will be T_E if eutectic forms in the interdendritic liquid. As indicated above there will be some composition below C_α where eutectic no longer forms, so that below this composition T_{base} increases. Therefore, ΔT_f goes through a maximum at some composition below C_α as shown on Fig. 9.47(b). It can be seen that ΔT_f is determined primarily by the phase diagram. Consequently, it should be clear from Eq. 9.50 and the above discussion that dendrite length is a function of two main factors:

1.   The freezing range of the alloy, ΔT_f. Alloys having a wide separation of liquidus and solidus will have large L.

2.   The heat transfer in the alloy. Variables that promote a small gradient in the array zone, $\bar{G}_A$, increase L. For example, L is increased by a sand mold versus a metal mold or by a low-freezing-temperature alloy versus a high-freezing-temperature alloy.

In general, most Al and Mg alloys have high values of L and steels have low values of L while Cu alloys overlap both of these. We will see that the dendrite length plays a strong role in controlling the structure, segregation, and porosity of cast metals.

C.   STRUCTURE

We discuss the structure of cast metals by considering the structure produced in ingots. Figure 9.48 presents a general schematic picture of the

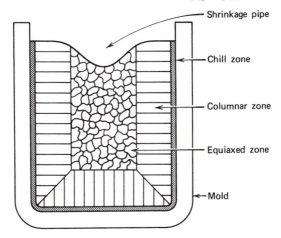

Figure 9.48 The three-zone structure of an ingot. (For actual photographs see Ref. 50, pp. 320–321)

type of structures obtained in ingots. Three rather distinct zones appear in the structure as shown.

1. Chill Zone. The structure in this zone consists of many small roughly equiaxed grains. Apparently nucleation usually occurs from very many sites along the mold wall causing this fine grain structure. However, this structure does not grow inward very far before the columnar structure emerges.

2. Columnar Zone. The grains in the chill zone have dendritic interfaces. Some of these grains will have their dendrites oriented perpendicular to the wall while others will have their dendrite axes at an angle to the mold wall as shown in Fig. 9.49(a) for grains I and II. By a process of competitive

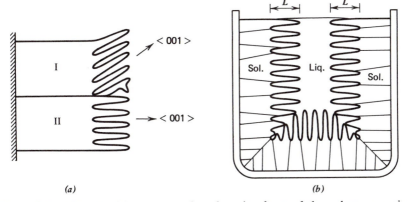

(a)   (b)

Figure 9.49 The dendrite array at the advancing front of the columnar grains.

growth the grains with dendrites perpendicular to the mold wall tend to crowd out the other grains, so that the columnar zone consists of relatively large long grains crystallographically oriented with their dendrite directions parallel to the heat flow direction. The columnar grains grow into the center of the ingot along the heat flow directions behind an advancing dendritic interface as shown in Fig. 9.49(b). In cubic metals dendrites grow in $\langle 001 \rangle$ directions so that the $\langle 001 \rangle$ directions lie parallel to the long axis of the columnar grains. This fact is put to use in producing iron alloys for magnets. The magnetic induction is anisotropic, being higher along the $\langle 001 \rangle$ directions. By placing a severe chill at the bottom of an ingot it is possible to freeze directionally from the bottom to the top, thereby producing an ingot that is virtually all columnar (unidirectional casting). In such an ingot all of the grains will have a $\langle 001 \rangle$ direction aligned and, consequently, this technique is widely used to produce high-grade magnets for such things as loudspeakers.

3. Equiaxed Zone. Within the liquid at the center of the ingot there are generally many small equiaxed grains suspended throughout. As freezing continues, these small grains begin to crowd together until finally they effectively block the inward motion of the columnar grains. This point is called the columnar to equiaxed transition. Again, the solid–liquid interfaces of these floating grains are dendritic.

It is possible by proper control to obtain either all columnar or all equiaxed grains in an ingot. To understand how this comes about we will focus our attention on the equiaxed grains. To obtain a columnar to equiaxed transition in an ingot it is apparent that the equiaxed grains must (a) be produced in the center of the ingot and (b) not melt in the hotter central region.

We now discuss each of these requirements separately.

a. Source of Equiaxed Grains. When liquid metal is poured into a mold the temperature of the liquid falls rapidly at the mold wall. This cold metal is denser than the hot metal at the center of the ingot and so it tends to sink while the hot metal rises, producing what is known as natural convection, see Fig. 9.50(a). Very large convection currents are present in the liquid metal shortly after pouring and they die out as the metal attains a constant temperature. The dendrites are very fragile, as may be seen from the sketch of Fig. 9.50(b), with the side branches often being narrower at their base as shown (see Ref. 33 for an explanation of this). It is currently thought that there are two main sources for the equiaxed grains.

1. The dendrites formed at the mold wall on the grains in the chill zone are simply torn off of these grains by the initial convection currents, and

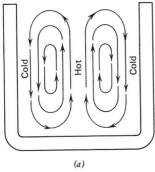

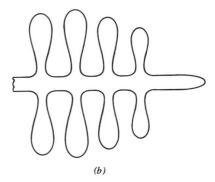

(a)   (b)

Figure 9.50 (a) Convection currents in the liquid metal cast into a mold. (b) Schematic view of a dendrite showing the narrowing of the side branches at their bases.

then the resulting dendrite fragments are swept into the center of the ingot by the convection currents to become the source of equiaxed grains.

2. After the columnar zone develops, dendrites may also be detached from these columnar grains by convection currents in the same manner as described above. However, it is also possible for the side arms to simply melt off the columnar dendrites due to temperature fluctuations, because the side arms have a higher solute content and, therefore, a lower melting temperature at their small bases. This mechanism is discussed in Ref. 33.

3. A third possible source of equiaxed grains was thought for some time to be the heterogeneous nucleation of grains in the center of the ingot due to constitutional supercooling ahead of the dendrite array. It is currently believed that this happens only rarely and that mechanism 1 is usually the predominate source of equiaxed grains. The solidification of an ingot would be a fascinating thing to observe if metals were transparent. This has been illustrated by experiments on H_2O-NH_4Cl mixtures that form a eutectic phase diagram. The progress of solidification is observed by making two sides of the mold out of plexiglass. Figure 9.51(a) illustrates the dendrite arrays leading the columnar grains as they grow out from the walls and up from the bottom of the cold mold. The source of the equiaxed grains is illustrated in Fig. 9.51(b), where small dendritic clusters are observed to fall out of the dendrite array at the walls, drop down to the bottom, and be carried out up into the liquid center by convection currents. See Refs. 33 and 50 for similar pictures and further discussion.

b. Survival of the Equiaxed Grains. In order to understand how the equiaxed grains survive in the supposedly hotter central region of the ingot, it is first necessary to understand a bit about heat flow in an ingot. When hot metal is poured into a cold mold one expects very steep temperature

Figure 9.51 NH_4Cl-H_2O solution cast into an aluminum mold chilled to liquid N_2 temperatures. (a) 1.7 min after pouring, (b) 4.2 min after pouring. (Courtesy of Wayne P. Bosze.)

gradients and high solidification rates. However, it may be seen from Eq. 9.9 that a high solidification rate requires a high solid temperature gradient but a low liquid temperature gradient. To maximize the solidification rate the system quickly adjusts the liquid temperature gradient downward, and this adjustment is aided by the convection currents in the liquid. Consequently, the temperature gradient in the liquid falls to near zero amazingly fast, as is illustrated in Fig. 9.52(a), which shows data for steel on the same experiment as Fig. 9.44(a). Notice that the steel was poured with a 155°F superheat, and after only $\frac{1}{2}$ minute this superheat has been virtually eliminated so that the temperature gradient in the liquid at the interface is nearly zero.

Referring again to Fig. 9.45(a) it may be seen that we can divide the casting into three regions, (1) solid, (2) dendrite array, and (3) liquid. Therefore, we can consider an average temperature gradient in each of these three regions, $\bar{G}_s$, $\bar{G}_A$, and $\bar{G}_l$. When a casting freezes, the value of $\bar{G}_l$ is quite small, being less than 1 or 2°C/cm after only a small amount of solid forms at the mold wall. The value of $\bar{G}_s$ and $\bar{G}_A$ are determined by the rate of heat extraction with $\bar{G}_s$ being larger than $\bar{G}_A$ as shown in Fig. 9.45(a).

Figure 9.52(b) presents a picture of an ingot with the temperature profile superimposed upon it at two different times. After a small length of columnar grains have formed, the temperature will have dropped to nearly a constant value in the liquid as shown. This liquid temperature must be nearly equal to the temperature of the solid–liquid interface advancing into

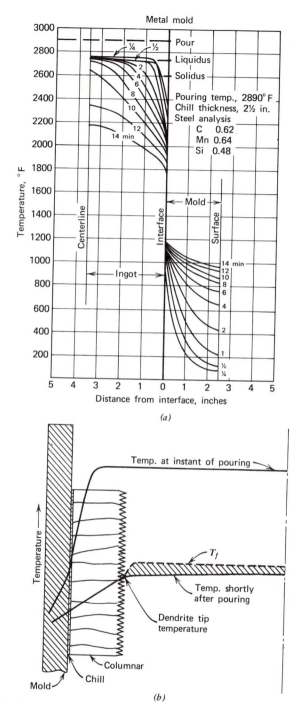

Figure 9.52 (*a*) Time–temperature profiles in a steel ingot poured with a 155°F superheat. Cast iron mold with a 2.5-inch wall thickness (from Ref. 31). (*b*) Temperature profiles at instant of pouring and shortly thereafter.

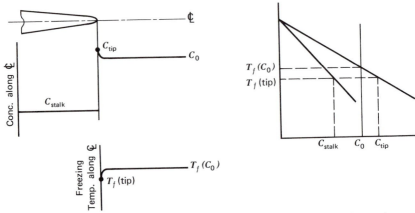

Figure 9.53 (a) Concentration profile along centerline of dendrite and corresponding freezing temperature profile. (b) Phase diagram locating initial alloy composition C_0.

it; thus, the liquid temperature will be essentially the temperature of the advancing dendrite tips on the columnar grains. Now consider the sketch of the solute profile at the tip of a dendrite that is growing into a liquid of composition C_0 as shown in Fig. 9.53. A small solute buildup will be present at the tip giving $C_{tip} > C_0$. The composition of the liquid in the center of the ingot will remain at C_0 because no macrosegregation can be produced to raise it above C_0 since the interface is dendritic (see p.293). Therefore, the freezing temperature at the dendrite tip will be a few degrees below the freezing temperature of liquid at $C = C_0$, as illustrated in the lower left figure of 9.53. Since the liquid temperature has dropped to the dendrite tip temperature, all of the liquid in the center is supercooled a little bit as indicated on Fig. 9.52(b). But, more importantly, any solid nuclei having $C < C_{stalks}$ will have a melting temperature above the central liquid temperature of $T_f(tip)$; see the phase diagram of Fig. 9.53. The dendrite fragments formed at the chill zone would have formed more rapidly from liquid of $C = C_0$ and would have $C < C_{stalks}$. Hence, if these fragments survived until the initial superheat had dissipated they would eventually become central equiaxed grains.

Those things that promote production and survival of dendrite fragments will increase the size of the equiaxed zone. For example,

1. Small superheat. A large superheat tends to melt the dendrite fragments.

2. Long freezing range alloy, big ΔT_f. This promotes longer dendrites that are more fragile and easily fragmented.

3. Sand mold rather than metal mold. Low $\bar{G}_A$ promotes longer fragile dendrites.

4. Low-melting-point alloys. Same as 3.

5. Rapid mixing. There are two effects here:

a. Increased fluid flow aids in fragmenting the dendrites.

b. The superheat is dissipated more quickly, thereby increasing the chance of survival of dendrite fragments.

At this point it is instructive to relate the above discussion to the more important commercial alloys. It is common to classify cast metals into two categories depending on their mode of solidification, (1) skin forming or (2) mushy forming. These are illustrated in Fig. 9.54(a) and 9.54(c) respectively, and the intermediate case is shown in 9.54(b).

4. Skin-Forming Alloys. An alloy will freeze in the skin mode if the dendrite length is small. Referring to Eq. 9.53 it is seen that this freezing mode is favored by alloys having a low freezing range, ΔT_f, and by a high temperature gradient in the dendrite array, $\bar{G}_A$. Examples of such alloys are low carbon and low alloy steels, aluminum bronze, and manganese bronze.

5. Mushy Forming Alloys. In this mode of freezing one never observes a substantial skin at the wall or an advancing dendrite array. The columnar to equiaxed transition occurs immediately at the wall. This mode of freezing is characteristic of alloys having long dendrites, namely, a high ΔT_f and low $\bar{G}_A$. One may consider that if the dendrite array did form at the wall and grow inward, the dendrites would be so long as to extent from the

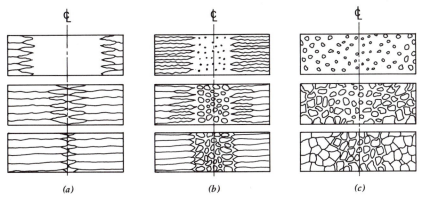

Figure **9.54** Schematic view of solidification from mold wall to centerline. (a) Skin forming, (b) an intermediate case, and (c) mushy forming (adapted from Ref. 35).

wall to the casting center. This, however, does not happen because the dendrites are very thin and fragile and they easily fragment because of liquid convection and thereby produce small solid nuclei throughout the casting shortly after pouring (teeming) as shown in Fig. 9.54(c). Examples of alloys freezing predominantly in the mushy modes are aluminum and magnesium alloys and some brasses.

It should be clear that whether a given alloy freezes in the skin or mushy mode depends on both the freezing range of the alloy and the heat transfer conditions. Hence, an alloy may freeze in the skin mode in a metal mold and in a mushy mode when cast into a sand mold. Specific examples for sand molds are:

Al and Mg alloys	Mushy mode
Cu alloys	
(a) Phos. bronze, Red brass	Mushy mode
(b) Al bronze, Yel. brass	Skin mode
Low alloy steels	Skin mode

An excellent graphical presentation of mushy and skin mode alloys is presented in Ref. 34.

D.   SEGREGATION IN CAST METAL ALLOYS

In Section 9.3 it was shown that when an alloy rod solidifies from one end with a flat solid–liquid interface a considerable segregation occurs along the length of the rod. In commercially cast metals the solid–liquid interface is dendritic, and this completely alters the type of segregation from the flat interface case of Section 9.3. It is shown in Fig. 9.55 that the dendrite morphology gives rise to lateral solute rejection and longitudinal solute rejection from the freezing solid. The longitudinal solute transport gives rise to a *macrosegregation* along directions parallel to the dendrite axes, and the lateral solute transport gives rise to a *microsegregation* along directions perpendicular to the dendrite axes.

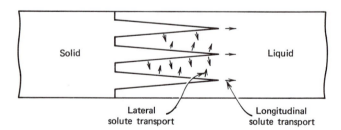

Figure 9.55   Schematic view of dendrite array showing directions of solute transport.

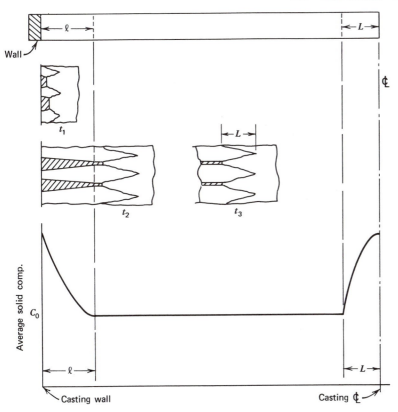

Figure 9.56 Establishment of the steady-state dendrite array and the corresponding inverse segregation produced.

1. Macrosegregation. During the columnar growth region it sometimes happens that the dendrites grow with a very nearly constant length. For instance, the dendrite length is nearly constant in the region of the ingot between 1 and 3 inches from the wall in the steel casting of Fig. 9.44(a). Under these conditions the dendrites have a nearly steady-state shape. Experiments have shown that when a dendrite array grows with nearly constant length there is very little macrosegregation in the longitudinal direction, even when there is severe convection in the liquid ahead of the advancing array. Essentially all of the solute is rejected laterally, so that one obtains only a microsegregation in the lateral direction. However, when the dendrite array is undergoing a change in length one does observe a macrosegregation in the longitudinal direction, as is illustrated in Fig. 9.56. A thin section from a casting is shown at the top of the figure. The dendrite array forms at the wall and as it advances to the right it continues

Table 9.5 Percent Volume
Change on Freezing, ΔV

Al	6.0%
Mg	5.1%
Cu	4.1%
Fe	2.2%

to increase in length (time t_1) untill the base of the array has moved a distance ℓ. After this point the dendrite array advances with constant length (steady state, times t_2 and t_3.) When the dendrite tips reach the centerline they encounter dendrites advancing in the opposite direction (assuming no equiaxed grains). Consequently, the dendrite array decreases in length over a region of length L on either side of the casting centerline. The macrosegregation observed as the array length changes is shown in the lower figure of 9.55 for an alloy of $k_0 < 1$. The composition is higher than the original composition C_0 in the region where the dendrites are lengthening at the wall. It is also higher than C_0 where the dendrites shorten at the centerline region, provided there is adequate liquid feeding into this centerline region, as is frequently the case. The segregation at the wall is called *inverse segregation* because it is reversed from the expected normal segregation that occurs with a flat interface. The segregation at the center is termed centerline segregation.

This phenomenon has been fairly well analyzed theoretically.[36] Metals undergo a volume contraction, ΔV, upon freezing. Consequently, when the metal at the base of the dendrite array freezes, fluid flows down the interdendritic channels to fill the ΔV of solidification. Analysis[36] shows that the segregation is a function of the ratio of this interdendritic flow velocity to the rate of interface advance. When the ratio is less than the steady-state value, a positive inverse segregation as shown in Fig. 9.56 is obtained. The amount of segregation is a function of both the volume contraction on freezing and the dendrite length. Table 9.5 lists the percent volume change on freezing for the major commercial metals. Hence, macrosegregation due to interdendritic fluid flow is mainly encountered in Al and Mg alloys. However, even in these alloys this effect is not very big, generally changing the concentration from C_0 by only a few weight percent.

2. Microsegregation. As discussed above, in dendritic solidification the solute is virtually all rejected in the lateral direction shown in Fig. 9.55. This results in a segregation in the lateral direction from the dendrite stalk to the interdendritic liquid, as shown in Fig. 9.46. This segregation is termed microsegregation because it extends over a length on the order of one-half the dendrite spacing, around 0.015–0.15 mm.

It may be seen from Fig. 9.46 that one way to characterize the amount of microsegregation would be to determine the volume fraction of second phase. However, in many alloys solid-state diffusion is sufficient to prevent formation of interdendritic eutectic; see the dashed curve of Fig. 9.46. Consequently, a common method of characterizing the amount of micro-segregation is to measure the segregation ratio, SR, defined as

$$SR = \frac{\text{Max. conc. (interdendritic region)}}{\text{Min. conc. (dendrite stalks)}}$$

This segregation ratio may be determined fairly easily with an electron microprobe. Dendrites in cubic materials often appear in transverse sections as small crosses as shown in Fig. 9.57(a). A typical microprobe trace across a dendrite stalk such as from A to B in Fig. 9.57(a) might give an intensity variation for the solute component as shown in Fig. 9.57(b). Since the intensity is proportional to concentration one may determine the value of SR from the peak-to-valley ratio. Some data on measured SR values in steels are presented in Table 9.6. These data were taken at different distances from the mold wall so they provide a measure of the effect of freezing rate and, therefore, dendrite spacing upon segregation. It may be seen that the SR does increase at lower rates but the increase is fairly small. Hence, the degree of microsegregation is fairly uniform in a cast metal.

From a practical point of view one often desires to eliminate the microsegregation in a cast alloy by homogenization at high temperatures. Consequently, diffusion must cause an average atom to move a distance of

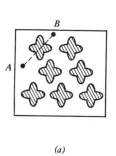

(a)

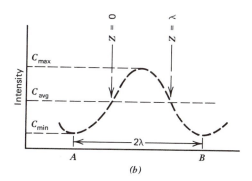

(b)

Figure 9.57 (a) Cross section of cubic dendrite stalks in a cast ingot. (b) The solute variation across dendrite from point A to B illustrating microsegregation.

Table 9.6 Variation of Segregation Ratio, SR, with Distance from Mold Wall. Dendrite Spacing Coarsens toward Center of Mold (see Table 9.8).

Dist. from Mold Wall (inches)	SR	Material
1.7–5.8	1.4–1.8	Mn in 4340 steel[37]
1.7–5.8	1.6–1.9	Ni in 4340 steel[37]
0.5–5	3.8–4.2	Cr in 52100 steel[38]
0.5–4	1.32–1.38	Ni in Fe-10% Ni[38]

λ, where 2λ is the dendrite spacing. From Eq. 6.49 we write

$$\lambda \approx 2.45 \sqrt{Dt} \qquad (9.52)$$

The time for homogenization may therefore be written as

$$t(\text{homog.}) \approx 0.167 \frac{\lambda^2}{D} \qquad (9.53)$$

This result shows that it is very important to minimize the spacing between dendrites because the homogenization time varies as the square of λ. A slightly more satisfying way of approaching this homogenization problem is to recognize that the microsegregation may frequently be approximated by a sine wave. For example, let the direction along the line A-B of Fig. 9.57(a) be called Z. Then we can write the equation for the microsegregation of Fig. 9.57(b) as

$$C(Z) = C_{\text{avg}} + A_0 \sin \frac{\pi Z}{\lambda} \qquad (9.54)$$

when A_0 is the amplitude of the segregation in the as-cast alloy, $A_0 = C_{\text{max}} - C_{\text{avg}}$. We have taken the position where $C(Z) = C_{\text{avg}}$, as the origin for $Z = 0$, as shown on Fig. 9.57(b). As diffusion occurs during annealing the amplitude of the sine wave will decay but the wave length will not change. (Consider why the wavelength does not change and whether, in fact, this statement requires an assumption). Consequently we may write two equations that form boundary conditions for the decay,

$$C(Z = 0, t) = C_{\text{avg}} \qquad (9.55)$$

$$\frac{dC}{dZ}\left(Z = \frac{\lambda}{2}, t\right) = 0 \qquad (9.56)$$

Equation 9.55 simply states that the concentration remains constant at C_{avg} at the position $Z = 0$, and Eq. 9.56 states that the peak of the sine wave remains at $Z = \lambda/2$ during the decay. It was pointed out on p. 142 that we need two boundary conditions and one initial condition to solve Fick's second law of diffusion, Eq. 6.9. Using Eqs. 9.55 and 9.56 as boundary conditions and Eq. 9.54 as the initial condition, the solution of Eq. 8.9 is found to be

$$C(Z, t) = C_{avg} + A_0 \sin\left(\frac{\pi Z}{\lambda}\right) e^{-\pi^2 Dt/\lambda^2} \tag{9.57}$$

From this equation it is a simple matter to show that the amplitude of the sine wave, $A = C(\lambda/2) - C(0)$, decays according to the simple relation

$$A = A_0 e^{-\pi^2 Dt/\lambda^2} \tag{9.58}$$

One can now ask: How long will it take for the amplitude of the microsegregation to decay to some arbitrary fraction of its original as-cast value, A_0? For example, you may show from Eq. 9.58 that the time required for the amplitude to decay to 1% of its original value is given as

$$t(99\% \text{ decay}) = 0.467 \frac{\lambda^2}{D} \tag{9.59}$$

It can be seen that this result agrees fairly well with the very simple approach that led to Eq. 9.53.

These results show that it is very important to minimize the spacing between dendrites because the homogenization time goes as the square of λ. Experiments have shown that the dendrite spacing is a function of solidification rate in much the same way as is eutectic spacing. Data[32] for many aluminum alloys are shown in Fig. 9.58. In this plot the spacing is for

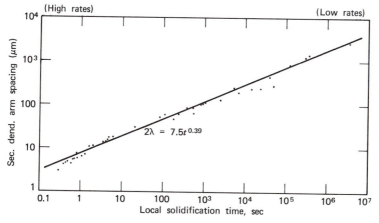

Figure 9.58 Variation of the secondary dendrite arm spacing as a function of the local solidification time (from Ref. 32).

secondary dendrite arms, which are much more abundant in the cast metal than primary arms. The local solidification time is the time for the dendrite array to pass any point in the casting and is therefore a measure of solidification rate. As expected, high rates produce smaller spacings. The final observed dendrite spacing is generally not the same as the spacing in the dendrite array that originally freezes from the liquid, although they are closely related. This is because as the solid cools slowly some of the larger dendrites tend to coarsen by solid-state diffusion at the expense of the smaller dendrites.

In conclusion, there are two important points. (1) The degree of microsegregation is only a weak function of solidification rate. (2) Homogenization of microsegregation is strongly dependent on dendrite spacing, which in turn may be significantly reduced at higher solidification rates.

E. POROSITY

There are two primary sources of porosity in cast metals. The first, cavity porosity, is due to improper feeding. The second, microporosity, is a result of the mushy mode of solidification in some alloys. We now discuss the cause of each of these forms of porosity.

1. Cavity Porosity. Suppose you wish to cast a tapered metal plate and you place a reservoir of metal (called a riser or feeder) at one end of the plate as shown in Fig. 9.59(a). As the metal freezes, it moves inward at roughly the same rate from each of the mold walls; see t_1. At time t_2 the advancing solid–liquid interfaces meet and form a bridge that traps a cavity of liquid metal as shown in Fig. 9.59(a). When this trapped liquid freezes it will form a shrinkage cavity due to the ΔV on freezing. Because the riser was placed at the narrow end of the taper the bridge was able to prevent the metal in the riser from feeding the solidification shrinkage. However, placing the riser at the fat end of the plate ensures that bridging does not occur between the feed metal in the riser and the solidification front; see

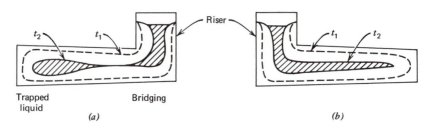

Trapped
liquid

Bridging

(a)

(b)

Figure **9.59**   (a) Improper riser design leading to bridging and cavity porosity. (b) Proper riser design.

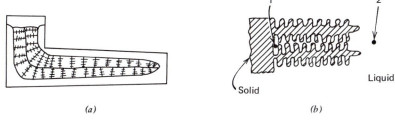

Figure 9.60 The two-phase, liquid plus solid region in cast metals.

Fig. 9.59(b). Consequently, we may define a general principle of proper feeding: *An open liquid channel must extend from the feed metal sources (risers) to all solidifying fronts.*

Cavity porosity results from improper feeding and it may be eliminated by proper riser design. This is a very interesting engineering design problem and it is one of the major topics of books devoted to cast metals.[39] A very good concise discussion of the many varied and fascinating problems involved in proper casting design is presented in Ref. 40.

2. Microporosity. We now reconsider the proper feeding design of Fig. 9.59(b) taking into account the fact that the solid–liquid interface is dendritic. When the advancing dendrite arrays meet at the center of the casting we are left with a final central section that consists of fine dendrites and interdendritic liquid; see Fig. 9.60(a). As this final interdendritic liquid freezes, feeding must occur through this two-phase region. To understand what is physically involved here consider Fig. 9.60(b), which shows a couple of long dendrites protruding into a liquid. As solidification occurs at point 1 it is necessary to have fluid flow from point 2 to point 1 in order to fill the shrinkage volume, ΔV, associated with this slidification. This means that the pressure at 1 must be less than at 2, and we may write

$$\Delta P = P(2) - P(1) \qquad (9.60)$$

To treat this problem it has been considered[41] that the final mushy region consists of n channels of radius R, and the analysis gives

$$\Delta P \propto \frac{(\Delta V \text{ on freezing})(\text{channel length})^2(\text{tortuosity factor})}{(\text{total channel area, } n\pi R^2)} \qquad (9.61)$$

where the tortuosity factor is an adjustable term that allows one to adjust the flow path length to account for the fact that the fluid must move through a circuitous channel, not a straight channel.

The pressure at point 2 will generally be close to atmospheric pressure, depending on the amount of liquid head. This means that the pressure at point 1 must be below atmospheric pressure, that is, a negative pressure,

$P(1) = P_2 - \Delta P$. Liquid metals normally contain a certain amount of dissolved gas. Consequently, when the pressure in the liquid falls below some nucleation pressure, say P^*, a gas bubble will nucleate. Therefore when $P(1)$ falls below P^* we obtain microporosity. Consequently to avoid porosity we require

$$P(1) = P_2 - \Delta P > P^* \tag{9.62}$$

This result illustrates that microporosity is a function of two main variables, (a) P^*, which is a function of the amount of dissolved gas in the metal, and (b) ΔP, which is a function of the variables of Eq. 9.61.

Microporosity is an inherent result of the nature of solidification of dendritic structures, and it may be present even with proper riser design. Equation 9.61 shows that microporosity is favoured in alloys having (1) a large volume change on freezing, ΔV, and (2) very long dendrites (this increases both channel length and tortuosity). Therefore we expect aluminum alloys to be quite susceptible to microporosity and steels less susceptible—see Table 9.5 and Section 9.5B—and this is found to be the case. Notice from Eq. 9.62 that microporosity may be reduced by casting under a high external pressure since this raises P_2. Also, vacuum degassing aids in avoiding this defect by lowering P^*.

Two characteristic types of microporosity are observed. In the first type the microporosity is uniformly dispersed throughout the section; we call this *dispersed microporosity*. In the second type the porosity is distributed in layers running across the section; we call this *layer microporosity*. Since these pores originate from interdendritic locations you might expect them to be quite small. Typical sizes are 5–10 μm for microporosity in columnar grains and 25 μm for pores in equiaxed grains. These sizes are so small that the pores are sometimes difficult to detect.

There are a number of other quite interesting sources of porosity in cast metals such as subsurface pinhole porosity in steels and blowholes in rimming steels. However, the above two sources of porosity are by far the most general causes of porosity in cast metals.

The student is urged to work Problems 9.10–9.13 to understand better the relationship of phase diagrams to the occurence of porosity and segregation in castings.

F. MECHANICAL PROPERTIES

After casting, a metal is generally used in one of four conditions: (1) as cast, (2) heat treated after casting, (3) mechanically worked after casting, and (4) worked and heat treated. In all cases the casting process has a strong influence upon the mechanical properties, but much more so in cases 1 and 2. There are four main casting variables that influence the

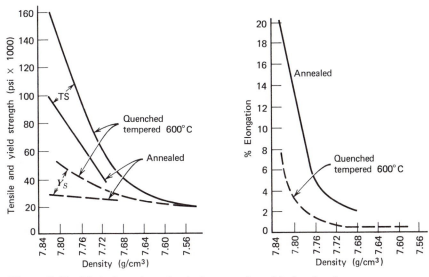

Figure **9.61**   Variation of mechanical properties with density. Investment cast to produce 0–3% shrinkage porosity. Low-alloy steel: 0.25 C, 0.3 Si, 0.5 Mn, 0.2 Ni, 3.1 Cr, 0.5 Mo (from Ref. 42).

mechanical properties of cast alloys, (1) porosity, (2) presence of second phase, (3) dendrite spacing, and (4) grain size.

1.   Porosity.   It is found that both cavity porosity and layer microporosity are more detrimental to mechanical properties than dispersed microporosity. The detrimental effect of porosity is due primarily to (1) stress concentration at the pores, which depends on pore shape, and (2) reduction in load-bearing area, which depends mainly on volume fraction porosity. Some experimental data taken from a study on a low alloy steel (similar to AISI 4130) are shown in Fig. 9.61. In this work the microporosity is measured by the density, that is, high density is equivalent to low porosity. It is seen from Fig. 9.61(*a*) that in both the annealed and heat-treated condition the tensile strength is decreased by the microporosity whereas the yield strength is only slightly affected. Ductility is quite strongly affected by the microporosity as shown in Fig. 9.61(*b*). The major effect of increased microporosity is to decrease ductility and tensile strength.

2.   Presence of Second Phase.   The effect of a second phase upon mechanical properties is illustrated by a study[43] on 7075 aluminum alloy (5.7 Zn, 2.3 Mg, 1.3 Cu, 0.18 Cr, 0.15 Ti). This is a high-strength age-hardenable alloy used in aircraft. The alloy was unidirectionally cast, which resulted in the presence of second phase, mainly $CuAl_2$, within the

Table 9.7 Variation of Mechanical Properties of AISI 7075 Aluminum Alloy as a Function of Volume % Second Phase (from Ref. 43)

Vol % 2nd Phase	UTS (psi)	YS (psi)	RA (%)	Elong. (%)
1.2	66,000	60,000	7	2
↓	↓	↓	↓	↓
0.1	84,000	72,000	32	13

interdendritic regions. The volume fraction second phase was then varied between 0.1% and 1.2% by thermal mechanical processing that involved heat treating, mechanical reduction, and a final heat treatment. The results are summarized in Table 9.7. It is seen that there is a significant increase in ductility as the second phase is removed. Notice also that in this case there is a significant improvement in both yield strength and ultimate tensile strength. This is apparently due to a strengthening of the matrix as the Cu from the $CuAl_2$ phase goes into solution. It is obvious that control of second phase is extremely important in optimizing the mechanical properties of some alloys.

3. Dendrite Spacing. A number of studies have recently pointed up the effect of dendrite spacing upon mechanical properties. As a typical example we consider a study[44] on the Al–Si casting alloy, A356 [Al–7% Si–0.5% Mg–0.2% Ti]. The alloy was unidirectionally solidified, solution treated at 540°C for 10 hours, quenched, and aged at 160°C for 3 hours. The dendrite spacing increased away from the chill wall. Specimens were taken at various distances from the chill wall to determine the effect of dendrite spacing upon mechanical properties and the results are briefly summarized in Table 9.8. It is apparent that the smaller dendrite spacing improves the ductility significantly. This is accompanied by an increase in ultimate tensile strength but with a constant yield strength, as found with

Table 9.8 Variation of Mechanical Properties with Dendrite Spacing in a Cast A356 Aluminum Alloy (from Ref. 44)

Dist. Wall (inches)	Dendrite Spacing (μm)	UTS (psi)	YS (psi)	Elong. (%)
8	100	46,000	42,000	1
↓	↓	↓	↓	↓
1	35	52,000	42,000	11

porosity. The improved mechanical properties in the samples having smaller dendrite spacing are due largely to the better annealing obtained in these samples as a result of the shorter wavelength of the periodicity of the microsegregation.

4. Grain Size. It is generally found that as one decreases the grain size the strength of a metal increases. There is a well-known relation called the Petch equation that shows that the strength is proportional to the reciprocal of the square root of the grain diameter (see p. 517). For cast metals, however, it is not always true that strength improves with decreasing grain size. As a general principle we may say: Small grain size will improve mechanical properties of a cast metal only if production of small grain size does not:

1. increase amount of microporosity
2. cause the microporosity to become layer type
3. increase volume percent second phase
4. increase dendrite spacing.

In general it is only possible to determine if any of these detrimental effects accompany a reduced grain size by a trial-and-error process. However, it is frequently found that small grain size does improve the mechanical properties of cast metals, although some bronzes and gun-metals [88 Cu–8 Sn–4 Zn] are exceptions to this rule, apparently because layer porosity is produced in the finer-grain-size material. The improved properties of fine-grain-sized castings are not thought to be due to a more effective blockage of dislocations by the higher density of grain boundaries. Rather, they are due to the finer distribution of microporosity and second-phase particles. An example of the effect of grain size in aluminum alloy 195 [Al–4.5 Cu] is presented in Fig. 9.62, and it may be seen that a significant increase in ductility and UTS is obtained at the smaller grain sizes. The grain size of the as cast metal may be decreased by increasing the size of the equiaxed zone. From the discussion of Section 9.5C we conclude that grain size can be reduced by

1. decreasing amount of supherheat
2. producing turbulence in the liquid
3. using a low thermal conductivity mold
4. adding grain refiners (called innoculents).

Since it is found[45] that too low superheat often reduces internal soundness in the casting, metals are frequently poured with moderate superheat and with the addition of innoculents. It should be realized that the final grain size may be changed from the as cast condition by heat treatment if the

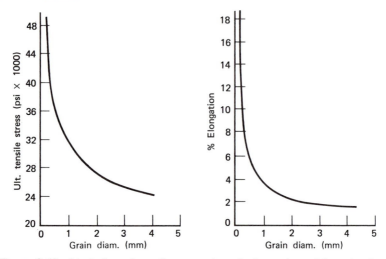

Figure 9.62 Variation of tensile strength and elongation with grain size in an Al–4.5 Cu alloy (from Ref. 45).

alloy undergoes a solid-state phase change (for example, steels) or if the material will be worked to allow recrystallization.

In conclusion it is apparent that we generally desire

1. small dendrite size
2. low porosity
3. small amount of second phase
4. small grain size if achieveable within above limitations.

The main beneficial effects of these features are improvement of ductility

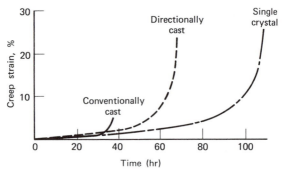

Figure 9.63 Creep properties of conventionally cast, directionally cast, and single-crystal samples of MAR-M200 tested at 1800°F and 30,000 psi (195 MN/m^2) (from Ref. 46).

and ultimate tensile strength, and these improvements are achieved through elimination of porosity and segregation defects.

There are some applications where one does not want a small grain size. For example, turbine blades in the hot stage of jet engines operate at very high temperatures under load. At high temperatures coherency across grain boundaries is reduced and the presence of grain boundaries tends to reduce the resistance to creep. Figure 9.63 presents creep data for a conventional turbine blade alloy, MAR-M200 [Ni + 9 Cr, 10 Co, 12.5 W, 2 Ti, 5 Al, 1 Nb, 1.5 C]. Data are also presented for single crystalline material as well as directionally solidified material. It is fairly easy to cast turbine blades unidirectionally so that the relatively large columnar grains grow along their length. As can be seen from Fig. 9.63 such techniques offer good possibilities of improving the performance of a jet engine using a given turbine blade alloy; see Ref. 46.

G. CAST IRONS

From a commercial point of view the most important cast alloys are cast irons. From a scientific point of view the nature of the solidification of cast irons is extremely interesting and not as well understood as we would like. Cast irons are essentially ternary alloys of Fe–C–Si of near eutectic composition. There are four main types of cast irons, (1) gray, (2) nodular, (3) white, and (4) malleable. A good concise summary of the chemistry and microstructure of these cast irons is presented in Chapter 18 of Ref. 47.

To understand the role of silicon in cast irons it is useful to first examine the binary Fe–C phase diagram shown in Fig. 9.64. Notice that liquid eutectic iron (4.3% C) may solidify as either austenite + graphite or as austenite + carbide (Fe_3C or cementite). The carbide forms at 1148°C, which is 6°C below the temperature of formation of the graphite. We call this temperature difference, ΔT_{G-C} as indicated on Fig. 9.64. When the eutectic forms with graphite we have a gray iron and when it forms with Fe_3C we have a white iron. The graphite is the thermodynamically stable phase whereas the Fe_3C is actually a metastable phase since it forms at a lower temperature. For kinetic reasons it is more difficult for the graphite to form than the Fe_3C. Consequently when one solidifies an iron–carbon eutectic alloy in a normal manner the eutectic solid contains Fe_3C. The major effect of silicon is to increase the value of ΔT_{G-C} as is shown in Fig. 9.65. As ΔT_{G-C} increases the thermodynamic driving force for forming graphite prior to forming Fe_3C is increased and, hence, the probability of forming a graphite eutectic rather than a carbide eutectic is increased. Hence, Si promotes the formation of graphite. See Fig. 9.71(a) for a typical gray cast iron microstructure.

To understand the nature of the solidification pattern of cast irons it is

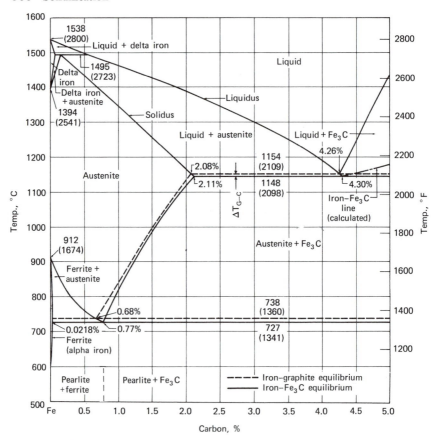

Figure 9.64 The iron–carbon phase diagram. (Reproduced by permission, from Metals Handbook, 8th ed., Vol. 8, American Society for Metals, Cleveland, 1973.)

first necessary to understand some characteristics of ternary phase diagrams.

1. Ternary Phase Diagrams.[48] In the ternary phase diagram the alloy compositions are plotted on a triangular grid as shown in Fig. 9.66(a). An alloy of any composition would be represented as a point on this grid. For example, the alloy represented by the point I on Fig. 9.66(a) has composition 40% A, 20% B, 40% C. The three binary phase diagrams A–B, B–C, and A–C are plotted along the edges of the triangle as shown in Fig. 11.60(b). Notice that A–B and A–C form eutectics whereas B–C forms continuous solid solutions. This is a rather simple ternary diagram with the interior formed as shown. Also note that whereas the liquidus of a binary alloy is a line, the liquidus of the ternary alloy is a two dimensional surface.

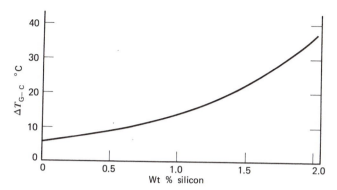

Figure 9.65   Difference of eutectic temperatures, ΔT_{G-C}, as a function of silicon content (adapted from Ref. 50, p. 189).

Figure 9.66(c) presents an isothermal cut through the phase diagram at temperature T_1. Consider now a liquid alloy of composition L_1, which when cooled to temperature T_1 is on the liquidus surface as shown. What is the composition of the solid that will freeze from this liquid? We know it must lie on the trace of the solidus surface with the T_1 isotherm, but unless we have the tie lines we do not know at which composition on this trace. Hence, in ternary diagrams we must be supplied with tie lines at each temperature, whereas in binary diagrams this difficulty is absent since the tie lines are automatically determined by specifying the temperature.

We will now consider what happens when we unidirectionally solidify an alloy of composition L_1 assuming a planar solid liquid interface. Referring to Fig. 9.66(b) it is seen that the first solid forms at α_1. As freezing progresses the liquid composition must change since the solid is freezing at a different composition. The liquid composition proceeds from L_1 to L_2 and the solid composition follows from α_1 to α_2. The liquid composition now lies on the trough. Liquids lying on the trough are in equilibrium with both the α and β phases. If we pass an isothermal plane through point L_2 on the trough, this plane intersects the α-phase boundary at α_2 and the β-phase boundary at β_2. Connecting these three points forms a "tie triangle" on the isothermal plane and the three connected compositions give the composition of the α and β phases in equilibrium with L_2 at this temperature. Consequently, once the liquid reaches composition L_2, the solid β_2 will begin to freeze out along with α_2 so that a two-phase eutectic forms just as at the eutectic point in binary diagrams. Only there is now a fundamental difference. In a binary eutectic the eutectic liquid always freezes at just one temperature, T_E, on the binary diagram. This is not true for this ternary eutectic. Text books on phase diagrams[48] show that the average composition of the eutectic, $\alpha_2 + \beta_2$, must lie somewhere on the tie line connecting

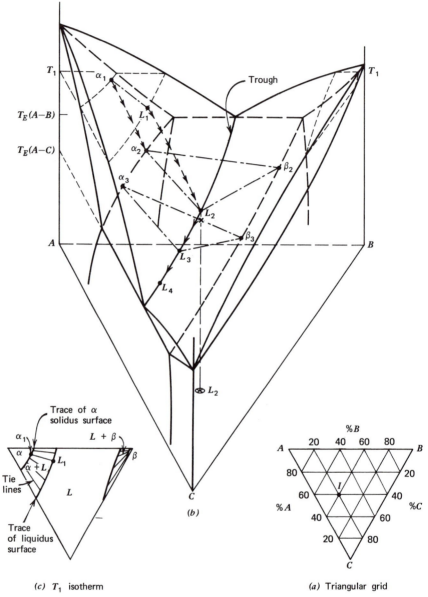

(b)

Trace of α
solidus surface

α₁

L + β

α

L₁

α + L

β

Tie
lines

L

C

Trace
of liquidus
surface

(c) T_1 isotherm

(a) Triangular grid

%B

A   20   40   60   80   B

80   20

60   *I*   40

%A   40   60   %C

20   80

C

Figure 9.66 Ternary phase diagram.

308

α_2 and β_2. This means that the average composition of the solid eutectic must necessarily be different than the composition of the liquid from which it is freezing, L_2. Consequently, as freezing progresses the liquid composition moves down the trough to a position such as L_3, where the α-phase composition in the solid eutectic has moved from α_2 to α_3 and the β-phase composition from β_2 to β_3. Since the eutectic temperature of the A–B binary $T_E(A$–$B)$, is higher than that of the A–C binary, $T_E(A$–$C)$ the trough line drops in temperature from L_2 to L_3. This means that *the eutectic freezes over a range in temperatures*. L_3 is positioned such that L_2 lies over the tie line between α_3 and β_3. Therefore, equilibrium freezing of a liquid at composition L_2 would give eutectic having α phase of composition α_3 and β phase of composition β_3 since these are the only two α- and β-phase compositions connected by a tie line containing L_2. Since alloys freeze in a nonequilibrium manner the final liquid will solidify from some composition beyond L_3 such as L_4. The main point of this discussion has been to show that in ternary eutectics in which the eutectic consists of two phases the freezing temperature is not fixed as in binary eutectics, but varies depending on the temperature slope of the trough on the ternary phase diagram.

The ternary Fe–C–Si phase diagram is very complex and difficult to visualize. Therefore, it is useful to present a kind of pseudo-phase diagram, Fig. 9.67, which simplifies interpretation and conveys the main ideas. The composition axis on this diagram is called the carbon equivalent, defined as $CE = \%\,C + \frac{1}{3}\%$ Si. Alloys with $CE < 4.3$ are termed hypoeutectic and those having $CE > 4.3$ are termed hypereutectic. Notice that the eutectic freezes out over a temperature range.

2. Gray Irons. Most gray cast irons are hypoeutectic and so we will discuss the manner in which these irons solidify. Figure 9.68(a) shows the progressive manner of freezing of a cast iron of $CE = 3.86$ that was solidified with a 285°F superheat in a 7-inch sand mold.[49] Consider a point in the mold 1.5 inches from the wall. After about 11 minutes the dendrite tips reach this point and they continue to grow in size until at around 55 minutes they reach a limiting size. Twenty minutes later at 75 minutes the interdendritic liquid begins to solidify as eutectic. The eutectic continues to freeze as the temperature drops until at about 93 minutes the solidification is completed at this point, 1.5 inches from the mold wall. This solidification procedure is shown schematically in Fig. 9.68(b), where it is illustrated that the solidifying ingot may be divided into three zones consisting of (I) completely solid austenite dendrites plus eutectic, (II) austenite dendrites plus liquid plus freezing eutectic, and (III) liquid plus freezing austenite dendrites. The austenite dendrites appear to reach the center of the ingot after only about 17 minutes. The interdendritic liquid at the center does

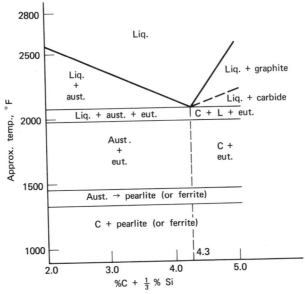

Figure 9.67   Schematic diagram showing approximate temperature ranges of various phase fields in cast irons. (From Ref. 47, Copyright McGraw-Hill Book Co., 1967. Used with permission of McGraw-Hill Book Co.)

not completely freeze until around 1.5 hours later so that we must consider cast irons as being long-range freezing alloys that freeze in more of the mushy mode than the skin mode.

A very interesting feature of cast-iron solidification involves the manner in which the interdendritic liquid forms eutectic solid. The eutectic does not freeze with a continuous front near the base of the austenite dendrites

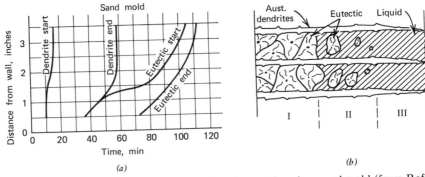

Figure 9.68   (a) Progress of solidification of a cast iron in a sand mold (from Ref. 49). (b) Schematic view of three stages in ingot from mold wall to centerline.

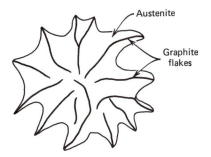

Austenite

Graphite flakes

Figure **9.69** A eutectic colony in gray cast iron [not to be confused with the colony structure described in Fig. 9.39].

in a manner similar to Fig. 9.43. Rather, the graphite actualiy nucleates directly out of the interdendritic liquid as shown schematically on Fig. 9.68(*b*). Recent research indicates that nucleation does not occur on austenite dendrite arms but it occurs within the interdendritic liquid in a manner that is not well understood. A spherical clump of eutectic then grows out from the graphite nuclei forming what are called, variously, eutectic cells or eutectic colonies. These cells are generally larger relative to the dendrites than shown on Fig. 9.68(*b*). A schematic illustration of a eutectic cell is shown in Fig. 9.69. The growth of this cell is an example of a faceted–nonfaceted eutectic growth front and it results in a highly irregular structure. Many experiments have shown that all of the graphite within the cell is interconnected. For excellent photographs see Ref. 50, pp. 185 and 195.

3. Kinetic Effects. As the rate of cooling is increased the graphite flakes become more finely dispersed until at sufficiently high rates the carbon solidifies as the carbide, cementite. The graphite flakes are classified progressively from the coarsest (type *A*) to the finest (type *D*), with type *A* usually called flake graphite and type *D* called undercooled graphite because it forms at a lower temperature. Excellent photographs illustrating these structures may be found in Ref. 51.

The kinetics of cast iron solidification may be understood from the plot of Fig. 9.70. The solidification of the cementite eutectic appears to occur in a coupled manner similar to the regular eutectic growth mode discussed in Section 9.4. The cementite curve gives the interface temperature as a function of the rate of freezing of the austenite–cementite eutectic. One has a similar curve for the graphite eutectic. Whichever eutectic form can grow at the higher temperature for a given rate is the one expected to be observed. Suppose, for example, that both eutectic forms were present at a given rate. The one that grew at the higher temperature would lead the

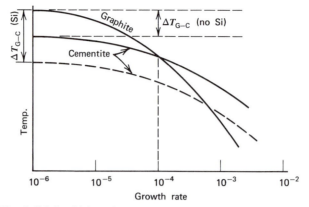

Figure 9.70 Solid–liquid interface temperature as a function of growth rate for growth of cementite eutectic and graphite eutectic.

other and, hence, crowd it out. Therefore, the crossover point on Fig. 9.70 gives the rate above which the cementite eutectic (white iron) forms. And it is seen from the dashed curve that by increasing ΔT_{G-C} the rate required to form white cast iron is raised. Hence, Si allows gray iron to form at higher rates. By modification of Eqs. 9.45 and 9.48 one may obtain an expression for the cementite curve on Fig. 9.70. It should be clear that the gray–white transition is a function of the kinetic equations describing the interface rates versus temperature for the two different types of eutectic. At present these equations are only qualitatively known (Ref. 50, p. 203).

Cast irons generally contain some sulfur impurities that have a strong influence upon the growth forms. It is found that the sulfur produces three obvious effects: (1) it produces coarser flakes, (2) it increases the undercooling, that is, it decreases the freezing temperature of the eutectic interface, and (3) it increases the colony density. These effects are well illustrated by the photographs of Ref. 50, p. 198 and Ref. 51, p. 590. The effects of sulfur are apparently connected to the fact that the equilibrium distribution coefficient of sulfur is approximately 0.02 in austenite and near 1 in graphite. Therefore, upon freezing the sulfur must be redistributed into the graphite to about the same extent as the carbon. This would result in a high liquid sulfur concentration in front of the austenite phase, which could influence the kinetics of the eutectic growth by affecting the interface kinetics, ΔT_K, the solid–liquid surface tension, or the diffusion field in the liquid. The relative importance of these three factors in controlling the effects of sulfur are at present unknown.

4. Nodular Irons. In 1948 it was reported by the International Nickel Company that addition of 0.02–0.1% Mg caused the graphite flakes to form in the shape of spheres rather than as flakes; see Fig. 9.71. In that

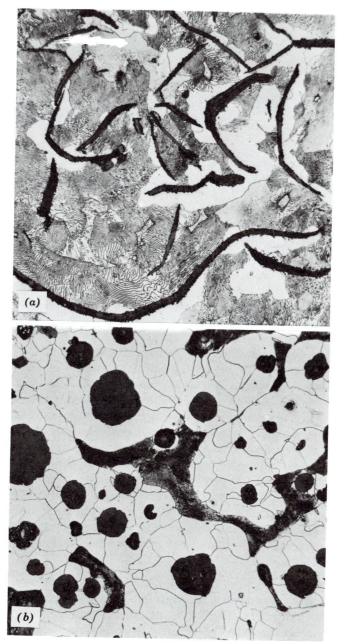

Figure 9.71 (*a*) Gray cast iron. Graphite flakes (black) surrounded by ferrite (white) in a pearlite matrix (250×). (*b*) Nodular cast iron (Mg added). Graphite nodules (black spheres) surrounded by ferrite (white) and some pearlite regions (black), (250×). (Courtesy Harlan Baker.)

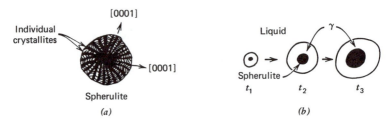

Figure 9.72   (a) A graphite spherulite in nodular cast iron. (b) Growth of a spherulite illustrating the surrounding austenite, γ, region.

same year it was reported by the British Cast Iron Research Association that addition of 0.2–0.4% Ce also produced spherical graphite shapes. At the present time industry uses the Mg process and the nodular iron so produced is usually called ductile cast iron. The ductile cast iron industry continues to grow at a fabulous rate, having grown from a tonnage of 18,000 in 1954 to an estimated 1.4 million tons in 1971.

The growth process of ductile cast iron can be thought of as being quite similar to gray cast iron except that the interdendritic liquid solidifies as spheres of graphite rather than eutectic colonies. These spheres are frequently called spherulites. The amount of supercooling for spherulite growth is larger than for eutectic cells and the number of spherulites per cubic centimeter is roughly 200 times greater than the eutectic cells of grey iron. The spherulite itself is essentially pure solid graphite, which is a spheroidal aggregate composed of a radially growing array of graphite crystallites originating at a common center. Figure 9.72(a) presents a schematic illustration of the appearance of the spherulites. The individual crystallites grow outward in such a manner as to maintain their C axis parallel to the growth direction (graphite has a hexagonal crystal structure). Hence the surface of the spherulite is composed of the basal planes of the graphite crystallites, and beautiful SEM pictures of these surfaces have recently become available.[52]

At the present time there are three areas in which we do not understand the solidification of ductile iron:

1.   Very little is known about how the spherulites nucleate.

2.   The flakes in grey iron are known to grow with their a axis parallel to the growth direction, whereas the graphite is growing parallel to the C axis in spherulite growth. The elements Mg and Ce are able to force this change in growth axis, but we do not understand how. Graphite grows with a faceted interface as shown in Fig. 9.10(a), and it is apparent that if solute atoms tended to absorb preferentially at the ledges they could poison this type of growth. It should be clear that our understanding of the graphite morphology depends on understanding the kinetics of the interface reactions that occur when graphite crystallizes from the melt.

3. When one examines as-cast nodules it is usually observed that they are surrounded by a layer of iron. Very good experimental evidence indicates that the spherulite is surrounded by a shell of austenite and that both the spherulite and shell grow simultaneously from the liquid as shown in Fig. 9.72(b). However, this mode of growth is hard to explain theoretically because all of the carbon added to the spherulite must diffuse through the austenite shell. This is a slow process and, also, one would expect tremendous pressures in the shell due to the volume increase as the graphite forms. New theories have been proposed to account for these difficulties,[50] but it is clear that much is still to be learned about one of our most common metals, cast iron.

5. **Mechanical Properties.** A gray cast iron is essentially a composite material consisting of graphite plus high carbon iron. Consequently the mechanical properties are strongly influenced by both the amount of graphite and its size, shape, and distribution. Gray iron with type-A flakes has a higher tensile strength than the finer type-D irons. The elastic properties of gray irons require special considerations because they differ somewhat from those of other engineering materials.[53] Gray cast irons are extremely brittle. Stress concentration at the tips of the graphite flakes causes local plastic flow at low external loads. The conventional 0.2% offset yield strength will give values that are quite close to the ultimate tensile strength. Consequently, the yield strength and the ductility of gray cast irons are not specified. The tensile strengths of common gray cast irons range from 20,000 to 60,000 psi (136–409 MN/m^2).

Changing the graphite morphology from flake to nodular has a tremendous influence on the mechanical properties. For example, in some irons this change in morphology can jump the tensile from 20,000 to over 100,000 psi with a concurrent *increase* in ductility. This is perhaps one of the most dramatic examples of the effect of microstructure upon the properties of a material. Most commercial ductile irons have properties in the following ranges:

Ductility: 2–30% elongation in 2 inches
Tensile: 55,000–120,000 psi (375–818 MN/m^2)
Yield: 30,000–90,000 psi (205–614 MN/m^2)

Malleable cast irons have not been discussed because they are obtained by heat treatment of white iron. Their mechanical properties are similar to ductile irons, however, and the student is referred to other texts[47,53] for a more complete discussion of cast irons.

REFERENCES

1. D. Turnbull and R. E. Cech, J. Appl. Phys. **21**, 805 (1950).

2. G. Colligon, J. Aust. Inst. Metals **10**, 277 (1965).

3. G. L. F. Powell, J. Aust. Inst. Metals **10**, 223 (1965).

4. D. Gomersall, S. Shiraishi, and R. Ward, J. Aust. Inst. Metals **10**, 220 (1965).

5. J. Marcantonio and L. Mondolfo, Met. Trans. **2**, 465 (1971).

6. K. Jackson, D. Uhlmann, and J. Hunt, J. Crystal Growth **1**, 1 (1967).

7. G. F. Bolling, J. J. Kramer, and W. A. Tiller, Trans. Met. Soc. AIME **227**, 1453 (1963).

8. W. G. Pfann, *Zone Melting*, Wiley, New York, 1966.

9. W.G. Pfann, Sci. American, **217**, 63 (December 1967).

10. J. A. Burton, R. C. Prim, and W. C. Slichter, J. Chem. Phys. **21**, 1987 (1953).

11. C. Elbaum, in *Progress in Metal Physics*, Vol. 8, B. Chalmers and R. King, Eds., Pergamon Press, New York, 1959, p. 203.

12. W. A. Tiller, K. A. Jackson, J. W. Rutter, and B. Chalmers, Acta Met. **1**, 428 (1953).

13. J. C. Warner and J. D. Verhoeven, Met. Trans. **4**, 1255 (1973).

14. R. F. Sekerka, J. Crystal Growth **3–4**, 71 (1968).

15. S. C. Hardy and S. R. Coriell, J. Crystal Growth **7**, 147 (1970).

16. L. R. Morris and W. C. Winegard, J. Crystal Growth **5**, 361 (1969).

17. R. J. Schaefer and M. E. Glicksman, Met. Trans. **1**, 1973 (1970).

18. F. R. Mollard and M. C. Flemings, Trans. AIME **239**, 1534 (1967).

19. J. D. Verhoeven and E. D. Gibson, Met. Trans. **4**, 2581 (1973).

20. R. W. Kraft, J. Metals **18**, 192 (1966).

21. F. S. Galasso, J. Metals **19**, 17 (1967).

22. G. A. Chadwick, in *Progress in Materials Science*, Vol. 12, No. 2, B. Chalmers and W. Hume-Rothery, Eds., Pergamon Press, New York, 1963.

23. L. M. Hogan, R. W. Kraft, and F. D. Lemkey, in *Techniques of* Chalmers and W. Hume-Rothery, Eds., Pergamon Press, New York,

24. J. L. Walter and H. E. Cline, Met. Trans. **4**, 1775 (1973).

25. K. A. Jackson and J. D. Hunt, Trans. Met. Soc. AIME **236**, 1129 (1966).

26. C. Zener, Trans. Met. Soc. AIME **167**, 405 (1946).

27. W. A. Tiller, in *Liquid Metals and Solidification*, ASM, Cleveland, 1958, p. 276.

28. H. W. Kerr and W. C. Winegard, Can. Met. Quarterly **6,** 55 (1967).

29. W. H. Weart and J. D. Mack, Trans. Met. Soc. AIME **212,** 664 (1958).

30. M. D. Rinaldi, R. M. Sharp, and M. C. Flemings, Met. Trans. **3,** 3139 (1972).

31. A. F. Bishop, F. A. Brandt, and W. S. Pellini, Trans. Am. Foundry. Soc. **59,** 439 (1951).

32. T. F. Bower, H. D. Brody, and M. C. Flemings, Trans. Met. Soc. AIME **236,** 624 (1966).

33. K. A. Jackson, J. D. Hunt, D. R. Uhlmann, and T. P. Seward, Trans. AIME **236,** 149 (1966).

34. H. F. Bishop and W. S. Pellini, Foundry **80,** 86 (February 1952).

35. R. W. Ruddle, Trans. Am. Foundry. Soc. **68,** 685 (1960).

36. M. Flemings and G. Nereo, Trans. AIME **242,** 41, 51 (1968).

37. H. Thresh, M. Bergeron, F. Weinburg, and R. K. Buhr, Trans. AIME **242,** 853 (1968).

38. M. Flemings et al., J. Iron and Steel Inst., **208,** 371 (April 1970).

39. R. Flinn, *Fundamentals of Metal Casting*, Addison-Wesley, Reading, Mass., 1963.

40. R. Flinn, Casting Techniques, in *Techniques of Metals Research*, Vol. 1, R. F. Bunshah, Ed., Wiley, New York, 1968, chapter 21.

41. J. Campbell, Cast Metals Res. J. **5,** 1 (March 1969).

42. S. J. Walker, Foundry Trade J. **127,** 943 (December 1969).

43. S. N. Singh and M. Flemings, Trans. AIME **245,** 1811 (August 1969).

44. S. F. Frederick and W. A. Baily, Trans. AIME **242,** 2063 (October 1968).

45. A. Cibula, *Grain Control*, University of Sussex Institute on Metallurgists, London, 1969, pp. 22–44.

46. F. L. Versnyder and M. E. Shank, Materials Sci. Eng. **6,** 213 (1970).

47. R. Heine, C. Loper, and P. Rosenthal, *Principles of Metal Casting*, McGraw-Hill, New York, 1967.

48. F. N. Rhines, *Phase Diagrams in Metallurgy*, McGraw-Hill, New York, 1956.

49. R. P. Dunphy and W. S. Pellini, Trans. Am. Foundry. Soc. **59**, 425 (1951).

50. *The Solidification of Metals,* Iron and Steel Institute, London, 1968, Publ. 110.

51. B. Lux, M. Grages, and D. Sapey, Prac. Metallog. **5**, 587 (1968).

52. B. Lux, W. Kurz, and M. Grages, Prac. Metallog. **6**, 464 (1969).

53. C. F. Walton, Ed., *Iron Castings Handbook,* Iron Founders' Society, Cleveland, 1971.

PROBLEMS

9.1 On p. 245 it is stated that if the solidus and liquidus are straight lines the equilibrium distribution coefficient k_0 is a constant independent of composition. Present a proof for this.

9.2 Present a derivation for Eq. 9.11.

9.3 (a) Derive Eq. 9.18. Hint: Do a mass balance on the solute within the zone as it advances a distance dZ. This will give you a differential equation that may be integrated to obtain Eq. 9.18.

(b) This equation is only valid from positions $Z = 0$ to $Z = L - l$, where L is the total length of the rod. Explain why the equation is not valid for all Z.

9.4

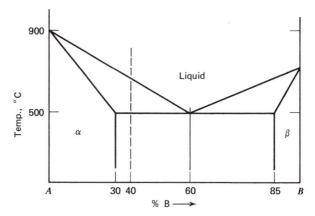

An alloy having the above phase diagram of overall composition 40% B is held in the molten state in a long boat and is slowly solidified from the left end. The temperature gradient is high enough so that the liquid–solid interface remains flat. The mixing in the liquid is very strong so that there is no solute buildup ahead of the interface in the liquid; hence, the liquid composition is uniform.

(a) What is the value of k_0 for this alloy? What is the value of k_e for this experiment?

(b) Using the normal freeze equation determine what fraction of the rod will be eutectic. Draw a picture of the rod after solidification and show where the eutectic is distributed.

(c) Complete equilibrium freezing is described with the lever law. Using the lever law, what fraction of the alloy would be eutectic after solidification if we had this equilibrium freezing?

(d) What is the answer to parts (b) and (c) if the initial composition of the alloy is only 5% B?

(e) Suppose a large casting were made of this alloy of 5% B. If you cut this casting open and looked at it under a microscope is there any chance you would see some eutectic? Based on the simple model of dendrite growth that was discussed in this chapter, what would be the volume fraction of the ingot occupied by β phase? Assume α and β have the same density.

9.5 Suppose you wished to make up a long rod of the alloy of Problem 9.4 having a uniform composition of 20% B. To do this you decide to start with a liquid of 20% B and freeze it. You have two choices:

1. You freeze from one end with a flat interface under conditions where k_e approaches 1 (say, 0.99).

2. You quench the alloy rapidly and obtain a dendritic structure. You then anneal the rod at the highest possible temperature to eliminate microsegregation.

(a) The usual value of δ/D is around 1000 sec/cm and the value of D around 10^{-5} cm^2/sec. Hence, choice 1 is not really practical. Explain why not. Hint: Consider stability of interface.

(b) The interdendritic spacing is inversely proportional to the rate of freezing. Suppose the average spacing between dendrite stalks is 100 μm. Estimate how long you should heat treat the rod in order to remove the microsegregation if you heat treat (a), 100°C below the eutectic temperature, and (b) 5°C below the eutectic temperature. The diffusion coefficient of B in A is given as $D = 0.085 \exp[-32,600/2T]$ cm^2/sec.

9.6

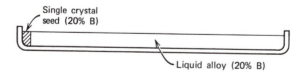

Single crystal seed (20% B)

Liquid alloy (20% B)

Suppose you wanted to grow a single crystal of the alloy of Problem 9.4 with a composition of 20% B. You do this by placing a small single crystal of the desired alloy at one end of a horizontal boat in contact with liquid alloy as shown above, and then slowly solidify the liquid. The liquid solidifies onto the seed crystal giving a single crystal of the same orientation as the seed.

(a) Calculate the temperature gradient in the liquid at the interface necessary to ensure that the interface remains flat throughout the solidification if the solidification rate were 1 cm/hour, and $D = 2 \times 10^{-5}$ cm^2/sec.

(b) Would you expect to get a single crystal if the solid–liquid interface were dendritic? Explain.

9.7   Use the phase diagram of Fig. 9.64 for this problem. To determine k_0 values approximate the liquidus and solidus curves with straight lines. Suppose you had a long rod of liquid steel having composition 0.4 wt % C. This melt is solidified from one end with a flat solid–liquid interface (no dendrites) at a rate of 10^{-3} cm/sec under conditions of complete mixing in the liquid.

(a) Your problem is to draw a graph showing composition versus distance along the rod. Calculate the composition at Z/L values of 0.0, 0.1, 0.2, 0.3, 0.4, 0.5, 0.6, 0.7, 0.8, 0.9, and 1.0. Also, give the fraction of the rod which is δ phase, γ phase, and eutectic, assuming that the δ phase and γ phase were to remain stable upon cooling. Note, you must assume negligible solid–state diffusion in order to use the normal freeze equation. Would you expect this assumption to hold here?

(b) Upon cooling, the δ phase and γ phase will undergo the solid-state transformations shown on the phase diagram. In some regions of the rod you will have pearlite plus intergranular Fe_3C. Locate this region on the rod. Also estimate the microstructure in the remaining regions. (If you are not familiar with pearlite formation see Section 12.2B.)

(c) Suppose there was a solute buildup present in the liquid and the δ/D value was 200 sec/cm. Calculate what fraction of the rod solidifies before γ phase begins to solidify from the liquid. Because the distribution coefficient k_0 is different between liquid/δ and liquid/γ, a transient may occur when the γ phase first begins to form. What will be the composition of the solid freezing from the liquid at completion of this transient?

9.8   (a) Suppose you repeat the above experiment with a steel of 1 wt % C under conditions of no mixing in the liquid. Taking the diffusion coefficient of C in liquid iron as $D = 0.016 \exp(-14000/RT)$ cm²/sec, calculate the temperature gradient required to prevent the interface from forming dendrites at a point where 50% of the bar has solidified. Use $R = 2$ cal/mol-°K.

(b) Repeat the calculation for conditions where the liquid is stirred to give a boundary-layer thickness of 0.01 cm.

(c) From your analysis discuss the effect of stirring on the stability of solid–liquid interfaces.

9.9   (a) Suppose the alloy of Problem 9.8 were cast into a rectangular mold having inside dimensions of 20×10 inches under conditions where no equiaxed grains formed. The columnar grains will grow across the narrow dimension and touch each other at the centerline plane. When the tips of the dendrites on these columnar grains first touch at the center, what is the length of the primary dendrite stalks if the temperature gradient is:

  1. 100°C/cm and

  2. 10°C/cm.

(b) What is the effect of dendrite length on (1) microporosity, and (2) the columnar to equiaxed transition.

(c) What effect would you expect a high solid-state diffusion coefficient to have on the dendrite length? Please explain your reasoning clearly.

9.10

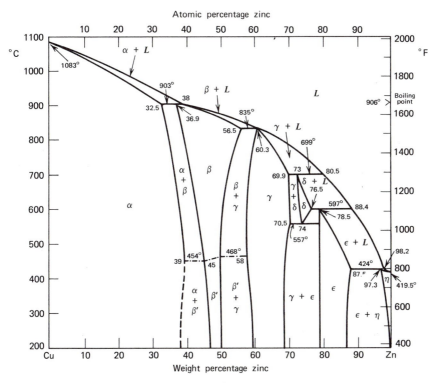

Copper–zinc phase diagram. [Reproduced by permission from *Metals Handbook*, American Society for Metals (1948).]

Cartridge brass is an alloy of 30 wt % Zn–70% Cu. See above phase diagram. Suppose you were to solidify a long rod of this alloy from one end. Assume:

1. The solid-liquid interface is flat (no dendrites).
2. There is complete mixing in the liquid ($k_e = k_0$).

(a) Make a sketch of the composition of the solid versus the fraction of the rod solidified. Clearly indicate the following *three* compositions on your plot, composition of initial solid to freeze, 32.5wt %, and 36.9wt %.

(b) Estimate the value of k_0 between α and liquid from the phase diagram. Now calculate the fraction of the rod that freezes from the liquid as pure α phase.

(c) Assume that the value of k_0 between β and liquid is 0.95. Also assume that upon cooling to room temperature the equilibrium phases form as shown on the phase diagram. Calculate what fraction of the rod would contain at least some β' phase after cooling to room temperature.

9.11

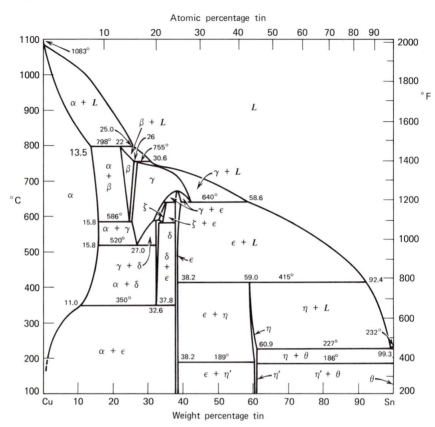

Atomic percentage tin

Copper–tin phase diagram. [Reproduced by permission from *Metals Handbook*, American Society for Metals (1948).]

Phosphor bronze is an alloy of essentially 10 wt % Sn–90 wt % Cu. See above phase diagram. Make assumptions 1 and 2 of problem above. Assume also that k_0 between α and liquid is a constant equal to 0.36 and that k_0 between β and liquid is constant at 0.87. Now calculate:

(a) The fraction of the rod that freezes directly from the liquid as pure α phase.

(b) The fraction of the rod that freezes directly from the liquid as pure γ phase.

9.12 Suppose that you were going to consider the two alloys of Problems 9.10 and 9.11 for a potential casting application. Which alloy would:

(a) give rise to more porosity problems?

(b) have more of a chance of containing second phase?

(c) be more prone to inverse segregation?

Explain your reasoning clearly, discussing what factors give rise to porosity, second phase, and inverse segregation, and explain how these two alloys differ as regards these factors.

9.13   Shown below is the location of the centerline of four dendrites, which are growing out of the page toward you.

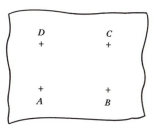

(a) Make a sketch of how you think the composition would vary from point A to point B for dendrites that form in the alloy of Problem 9.10.

(b) Repeat for the alloy of Problem 9.11

(c) Now make a sketch of what phases would be present in an as-cast alloy of Problem 9.10 along the line from point A to B. Assume the alloy is at room temperature and carefully consider the phase diagram that shows some solid-state phase transformations.

(d) Repeat part (c) for problem 9.11. The transformations shown on the Cu-rich side of the Cu–Sn phase diagram below 500°C do not occur in cast metals. (They are too sluggish and occur only upon extended annealing.) Therefore, you may obtain an operational phase diagram by simply extending all vertical phase boundaries at 500°C straight down to room temperature.

(e) You may check your answers to parts (c) and (d) by comparing figures 60 and 66 in the book, *Annotated Metallographic Specimens*, A. R. Bailey and L. E. Samuels, Metallurgical Services, Surrey, England 1971.

9.14   Make a guess at how you think the following variables would affect the dendrite zone length, and then explain the reasoning for your guess.

(a) An increase of alloy composition.

(b) A decrease of thermal conductivity of the mold material.

(c) An increase of the latent heat of the alloy.

(d) An increase of convective mixing within the liquid ahead of the interface.

9.15

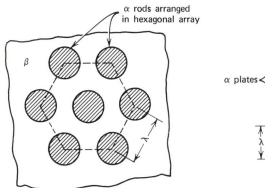

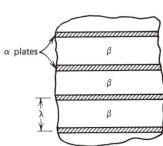

Suppose a particular eutectic might exist as either a rod-type or a lamellar-type structure.

We will assume that the above photomicrographic representations lie perpendicular to the rods and lamellae and that the distance of separation is λ in both cases.

(a) Compute the α-β interface area per square centimeter of photomicrograph for both cases. You will find that the α-β interface area is a function of only λ for the lamellar case. For the rod case, however, it is a function of both λ and the area fraction of the photomicrograph occupied by α rods. Your answer for the rod case should come out to be $3.809 \times \sqrt{A}/\lambda$ where A is area fraction of α rods.

(b) Take the area fraction of α on the photomicrograph as equal to the volume fraction of α in the eutectic. Then, based on your answer to part (a), explain why you might expect a lamellar morphology if the volume fraction α were above 27.6% and a rod type if it were below 27.6%. Explain any assumptions. [Note: We have taken λ the same for both rod and lamellar cases. This assumption may be eliminated and instead of 27.6% one obtains 31.8%; see J. D. Hunt and J. P. Chilton, J. Inst. Metals **91**, 342 (1962–63).]

9.16 Consider an alloy of composition 10% B having the phase diagrams of problem 9.4. A long rod of this alloy is solidified slowly from one end under conditions where the normal freeze equation applies, $X_s = X_0 k_0 (1-f)^{k_0-1}$, where f is the fraction of the rod solidified and X_s and X_0 are fractional concentrations (as opposed to the volume concentrations used in derivation of Eq. 9.17).

(a) Show that after the fraction f of the rod has solidified the *average* solid composition is

$$\bar{X}_s = \frac{X_0}{f} [1 - (1-f)^{k_0}]$$

(b) As the rod solidifies the composition of the liquid increases. This in turn lowers the freezing temperature of the liquid. Show that the freezing temperature of the liquid varies with the fraction of solidified rod as

$$T_f = T_p - m_l X_0 (1-f)^{k_0-1}$$

where m_l is the slope of the liquidus line on the phase diagram, and T_p is the freezing temperature of the solvent (component A on phase diagram).

(c) Make a sketch of the phase diagram. On this diagram plot the average composition of the solid, $\bar{X}_s$, for conditions where the liquid freezing temperatures, T_f, are 750, 700, 600, 500°C. What fraction of the rod will freeze as eutectic?

(d) Nonequilibrium freezing is often described by plotting $\bar{X}_s$ on the equilibrium diagram as you have done in part (c). For examples, see pp. 42 and 86 of Ref. 48.

CHAPTER 10
RECOVERY AND RECRYSTALLIZATION

We will now proceed to examine a number of different types of solid–solid phase transformations that allow us to control the structure of alloys. The first of these is recrystallization. The general phenomenon of recrystallization may be described as follows. First, we plastically deform a metal by a significant amount. This causes the grains to become elongated as shown at the left of Fig. 10.1. Now we heat the metal up to a temperature on the order of one-half of its melting temperature and hold it at this temperature. A very interesting sequence of events occurs as shown in Fig. 10.1. If we were to observe the structure with a hot-stage optical microscope we would find that nothing would happen until at time t_1 new grains would begin to nucleate within the cold-worked grains of the matrix. These grains would then grow rather quickly until all of the cold-worked grains were gone at time t_2. These new grains would then grow at a slower rate until at time t_3 a terminal grain size would be obtained. Although nothing is observed to happen up to time t_1 in an optical microscope, we will find that much is happening during this time interval and it is called recovery as shown on

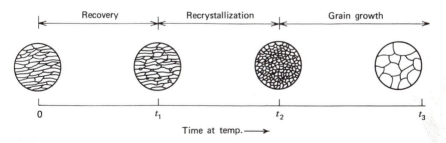

Figure 10.1 Schematic representation of the recovery–recrystallization–grain growth sequence.

325

Fig. 10.1:

Recovery: All annealing phenomena that occur before the appearance of new strain-free grains.

The process that occurs from time t_1 to t_2 is termed recrystallization and is defined as follows.

Recrystallization: The nucleation of the new strain free grains and the gradual consumption of the cold worked matrix by growth of these grains.

The driving force for the growth of the new strain-free grains from time t_1 to t_2 is the stored energy of the cold-worked matrix. After time t_2 the driving force for growth becomes only the curvature of the grain boundaries. The difference in the nature of these two driving forces and how they affect grain boundary motion was discussed in Section 7.5. The process occurring after time t_2 is generally referred to simply as grain growth; see Fig. 10.1. Consequently, it is seen that the processes that occur upon annealing of a cold-worked metal are subdivided into three stages, recovery, recrystallization, and grain growth.

When we plastically deform a metal, a considerable amount of energy is expended. Most of this energy goes into heat, but a small fraction of it remains in the metal as stored energy; see Fig. 10.2. This stored energy gives rise to the two relaxation stages upon heating, recovery, and

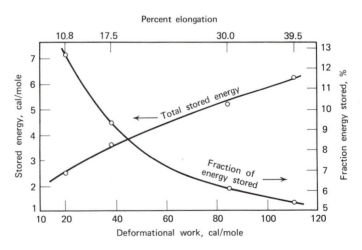

Figure 10.2   Amount of energy stored during deformation of Cu (data from Ref. 1).

recrystallization. The third stage, grain growth, is only indirectly related to these two stages and it has been considered in Chapter 7.

Strictly speaking, the driving force for these two relaxation processes is the Gibbs free energy per atom, $\bar{G}$, the chemical potential discussed on p. 151. By definition we have

$$\bar{G} = \bar{H} - T\bar{S} = \bar{E} + P\bar{V} - T\bar{S} \tag{10.1}$$

and for processes occurring at constant temperature and pressure one obtains

$$\Delta\bar{G} = \Delta\bar{E} + P\,\Delta\bar{V} - T\,\Delta\bar{S} \tag{10.2}$$

The term $P\,\Delta\bar{V}$ usually only amounts to a few percent of $\Delta\bar{E}$ as is shown in Problem 10.1. This result generally holds for solid-state processes at atmospheric pressure because $\Delta\bar{V}$ is small in the solid state. It is apparent from Eq. 5.2 that in order to evaluate the entropy term we must determine the degree of disorder produced in the lattice by plastic deformation. As we shall see, the major defects produced by the plastic flow are dislocations. One may estimate the entropy term from this source, and the $T\,\Delta\bar{S}$ term turns out to be only a few percent of $\Delta\bar{E}$ (Ref. 2, p. 15). Hence, to quite a good approximation the driving force for recovery and recrystallization may be simply taken as the stored energy $\Delta G \approx \Delta E = E_s$.

10.1   STORED ENERGY

When an alloy is plastically deformed many defects are forced into the crystal lattice, and these defects along with elastic strains serve as mechanisms for the storage of energy in the alloy.

A.   MECHANISMS OF ENERGY STORAGE

1.   Elastic Strain.   It was shown on p. 95 that if a lattice is strained by an amount ε it will possess a strain energy per unit volume given as $\varepsilon^2\bar{E}/2$, where $\bar{E}$ is Young's modulus. Lattice strain will produce a line shift in the x-ray pictures of a metal so that one may estimate the strains and, hence, the strain energy from x-ray examination of metals. Experiments show that elastic strain energy can account for only 5–10% of the total stored energy.

2.   Lattice Defects.   Plastic deformation will produce the following defects in the crystal lattice: dislocations, vacancies, interstitial atoms, stacking faults, and twin boundaries. The fraction of the stored energy produced by each of these defects depends on two things, the energy per defect and the density of defects produced by deformation. The two most

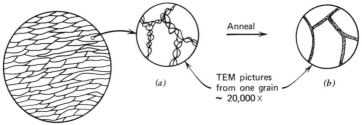

Optical micro. of cold—worked Al ~ 100×

Figure 10.3   Schematic representation of grain structure and subgrain structure in aluminum.

important defects produced by room-temperature deformation are dislocations and vacancies. The vacancies only account for a small fraction of the total stored energy so that the major portion of the stored energy, generally around 80–90%, is due to the generation of dislocations.

As mentioned on p. 118, severe cold working of an annealed metal will increase the dislocation density from around 10^7 to 10^{11} dislocations per square centimeter. The distribution of dislocations in the cold-worked matrix may be observed with transmission electron microscopy; see p. 329 of Ref. 3 for examples. It is found that if the dislocations have a low mobility at the temperature of deformation, they appear as a fairly random array in the deformed metal. However, if the dislocations are able to cross slip (high stacking-fault energy), they immediately begin to condense into tangles so that the metal contains regions of high and low dislocation density as shown in Fig. 10.3(a). The regions of low dislocation density are called variously cells or subgrains. Aluminum deformed at room temperature forms a distinct subgrain or cell structure because its high stacking-fault energy does not inhibit cross slip. Consequently, if one were to examine a cold-worked grain of aluminum in a transmission electron microscope (TEM) the grain would not appear as a small single crystal, but would be filled with subgrains as shown in Fig. 10.3(a). In general, it is observed that if distinct dislocation tangles are not already present in a metal after deformation, annealing will cause them to form, and additional annealing will condense the tangles to form clearly defined subgrain boundaries as shown in Fig. 10.3(b). These results show that the dislocation network can lower its energy by forming tangles rather than random arrays. It is very difficult to calculate accurately the energy of the dislocation arrays produced by deformation because of the complexity of the arrays and the uncertainty of the interaction energies between dislocations. It is important to realize that cold-worked grains that appear as perfect little single crystals under an optical microscope may consist of millions of tiny subgrains of slightly different crystallographic orientations separated by cell walls composed of condensed dislocation tangles.

B. VARIABLES AFFECTING THE AMOUNT OF STORED ENERGY

1. Purity. The addition of impurity atoms to a metal increases the amount of stored energy at a given strain. Apparently, the impurity atoms hinder dislocation motion and thereby produce an enhanced dislocation multiplication.

2. Deformation. More complex deformation processes produce higher stored energy. Simple tension may activate slip on only two slip planes in an fcc metal, whereas more complex deformation such as extrusion will generally activate slip on all four possible sets of slip planes. In the latter case dislocation intersection will be much more frequent, giving rise to higher dislocation densities.

3. Temperature. Deformation at lower temperatures increases the amount of stored energy. This is because there is less thermal energy to assist in the release of energy and to reduce the interaction between defects during deformation.

4. Grain Size. The amount of stored energy increases as the grain size decreases. Consider a large grain that is deformed a given amount. Now imagine this same volume to be divided into many small grains and then deformed the same amount. In this second case the deformation would result in many more grain-boundary-dislocation interactions. Since the grain boundaries are effective in blocking dislocations, the smaller grain size promotes dislocation interaction and multiplication. The dislocation density produced by strain has been shown to be inversely proportional to grain size (Ref. 3, p. 124).

10.2 RELEASE OF STORED ENERGY DURING ANNEALING

There are a number of different experimental techniques for measuring the amount of stored energy in a metal.[2] As a cold-worked metal is heated, the stored energy is released when the temperature becomes sufficient to allow the relaxation processes to occur. This energy release may be measured by comparing the annealing behavior of a cold-worked and an annealed specimen. One technique involves heating a cold-worked and an annealed specimen while measuring the power difference, ΔP, required to increase the temperature of both specimens at equal rates. Using this technique

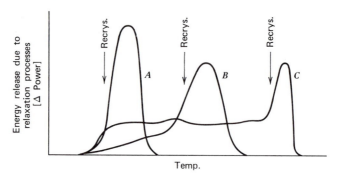

Figure **10.4**   Three types of energy release curves.

various types of curves are obtained as is shown in Fig. 10.4. Three points are significant concerning these curves:

1. In each case the recrystallized grains first appear at the onset of the large power peak.
2. The fraction of the stored energy released during recovery is small for type *A* and large for type *C*.
3. Type *A* curves are usually obtained for pure metals and type *B* or *C* for impure metals.

Let *S* be the total stored energy and *Sr* be the energy released during recovery, so that *Sr/S* is the fraction of the stored energy released during recovery. In one study[3] this fraction has been found to vary from 0.03 for high-purity metals to as high as 0.7 for some alloys. It is apparent that impurity atoms inhibit the nucleation of recrystallized grains and thereby allow a larger fraction of the stored energy to be released by recovery processes prior to the onset of recrystallization.

A. PROPERTY CHANGES

The changes in the physical properties that occur upon annealing are illustrated in Fig. 10.5 for a polycrystalline metal, where the energy release curve is shown at the bottom as a reference.

1. Hardness.   There is usually only a small change in hardness during recovery, about one-fifth of the total change. Since hardness (and strength) are decreased with decreasing dislocation densities, Fig. 10.5 indicates only a small drop in dislocation density during recovery but a large drop during recrystallization. We expect the latter result because the new recrystallized grains are essentially strain free.

2. Resistivity.   The electrical resistivity of a metal is a measure of the resistance offered by the metal lattice to the flow of electrons produced by an electric field. Defects in the lattice may act as scattering sites for the moving electrons and thereby increase resistivity. Point defects such as vacancies and interstitial atoms are more effective at scattering than dislocations, so that changes in resistivity reflect changes in vacancy and interstitial concentrations. The drop in resistivity during recovery shown on Fig. 10.5 means that a significant decrease in point-defect concentration must occur during recovery.

3.   Density.   The density of a cold-worked metal is decreased due to the generation of vacancies. Edge dislocations also make a small contribution to the lower density because of the dilation of the lattice about this type of dislocation. The density data of Fig. 10.5 show that the recovery mechanisms must involve a decrease in vacancy concentration and probably some decrease in edge dislocation density.

4.   Cell Size.   The data on cell size in Fig. 10.5 are self-explanatory. The cell size grows only slightly in the first portion of the recovery stage but shows a definite increase just prior to recrystallization.

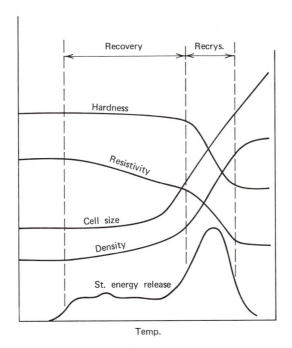

Figure 10.5   Variation of several physical properties during recovery and recrystallization.

B. RECOVERY MECHANISMS

Many experiments[2-4] have been done in an effort to determine what mechanisms are operating during the recovery stage. A summary is presented in Table 10.1. The mechanisms operating in the low temperature range involve vacancy motion; those operating in the intermediate temperature range involve dislocation motion without climb; those in the high temperature range involve dislocation motion with climb. This order of occurrence is a reflection of the relative thermal activation required for the different mechanisms; for example, as discussed in Chapter 4, dislocation climb is more difficult than glide. Table 10.1 illustrates the extreme complexity of the recovery stage.

1. Subgrain Growth. The origin of subgrains was discussed in connection with Fig. 10.3. After deformation the dislocation tangles isolate cellular regions of relatively low dislocation density as shown in Fig. 10.3(a). These cells are slightly misoriented with respect to each other (a few degrees) and are in a size range on the order of 0.1–1 μm. Upon annealing the dislocation tangles condense into sharp two-dimensional boundaries and the dislocation density within the cells decreases; see Fig. 10.3(b). Near the end of the recovery stage these subgrains begin to increase in size (see Ref. 3, p. 331 for pictures).

2. Subgrain Coalescence. Electron microscope studies have shown that in some cases the boundaries between subgrains simply disappear during the recovery stage. The process is illustrated schematically in Fig. 10.6. By processes that are not clear, the orientation mismatch between two neighboring grains disappears. This is probably accomplished by movement of interface dislocations involving climb so that diffusion is required.

Table 10.1   Recovery Mechanisms

Temp.	Mechanisms Operating
Low	1. Migration of point defects to sinks (grain boundaries, dislocations, etc.)
	2. Combination of point defects
Intermediate	1. Rearrangement of dislocations within tangles
	2. Annihilation of dislocations
	3. Subgrain growth
High	1. Dislocation climb
	2. Subgrain coalescence
	3. Polygonization

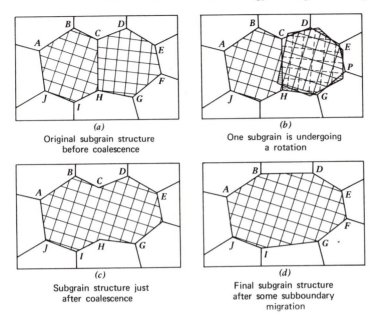

Figure 10.6 Schematic representation of subgrain coalescence by subgrain rotation. [From J. C. M. Li, J. Appl. Phys. **33**, 2959, (1962).]

3. Polygonization. It was discovered using x-ray analysis that when a single crystal is bent slightly as shown in Fig. 10.7(a) and then annealed, it breaks up into small single-crystal blocks as shown in Fig. 10.7(b). The Laue pattern from a crystal such as Fig. 10.7(a) shows a smearing of the diffraction spots due to the curved crystal planes. Upon annealing each smeared spot breaks up into a series of discrete smaller spots indicating the polygonized structure of Fig. 10.7(b). After the single crystal has been bent, an excess of positive edge dislocations is generated as shown in Fig.

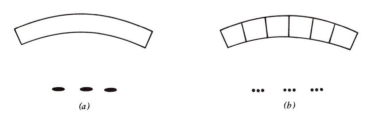

Figure 10.7 (a) A bent single crystal and the corresponding Laue spot pattern. (b) The polygonized crystal and its corresponding Laue spot pattern.

Figure 10.8 (a) The excess of edge dislocations produced by bending. (b) Alignment of the edge dislocations after polygonization.

10.8(a). Annealing causes these edge dislocations to line up over one another in small-angle tilt boundaries as shown in Fig. 10.8(b). This process has been verified using etch-pit techniques to locate the dislocation arrangements during the process. A classical set of experiments using the etch-pit technique to follow the dislocation arrangement in a Fe–Si alloy may be found in Ref. 5. It should be clear that in order for the dislocations to become lined up they must undergo both glide and climb. Polygonization is not as well defined in polycrystalline materials because of the complexity of the deformation. However, we may say that, in general, (a) polygonization requires an excess of edge dislocations, (b) it is only produced at higher recovery temperatures because dislocation climb is involved, and (c) it produces subgrains that are roughly 10 times larger than those produced via dislocation tangle condensation.

10.3   KINETICS OF RECOVERY

It is possible to obtain information concerning the mechanisms of recovery from analysis of the kinetics of recovery. Let P be some physical property that changes during the recovery stage (e.g., resistivity). We may write

$$P = P_0 + P_d \qquad (10.3)$$

where P_0 is the value of the physical property in the annealed state prior to deformation, and P_d is the increment in the physical property due to the defects produced by deformation. We will assume that P_d is proportional to the volume concentration of defects, C_d, produced by deformation, so that we may write

$$P = P_0 + \text{const} \cdot C_d \qquad (10.4)$$

We are interested in the time rate of change of the physical property, so we may write

$$\frac{d(P - P_0)}{dt} = \text{const} \cdot \frac{dC_d}{dt} \qquad (10.5)$$

The decay rate of the defects is a function of the concentration of defects and the mobility of these defects. This problem may be treated as in chemical reaction rate theory and we obtain[6]

$$\frac{dC_d}{dt} = -K(C_d)^n e^{-Q/kT} \qquad (10.6)$$

where Q is the activation energy for the process of defect anihilation occurring and n is an integer, being 1 for first-order kinetics, 2 for second-order, and so on, and K is a constant. Combining Eqs. 10.4, 10.5, and 10.6 we obtain

$$\frac{d(P-P_0)}{(P-P_0)^n} = -Ae^{-Q/kT}\, dt \qquad (10.7)$$

where A is a constant. For first-order kinetics we obtain

$$\ln(P-P_0) - \text{const} = -Ae^{-Q/kT}t \qquad (10.8)$$

where we have introduced an integration constant. Equations 10.7 and 10.8 describe the time-dependent decay of physical properties during recovery; they show that for first-order kinetics an exponential type of decay is expected. From analysis of kinetic data on recovery using these equations it is sometimes possible to obtain meaningful data on the activation energy for the defect causing recovery.

An excellent example on the recovery of Zn single crystals[2] will be presented. As shown in Fig. 10.9(a) the Zn single crystals were strained in

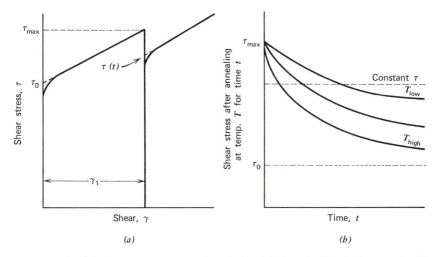

Figure 10.9 (a) The shear stress strain relationship for the Zn single crystals. (b) Time–temperature variation of yield stress during recovery.

pure shear by some amount γ_1. This increased the yield stress from τ_0 to τ_{max}. Recovery causes the yield stress to decay from τ_{max} back toward τ_0, and the rate is a function of time and temperature as shown in Fig. 10.9(b). The curves of Fig. 10.9(b) are obtained after annealing the crystals following the initial strain for various times at given temperatures. The recovered yield stress is then determined on each specimen by reloading to determine $\tau(t)$ as shown in Fig. 10.9(a). The general solution of Eq. 10.7 can be written as

$$f(P-P_0)=-Ae^{-Q/kT}t \qquad (10.9)$$

where f is an unspecified function depending on n of Eq. 10.6. If we now take τ constant as shown on Fig. 10.9(b), the left-hand side of Eq. 10.9 is constant so we have by taking logs,

$$\ln t = \text{const} + \frac{Q}{kT} \qquad (10.10)$$

From a plot of $\ln t$ versus $1/T$ at a constant value of τ as may be obtained from Fig. 10.9(b) we determine a value of Q from the slope. The data on Zn gave a linear relation between $\ln t$ and $1/T$ that indicated that only one activation energy and, hence, only one defect anihilation process was involved. The value of Q was found to be approximately the same as the activation energy for self-diffusion. From the discussion on p. 162 we may write

$$Q(\text{self-diffusion}) = \Delta E(\text{vacancy migration})$$

$$+ \Delta E(\text{vacancy formation}) \quad (10.11)$$

Consequently, we conclude that both vacancy formation and migration are involved in this recovery process. It is therefore believed that the recovery process involves dislocation climb.

Interesting experiments on iron[2] have given a value of Q close to ΔE(vacancy migration) at short recovery times and close to Q(self-diffusion) at long recovery times. This result indicates that the recovery mechanism was time dependent, involving vacancy migration at short times and dislocation climb at long times.

CONCLUSIONS

1. Recovery usually occurs exponentially with time.

2. Proper analysis of kinetic data allows one to determine Q in some cases.

3. Generally, more than one recovery mechanism operates so that Q is not a constant.

10.4 NUCLEATION MECHANISMS FOR RECRYSTALLIZATION

In Chapter 8 we discussed the classical theory of nucleation in which a nucleus of the new phase having some critical size for growth is formed by a statistical fluctuation in the cluster size distribution within the parent phase. Experiments have shown that this theory cannot apply to the nucleation of the new recrystallized grains, primarily because the radius of the critical sized cluster predicted by Eq. 8.2 is too large compared to experiments (see Ref. 7, p. 720). Two mechanisms have been observed for the nucleation event in recrystallization depending on the metal and the degree of deformation. The deformed metal contains two main types of interfaces, preexisting grain boundaries and the subgrain boundaries resulting from the deformation. "Nucleation" has been found to originate by the sudden growth of either of these two types of boundaries.

1. Sudden Growth of Preexisting Grain Boundaries. This mechanism has been observed with both the electron microscope and x-rays. The boundary between an original grain of high dislocation density and an original grain of low dislocation density is observed to suddenly grow out as shown in Fig. 10.10(a) (see Fig. 14.4 of Ref. 7). Hence, the "nucleation" event is essentially a growth phenomenon. A model based on electron microscope observation assumes that the mobile boundary is pinned at two points as it bulges out; see Fig. 10.10(b). Let the volume change in going from position I to II be dV and the overall free-energy change per volume be ΔG. The surface free energy of the boundary is taken as γ, and the stored energy per volume in the cold-worked grain is E_s. For the boundary moving from position I to II a free-energy balance gives

$$\Delta G = -E_s + \gamma \frac{dA}{dV} \qquad (10.12)$$

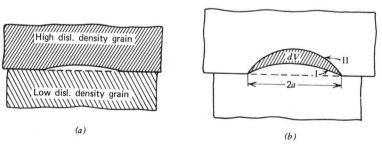

(a)   (b)

Figure 10.10   (a) Sudden growth of a boundary into a high dislocation density grain. (b) Model to describe this "nucleation" event.

where the stored energy of the new grain formed behind the moving boundary is taken as zero. You may show yourself with a simple geometrical model that for the motion of any surface of area A,

$$\frac{dA}{dV}=\frac{1}{r_1}+\frac{1}{r_2} \tag{10.13}$$

where r_1 and r_2 are the principle radii of curvature defined on p. 204. Hence, for spherical interfaces we have

$$\Delta G=-E_s+\frac{2\gamma}{r} \tag{10.14}$$

In order to have growth, the value of ΔG must be negative, so that E_s must be larger than $2\gamma/r$. Since the boundary is pinned at a distance of $2a$ it follows from the discussion of the Frank–Read generator (p. 119) that the radius of curvature will go through a minimum, $r_{min}=a$. Therefore, in order to have growth of the bulging boundary it is necessary that

$$E_s>\frac{2\gamma}{a} \tag{10.15}$$

In this model, then, "nucleation" occurs along a preexisting boundary at points where a bulge of length $2a$ protrudes out a distance greater than a. The incubation time would be the time required for the bulge to grow out to the critical position of $r_{min}=a$. The velocity of the boundary can be written from Eq. 7.32 as

$$V=B\frac{E_s-2\gamma/r}{\lambda} \tag{10.16}$$

It is apparent from this equation that "nucleation" by this growth mechanism will occur at boundaries having a high grain-boundary mobility, for example, high-angle boundaries or perhaps the special coincident boundaries.

2.   Sudden Growth of Subgrain Boundaries.   There appear to be essentially two mechanisms that lead up to the sudden growth of subgrain boundaries. In the first, the subgrains grow either by the coalescence mechanism previously discussed or by subgrain boundary migration. Eventually a high-mobility boundary forms (probably a high-angle boundary), which then suddenly grows out if a condition such as Eq. 10.15 is satisfied. A review of the detailed experimental evidence for this mechanism may be found on pp. 79–85 of Ref. 2. The details of the second mechanism are obscure. It appears from transmission electron microscope studies using cine[8] films that adjustments within the subboundaries occur on an atomic scale. These adjustments modify an existing high-mobility boundary, which then grows out forming the "nucleation" event. This mechanism is more common in highly deformed metals, probably because it requires a preexisting high-mobility (high-angle) boundary. A large deformation is required to produce large misorientations between subgrains with consequent high-angle subboundaries.

GENERAL CONCLUSIONS

Nucleation of grains results from the sudden growth of a high-mobility boundary. These boundaries are either:

1. An original high-angle boundary.
2. (a) A high-angle subgrain boundary formed by a subgrain enlargement mechanism.

(b) An existing high-angle subgrain boundary modified by unknown atomic rearrangements.

Mechanisms 1 and 2(a) are more common in lightly deformed metals and mechanism 2(b) is more common in highly deformed metals.

The "nucleation" event appears to be a growth phenomenon so that variables affecting growth will also affect "nucleation" in a similar way. Even though the new grains do not spontaneously nucleate in the sense of the classical nucleation picture, we may still treat the recrystallization transformation as being composed of nucleation and growth stages.

10.5 KINETICS OF RECRYSTALLIZATION

The recrystallization stage occurs by the nucleation of new strain-free grains that grow and consume the cold-worked matrix as indicated on Fig. 10.1. The rate at which the volume will be transformed into new grains is a function of both the nucleation rate and the growth rate of these new grains.

A. NUCLEATION AND GROWTH RATE EQUATIONS

Consider a cold-worked metal annealed isothermally at some temperature. After a time τ a new grain nucleates at some point in the cold-worked matrix and begins to grow. Frequently, this new grain will increase in size at a constant rate until it impinges upon a neighboring grain. Consequently, its radius varies with time as shown in Fig. 10.11. During the linear portion we may take the radius of a nucleus as

$$R = G(t-\tau) \qquad (10.17)$$

where G is the growth rate defined as dR/dt. If we take the nucleus as a

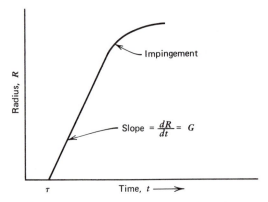

Figure 10.11   Time variation of the radius of a new grain.

sphere we can write

$$\text{Transformed volume per nuclei} = \tfrac{4}{3}\pi G^3(t-\tau)^3 \qquad (10.18)$$

This equation provides an expression for the volume transformed per nuclei so that now we must determine the number of nuclei. As on p. 219 we define a nucleation rate

$$\dot{N} = \frac{\text{number nuclei formed/unit time}}{\text{untransformed volume}} \qquad (10.19)$$

The number of nuclei formed in the time interval dt is $\dot{N} \cdot dt \cdot V_u$, where V_u is the untransformed volume. The value of V_u will be a function of time and therefore difficult to determine. Suppose that we consider the function $\dot{N} \cdot dt \cdot V$ where V is the total volume of our sample. This function counts the number of nuclei formed in both the untransformed and the transformed volumes of the sample. Since nuclei may not form in the already transformed volume, we call these *phantom nuclei*, as shown on Fig. 10.12. We then define an imaginary number of nuclei, n_{imag}, as the sum of the real

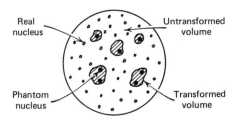

Figure 10.12   The real and phantom nuclei in a transforming matrix.

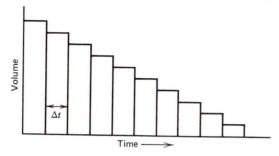

Figure 10.13   Volume associated with nuclei formed during specific time intervals.

nuclei, n_r, and the phantom nuclei, n_p,

$$n_{imag} = n_r + n_p \qquad (10.20)$$

The volume associated with the imaginary nuclei may be expressed as follows:

$$\text{Imag. vol. transformed} = \int_0^t \underbrace{\tfrac{4}{3}\pi G^3(t-\tau)^3}_{\text{vol/nuclei}} \cdot \underbrace{\dot{N}V \, dt}_{\substack{\text{No. imag.} \\ \text{nucl. formed}}} \qquad (10.21)$$

To see how this equation is obtained consider Fig. 10.13. The time during which transformation has occurred is arbitrarily divided into intervals of duration Δt. The number of nuclei formed during each of these time intervals is $\dot{N}V\,\Delta t$. The volume of the nuclei formed in the last interval is $[\dot{N}V\,\Delta t] \cdot 4\pi G^3(\Delta t - \tau)^3/3$. The volume of the nuclei formed in the first time interval is $[\dot{N}V\,\Delta t] \cdot 4\pi G^3(N\,\Delta t - \tau)^3/3$, where N is the total number of time intervals considered. Letting n_i be the number of the time interval, the imaginary volume transformed may be written

$$\text{Imag. vol. transformed} = \sum_{n_i=1}^N \tfrac{4}{3}\pi G^3(n_i\,\Delta t - \tau)^3 [\dot{N}V\,\Delta t] \qquad (10.22)$$

This result then leads directly to Eq. 10.21.

It turns out to be more convenient to work with the fraction of the volume transformed, which we will call X,

$$X_{imag} = \frac{\text{Imag. vol. trans.}}{\text{Total vol.}} = \int_0^t \tfrac{4}{3}\pi G^3(t-\tau)^3 \dot{N} \, dt \qquad (10.23)$$

We would now like to relate X_{imag} to the real volume fraction transformed, X_r. The real and phantom nuclei formed in any time interval dt will have the same volume per nuclei, $\tfrac{4}{3}\pi G^3(t-\tau)^3$, so that we may write

$$\frac{dn_r}{dn_{imag}} = \frac{dV_r}{dV_{imag}} = \frac{dX_r}{dX_{imag}} \qquad (10.24)$$

Let the number of nuclei formed per volume in a time dt be called $d\rho$, so that we have $dn_r = V_u \cdot d\rho$ and $dn_{\text{imag}} = V \cdot d\rho$. We now assume that the nuclei are formed randomly throughout the matrix. This means $d\rho$ will be independent of position, so we may write

$$\frac{dn_r}{dn_{\text{imag}}} = \frac{V_u}{V} = \frac{V - V_{\text{tran}}}{V} = 1 - X_r \qquad (10.25)$$

Combining Eqs. 10.24 and 10.25,

$$\frac{dX_r}{dX_{\text{imag}}} = 1 - X_r \qquad (10.26)$$

This simple differential equation is solved to give

$$X_r = 1 - e^{-X_{\text{imag}}} \qquad (10.27)$$

where the integration constant is found to be zero since both X_r and X_{imag} are zero at time zero. If we assume that both G and $\dot{N}$ are constant and that τ is negligibly small we may integrate Eq. 10.23 to obtain

$$X_{\text{imag}} = \frac{\pi}{3} \dot{N} G^3 t^4 \qquad (10.28)$$

Combining Eqs. 10.27 and 10.28 we obtain

$$X_r = 1 - \exp\left(-\frac{\pi}{3} \dot{N} G^3 t^4\right) = \text{vol. frac. trans.} \qquad (10.29)$$

This equation is frequently referred to as the Johnson–Mehl[9] equation and it is applicable to any phase transformation subject to the four restrictions of random nucleation, constant $\dot{N}$, constant G, and small τ. Many solid-state phase transformations nucleate at grain boundaries so that a random distribution of nucleation does not occur. Appropriate modifications of Eq. 10.29 for these cases have been considered (Ref. 10 and Ref. 7, Chapter 12). For an alternative derivation of Eq. 10.29, see Ref. 10, p. 134.

A plot of Eq. 10.29 is shown in Fig. 10.14 for different values of the

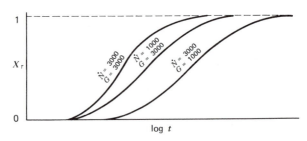

Figure 10.14   A plot of the Johnson–Mehl equation (Eq. 10.29) for various values of G and $\dot{N}$.

growth rate G and the nucleation rate $\dot{N}$. These curves demonstrate the obvious fact that the volume fraction transformed is a much stronger function of G than of $\dot{N}$. These curves are said to have a "sigmoidal" shape, and this shape is characteristic of nucleation and growth transformations.

It is often found that the growth rate G is a constant but the nucleation rate $\dot{N}$ is usually not constant in solid–solid phase transformations, so that one would not expect Eq. 10.29 to apply strictly. Avrami considered the case where the nucleation rate decayed exponentially with time. In this case the above analysis with variable $\dot{N}$ leads to Eq. 10.30, where k and n are constants. For rapid decay the results predict $n = 3$ and for slow decay $n = 4$ (see Ref. 7, Chapter 1).

$$X_r = 1 - e^{-kt^n} \qquad (10.30)$$

Equation 10.30 is often referred to as the Avrami equation. Most solid–solid phase transformation data may be correlated with this equation, but the value of n is not always found to be between 3 and 4. Hence, Eq. 10.30 is a good phenomenological equation for the description of the transformation kinetics of solid–solid phase transformations. The values of n are determined from the slope of a plot of $\log \log (1 - X_r)$ versus $\log t$. The values of n may be compared with theoretical models such as the Johnson–Mehl and the Avrami models presented above or with more recent theories[7,10]; see p. 441.

B. EXPERIMENTAL DETERMINATION OF G AND $\dot{N}$

The determination of the growth rate of the recrystallized grains G and their nucleation rate $\dot{N}$ is a very tedious process requiring much metallographic work. A series of identical samples are strained the same amount and then annealed at some temperature. The samples are then individually quenched after varying times. From metallographic examination the radii of the largest unimpinged grains are determined in each sample and plotted versus the time during which that sample was annealed, as shown in Fig. 10.15. It is assumed that the largest grains nucleated first; the slope of the curve in Fig. 10.15 gives the growth rate and the intercept on the time axis gives the incubation time. Note that the growth rate was a constant in this study.[11]

From the metallographic examination of the quenched samples one also obtains the number of new recrystallized grains, N_s, as a function of time as shown in Fig. 10.16(a). The slope of this curve then determines a surface nucleation rate, $\dot{N}_s$ = number grains/area-sec, as is shown in Fig. 10.16(b). Notice that in this study[11] the nucleation rate was far from constant as is required by the Johnson–Mehl equation, and it increases with time rather than decays as required by the Avrami model.

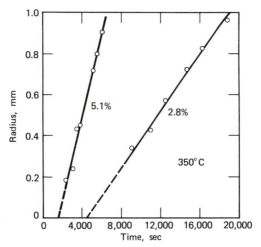

Figure 10.15   Radius of largest grain versus time of heating at 350°C, for 2.8 and 5.1% elongation in aluminum (from Ref. 11).

In metallographic examination we obtain a two-dimensional view from which we would like to determine the three-dimensional picture. In the present case we measure a surface density N_s from which we would like to infer the volume density N_v. Suppose we assume that (1) all of the grains are spheres and (2) all of the grains are of the same size. From metallographic examination we determine the largest radius of the grains and if a good statistical sample of grains is present this radius, r_{max}, will equal the true radius of the spherical grains. Consider a unit area of surface as shown

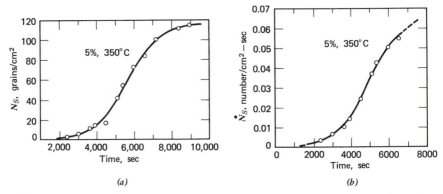

(a)

(b)

Figure 10.16   Aluminum annealed at 350°C after 5% elongation. (*a*) Time variation of surface grain density. (*b*) Time variation of surface nucleation rate (from Ref. 11).

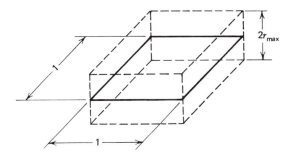

Figure **10.17**   A volume for analysis of grain density.

in Fig. 10.17 that is examined metallographically. Any grain whose center lies within r_{max} of this surface will appear on the surface. Therefore, all grains whose centers lie within the volume shown, $1 \cdot 1 \cdot 2r_{max}$, will touch the surface, so that we may write

$$N_v[1 \cdot 1 \cdot 2r_{max}] = N_s[1 \cdot 1] \tag{10.31}$$

From this result we may determine volume density of grains, N_v, from the measured surface density N_s. In general, the grains will not be spherical for reasons discussed in Section 7.3 and, also, the grain sizes will have some distribution. Consequently, Eq. 10.31 is only a first-order approach to this problem. A more complete consideration of this important problem of how one relates the two-dimensional features of metallographic samples to the corresponding three-dimensional picture may be found in books on quantitative metallography.[12,13]

C. EFFECT OF STRAIN, PURITY, GRAIN SIZE, AND TEMPERATURE ON G AND $\dot{N}$

1. Growth Rate.   The growth rate of a recrystallized grain can be described by writing Eq. 7.32 for the boundary between the recrystallized grain and the cold-worked matrix,

$$G = B \cdot \frac{\Delta \mu}{\lambda} \tag{10.32}$$

In this case the chemical potential difference is essentially the stored energy per mole, E_s, as discussed in Section 10.1, and we write

$$G = \frac{D_B}{kT} \cdot \frac{E_s}{\lambda} \tag{10.33}$$

where we define E_s per unit mole rather than per unit volume as in Eq. 10.12 and D_B is the grain-boundary self-diffusion coefficient. The effect

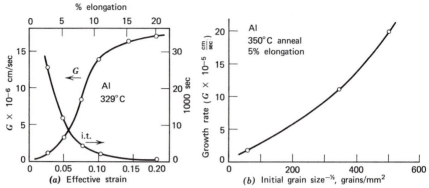

Figure 10.18   Variation of growth rate G during recrystallization of aluminum as a function of (a) prior strain and (b) original grain size (from Ref. 11).

of different variables upon growth rate may be understood by using Eq. 10.33. Although the growth rate was observed to be a constant in the work of Fig. 10.15, it is sometimes found to vary with time.[3] The reason is thought to be that during the recrystallization anneal, recovery processes will occur thereby changing the stored energy E_s and hence the growth rate; see Eq. 10.33.

It is clear from Eq. 10.33 that increasing the stored energy will increase the growth rate. Therefore, the growth rate is increased by increasing strain or by decreasing grain size, both of which increase E_s. Data illustrating these effects are shown in Fig. 10.15 and 10.18. That the growth rate does not increase much at strains beyond 15% is a reflection of the fact that only a small increase in stored energy was obtained beyond 15% strain. Also shown in Fig. 10.18(a) is the incubation time as a function of strain. The incubation time drops sharply with strain to about zero at 15% strain, which indicates that nucleation is easier at high strains.

The effect of purity upon the growth rate of the recrystallized grains is very strong. This is illustrated in Fig. 7.41 in which it may be seen that adding only 60 ppm Sn to Pb decreases the interface growth rate by a factor of about 5000. As discussed in Section 7.5, impurities have a strong influence upon the grain-boundary mobility, apparently due to a drag effect.

The temperature dependence of the growth rate is found to follow an Arrhenius equation.

$$G = G_0 e^{-Q_g/RT} \tag{10.34}$$

The only term in Eq. 10.33 strongly temperature dependent is D_B, which we expect to go as $D_0 e^{-Q/RT}$. Hence, we expect the measured Q_g values for

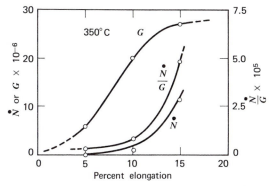

Figure 10.19   Variation of growth rate G and nucleation rate $\dot{N}$ during recrystallization of aluminum at 350°C as a function of prior elongation (from Ref. 11).

grain growth (Eq. 10.34) to correspond to the activation energy for grain-boundary self-diffusion. This is found to be the case only for very high-purity metals. The measured Q_g values in nominally pure metals (a few 100 ppm impurity) are much higher than the grain-boundary self-diffusion activation energies. Apparently, this results from a temperature dependence of the impurity drag effect.

2.   Nucleation Rate.   The nucleation rate of recrystallization is found to increase with strain. The results from one study are shown in Fig. 10.19, which also shows that a critical amount of strain is necessary in order to cause recrystallization to occur. These results are generally true for metal systems.

The effect of grain size upon nucleation rate is quite pronounced. Data from the study of Anderson and Mehl[11] are presented in Fig. 10.20 to illustrate this point. It is seen that the smaller grain size causes a much higher rate of nucleation, and it also causes nucleation to occur much sooner. The smaller grains produce a more complex stress pattern that results in higher local deformations, which enhances the nucleation rate.

The effect of purity upon nucleation rate does not seem to have been well studied. However, since impurities increase the amount of stored energy for a given strain one expects the nucleation rate to be increased by the presence of impurities.

The temperature dependence of the nucleation rate has been studied and it is found that the nucleation rate also follows an Arrhenius equation,

$$\dot{N} = N_0 e^{-Q_n/RT} \qquad (10.35)$$

The study of Anderson and Mehl[11] found the activation energy for $\dot{N}$ and G to be roughly the same except at low strains, on the order of 5%.

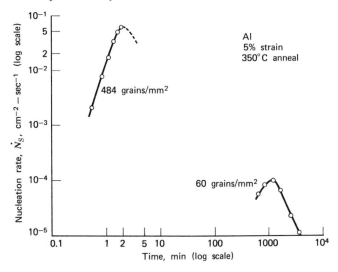

Figure 10.20 Time variation of nucleation rate during recrystallization of two aluminum samples with different original grain sizes, both strained 5% (from Ref. 11).

10.6 CONTROL OF RECRYSTALLIZATION TEMPERATURE AND GRAIN SIZE

From a practical point of view we are mainly interested in controlling the recrystallization temperature and the recrystallized grain size. In order to estimate how these are affected by the parameters which we may control, the Johnson–Mehl equation (Eq. 10.29) will be used to describe the recrystallization kinetics.

A.   RECRYSTALLIZATION TEMPERATURE

The recrystallization temperature is defined as the temperature at which recrystallization occurs within some specified time, frequently 1 hour. From the Johnson–Mehl equation you may show that the time required to obtain 95% of the volume transformed, $t_{0.95}$, is given by

$$t_{0.95} = \left[\frac{2.85}{\dot{N}G^3}\right]^{1/4} \tag{10.36}$$

For our present purposes we will consider 95% transformation as complete transformation. Since we know that $\dot{N}$ and G both increase with temperature, we expect Eq. 10.36 to vary with temperature as shown in Fig. 10.21. From the plot the 1-hour recrystallization temperature may be determined.

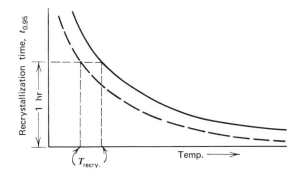

Figure 10.21   Definition of the 1-hour recrystallization temperature.

Suppose that somehow we were able to cause $\dot{N}$ and G to be increased. This would decrease the recrystallization time curve as shown by the dashed curve of Fig. 10.21, which in turn decreases the 1-hour recrystallization temperature as shown. Hence, *there is an inverse relationship between* T_{Recry} *and the rates* G *and* $\dot{N}$; see Problem 10.4.

B.   RECRYSTALLIZED GRAIN SIZE

Using the Johnson–Mehl equation it may be shown that the recrystallized grain size d varies as

$$d = (\text{const})\left[\frac{G}{\dot{N}}\right]^{1/4} \tag{10.37}$$

see Problem 10.3. This equation shows what one qualitatively expects: A small recrystallized grain size is promoted by a high nucleation rate and a slow growth rate; that is, the grains nucleate at a high density and grow very little.

The predominant factors that affect $\dot{N}$ and G are (1) amount of prior strain, (2) temperature of anneal, (3) prior grain size, and (4) purity. We will now use Eqs. 10.36 and 10.37 to discuss how these factors influence (a) the value of T_{recry}, and (b) the value of d.

1.   Prior Strain.   The effects of prior strain are shown in Fig. 10.19. (a) T_{recry} is decreased by high strains since high strains increase both $\dot{N}$ and G. This effect is illustrated in Fig. 10.22(a), which presents the 1-hour recrystallization temperature for iron and aluminum. (b) High strains decrease recrystallized grain size, d, because $\dot{N}/G$ increases with strain as illustrated in Fig. 10.19. Data illustrating this effect in α brass are presented in Fig. 10.22(b).

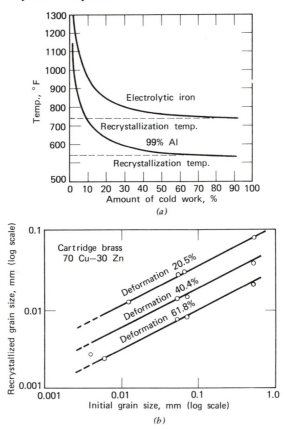

Figure 10.22   (*a*) Variation of 1-hour recrystallization temperature with amount of deformation (from Ref. 14). (*b*) Variation of recrystallized grain size as a function of initial grain size for various prior deformations (from Ref. 15).

2. Temperature of Anneal.   (a) It is not meaningful to talk of the effect of the anneal temperature on recrystallization temperature, but one can say that higher anneal temperatures produce faster recrystallization since both $\dot{N}$ and G increase with temperature. (b) The effect of anneal temperature upon recrystallized grain size does not seem to be well documented. Since $\dot{N}$ and G both follow an Arrhenius equation with Q nearly the same, one expects $\dot{N}/G$ to be nearly constant with temperature. Hence, d is probably a weak function of anneal temperature.

3. Prior Grain Size.   At a given strain one expects more stored energy in a fine-grained metal because of the more complex deformation required by the fine grains. (a) T_{recry} is lowered by fine grains since both $\dot{N}$ and G increase due to increased stored energy. (b) d is decreased by fine grains as is shown in Fig. 10.22(b) for α brass. The smaller grains cause increased local deformation, which increases $\dot{N}$ faster than G as illustrated on Fig. 10.19.

4. Impurities. Addition of impurities increases the amount of stored energy at a given strain, so that one might expect the growth rate to increase; see Eq. 10.33. However, impurities are extremely effective at decreasing the interface mobility. As we have seen (Fig. 7.41), the addition of only 60 ppm Sn to Pb decreases the interface mobility by a factor of 5000. This latter effect dominates so we expect impurities to decrease the growth rate and nucleation rate. (a) T_{recry} is increased by addition of impurities. This effect is illustrated in Table 10.2. The data of this table also illustrate that the recrystallization temperature is on the order of one-half the melting point. (b) For a given deformation the presence of impurities allows a higher stored energy. Since $\dot{N}/G$ increases with the amount of stored energy, one expects impurities to promote a smaller recrystallized grain size.

5.   Temperature of Prior Strain. Straining at higher temperatures allows more recovery to occur so that one expects less stored energy to drive both $\dot{N}$ and G. Hence one expects: (a) Increased straining temperatures will

Table 10.2   Approximate Values of 1-Hour Recrystallization
Temperatures

	Material	Recrystallization Temperature[a]
Copper[14]	(99.999%)	250 (120)
	(OFHC)	400 (210)
	5% Zn	600 (320)
Aluminum[14]	(Zone refined)	50 (10)
	(99.999%)	175 (85)
	(99.0+%)	550 (240)
	Alloys	600 (320)
Nickel[14]	(99.99%)	700 (370)
	(99.4%)	1100 (630)
Tungsten[26]	(High purity)	2200–2400 (1200–1300)
	(Containing micropores)	2900–4200 (1600–2300)
Tin[14]		25 (−4)

[a] Values in °F; values in parentheses, °C.

increase T_{recrys} because G and $\dot{N}$ will decrease. (b) Increased straining temperature will lower $\dot{N}/G$ due to decreased local strains which will increase d.

6. Strain-Anneal Technique. Recrystallization may be used as a technique to produce single crystals of metals. In this case one is interested in a maximum d, hopefully equal to several centimeters. It is clear from Fig. 10.19 that to minimize $\dot{N}/G$ one must strain at just the critical amount required to produce nucleation. This is done by a trial-and-error technique and strains on the order of 2–5% are usually required to optimize the grain size. Some specific results are presented in Ref. 16 and a good discussion of the use of this strain-anneal technique for many different metals is presented in Ref. 17.

In summary it can be seen that recrystallization offers a means of controlling grain size. Both the recrystallized grain size and the recrystallization temperature are controlled by $\dot{N}$ and G. From a knowledge of how physical variables affect $\dot{N}$ and G one can determine how T_{recry} and d are affected by means of Eqs. 10.36 and 10.37.

10.7   RELATED TOPICS

A.   HOT WORKING

If a rod of soft solder is bent back and forth several times it is found that the rod remains soft even after quite a few bends. The same operation done on a rod of annealed copper results in a rapid work hardening so that the copper can no longer be bent by hand after only a few bends. We discussed work hardening on p. 121 and pointed out that it is due to dislocations becoming immobilized by various mechanisms. The question of interest here is why the solder does not work harden. The reason is that the recrystallization temperature of solder is sufficiently low to allow recrystallization to occur during the bending process. The new strain-free grains have a low dislocation density and thereby prevent the work hardening that would occur in their absence. The simultaneous occurrence of recrystallization and deformation occurs during hot working and is illustrated in Fig. 10.23.

B.   TEXTURE

It was pointed out on p. 69 that when a metal is severely cold worked the individual grains will all tend to rotate so as to produce a preferred crystallographic orientation in the direction of deformation. This preferred orientation among the grains is generally referred to as a *deformation texture*.

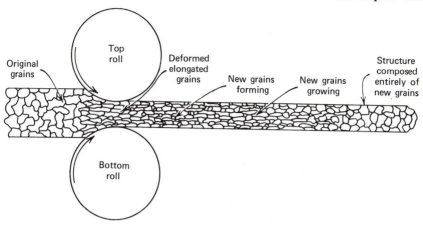

Figure 10.23 Recrystallization during hot rolling. (From *Making, Shaping and Treating of Steel*, 7th ed. Copyright 1957, United States Steel Corp., p. 386.)

When a metal containing a deformation texture is annealed to produce recrystallization one might expect the new grains to be randomly oriented. This is frequently not the case and one often obtains what is termed an *annealing texture* or a *recrystallization texture*. This result means that the new recrystallized grains must form with a preferred crystallographic orientation relative to the deformed grains. The preferred orientation of the recrystallized grains must be due to one or both of two causes. (a) The nuclei form with a preferred orientation. (b) Only those grains with the preferred direction grow to an appreciable volume; the mobility of other grains is too low and they are suppressed by competitive growth. There is some debate on the relative merits of these mechanisms; for further discussion one is referred to Refs. 4 and 21. Current experimental work indicates that mechanism (b) is the predominant but not the sole cause of annealing textures (Ref. 21, p. 951). It was pointed out in Chapter 7 (p. 209) that certain orientation mismatches form grain boundaries that have unusually high mobilities (see Fig. 7.41). These special boundaries are termed coincident boundaries. Experimental evidence[4,21] suggests that the annealing texture results from preferred growth of those grains forming coincident boundaries with the grains of the primary texture. The high mobility of these particular grains would account for their preferred growth, and their special orientation relation to the deformation grains would account for the recrystallization texture.

The cube texture is often obtained as the recrystallization texture in fcc metals. A particular example where this texture is advantageous is in Fe–Si alloys. Because of the anisotropy of the magnetic properties of iron the cube texture makes a much superior transformer sheet material.

C. SECONDARY RECRYSTALLIZATION

The type of recrystallization that we have discussed up to this point and that is shown in Fig. 10.1 is frequently referred to as primary recrystallization. In some cases during the grain growth stage shown on Fig. 10.1 a few large grains will form that grow preferentially and thereby consume the smaller recrystallized grain structure. This phenomenon is generally called secondary recrystallization. Its appearance is very similar to primary recrystallization, as illustrated schematically in Fig. 10.24. An excellent series of photographs may be found on p. 497 of Ref. 4.

Secondary recrystallization is found to occur in metals that have achieved a relatively stable small grain size. Such a condition is obtained in three common ways. (1) The metal is worked sufficiently to obtain a deformation texture and then a recrystallization texture is formed after primary recrystallization. Since the recrystallized grains have a similar orientation there is an absence of high-angle boundaries. Hence, boundary mobilities are low and the grain structure is relatively stable. (2) The metal contains inclusions that result in a limiting grain size as discussed in Chapter 7. It was shown on p. 210 that the restraining force per area of interface produced by particles of radius r having volume fraction f would be given as $(3f/2\pi r^2) \cdot \pi r \gamma (1+\cos \alpha)$. From Eqs. 7.32 and 7.38 we can write the boundary velocity as

$$v_{gb} = \frac{B \bar{V} \Delta P}{\lambda} \qquad (10.38)$$

where ΔP is the pressure difference across the boundary. The pressure difference due to the boundary curvature, $2\gamma/r$ (see Eqs. 7.38 and 7.39), that causes grain growth is opposed by the restraining force per unit area so

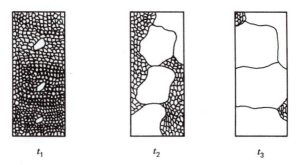

$$t_1 \qquad\qquad t_2 \qquad\qquad t_3$$

Figure **10.24** Schematic illustration of secondary recrystallization.

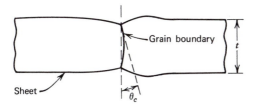

Figure 10.25   A vertical grain boundary migrating to the right in a sheet specimen.

that we have

$$v_{gb} = \frac{B\bar{V}}{\lambda}\left[\frac{2\gamma}{r} - \frac{3f}{2r}\gamma(1+\cos\alpha)\right] \tag{10.39}$$

and a limiting grain size is achieved when the term in brackets is zero. (3) The metal is in thin sheet form. The boundaries intersecting the surface tend to form grooves at the surface intersections and these grooves retard the boundary motion. An analysis of this problem (Ref. 4, p. 466) has shown that in order for a boundary, such as that shown in Fig. 10.25, to migrate to the right the angle of the boundary with the surface normal must be greater than some critical value θ_c. For steady-state migration (constant velocity) this angle must equal θ_c and the expression for the grain-boundary velocity is found to be,

$$v_{gb} = \frac{B\bar{V}}{\lambda}\left[\frac{\gamma}{r_1} - \frac{2\theta_c\gamma}{t}\right] \tag{10.40}$$

where r_1 is the radius of the grain boundary in the plane of the sheet and t is the sheet thickness. [The value of θ_c is found to be $\gamma/(6\gamma_s)$, where γ_s is the surface tension of the vapor–solid interface, and we have taken γ_s to be the same for both grains here.] Equation 10.40 shows that the restraining force per unit area due to the surface intersection is proportional to θ_c/t, and a limiting grain size will therefore be formed in thin sheets when this force balances the driving force due to curvature. The radius of curvature of the limiting grain size and, hence, the limiting grain diameter is proportional to t/θ_c.

If a metal containing the relatively stable grain structure produced by one of the above three processes is heated to a higher temperature, exaggerated growth of a few select grains (secondary recrystallization) often occurs. A discussion of the mechanism by which these select grains *nucleate* in a dispersion stabilized matrix is presented on p. 980 of Ref. 21. As may be seen in Fig. 7.33 the large grains will be concave outward and the driving force for their *growth* will be interface curvature just as for the normal grain growth discussed on pp. 204 and 209. These select grains grow faster than the smaller stabilized grains because they are everywhere concave outward and, also, in some cases because they have a higher mobility. This latter condition would be particularly true in the case of texture stabilized grains. The student is referred to Chapter 10 of Ref. 4 for a more detailed discussion of this phenomenon.

D. EXAMPLE

A fascinating example that involves many of the ideas of this chapter is found in nonsag tungsten filament wire. In Chapter 7 (p. 210) it was pointed out that by adding thoria particles to tungsten wire the limiting grain size could be significantly reduced. The smaller grain size prevents grain boundaries from extending clear across the filament diameter and thereby reduces filament failures due to grain-boundary offsets such as illustrated in Fig. 7.43. Therefore, thoria-dispersed tungsten filaments have a much better creep resistance than undoped tungsten, as is shown in Fig. 10.26. However, nonsag tungsten filaments having much better creep resistance than thoria-dispersed filaments have been produced for around 40 years by the addition of small amounts of K, Al, and Si impurities (see

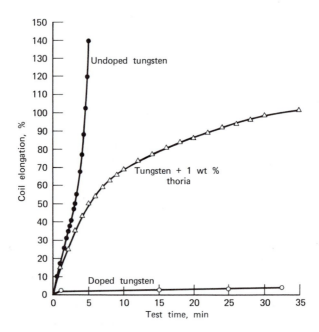

Figure 10.26 Creep deformation of 225- μm-diameter tungsten wires at 2500°C (from Ref. 22. Copyright 1972 by Technical Publishing Company).

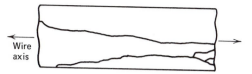

Figure **10.27**   Elongated grain boundary structure in doped tungsten wires.

Ref. 20, p. 139). The effects of these impurities is to produce a recrystallized grain structure consisting of large grains elongated along the wire axis as shown in the sketch of Fig. 10.27. The grains do not offset as shown in Fig. 7.43 because they are interlocked along the wire axis. Since creep occurs predominantly by grain-boundary sliding, these larger-grained filaments exhibit much better creep resistance than the fine-grained thoria-dispersed filaments; see Fig. 10.26.

The action of the dopants in producing the elongated interlocking grain structure is unknown and has long been disputed, but recent experiments present strong evidence for the following picture. The dopants are added as oxides and during the sintering treatment over 90% of the dopants are volatilized.[23] By a mechanism not well understood, the volatilization of the dope produces arrays of small bubbles in the ingot. Recent experiments have shown that the bubbles are present in the sintered tungsten ingot and they become elongated as the ingot is worked down. During annealing the elongated bubbles change their shape. At temperatures above 1600°C the elongated bubbles break up into strings of smaller bubbles that lie along the axis of the wire with sizes ranging down to 100 Å and less.[23] The bubbles act as barriers to grain-boundary motion just as do second-phase particles. Because they are aligned along the wire axis, they allow grain growth parallel to the wire axis but inhibit grain growth normal to the axis. A limiting grain size is produced by the restraining action of the bubbles, and this is followed by secondary recrystallization that produces the large elongated and interlocked grains. The secondary recrystallization occurs at temperatures around 2100°C and the primary recrystallization occurs in the temperature range of 1400–1600°C. The tungsten filament wire has a $\langle 100 \rangle$ fiber texture and apparently the recrystallization texture is also $\langle 100 \rangle$. Recent evidence[24] indicates that secondary recrystallization occurs by passage of coincidence boundaries into the $\langle 100 \rangle$ fiber grains. As explained on p. 209 with reference to Fig. 7.41, these coincidence boundaries have a much higher mobility due to a lower solute segregation. Thus, they are able to grow preferentially to produce the secondary recrystallization.

There are a number of other cases where submicroscopic bubbles (pores) have been found to be very effective at inhibiting nucleation. For example, if nickel sheet, dispersion hardened with thoria particles, is deformed 40%

by rolling, it can then be heated to its melting point without recrystallizing.[25] It is found that metals containing a dispersed hard second phase (such as thoria in nickel or, interestingly, carbides in tempered martensites) will form small pores around the dispersed particles upon deformation. These pores inhibit nucleation and also reduce grain growth. Similar effects have been observed in Cu, Ni, Al, and Al alloys,[26] where the pores have been introduced by nuclear irradiation or by the metal preparation process.

REFERENCES

1. P. Gordon, Trans. AIME **203**, 1043 (1955).

2. J. G. Byrne, *Recovery Recrystallization and Grain Growth*, Macmillan, New York, 1965.

3. L. Himmel, Ed., *Recovery and Recrystallization of Metals*, Gordon and Breach, New York, 1963.

4. *Recrystallization, Grain Growth and Texture*, American Society of Metals, City, 1966.

5. W. Hibbard and C. Dunn, Acta Met. **4**, 306 (1956).

6. A. Damask and G. Dienes, *Point Defects in Metals*, Gordon and Breach, New York, 1963.

7. J. Christian, *The Theory of the Transformations in Metals*, Pergamon, New York, 1965.

8. L. Michels and B. Ricketts, Trans. AIME **239**, 1841 (1967).

9. W. Johnson and R. Mehl, Trans. AIME **135**, 416 (1939).

10. J. Cahn and W. Hagel, in *Decomposition of Austenite by Diffusional Processes*, V. Zackay and H. Aaronson, Eds., Interscience, New York, 1962, p. 131.

11. W. Anderson and R. Mehl, Trans. AIME **161**, 140 (1945).

12. R. Dehoff and F. Rhines, *Quantitative Microscopy*, McGraw-Hill, New York, 1968.

13. E. E. Underwood, *Quantitative Stereology*, Addison-Wesley, Reading, Mass., 1969.

14. A. Guy, *Elements of Physical Metallurgy*, Addison-Wesley, Reading, Mass., 1960.

15. S. Channon and H. Walker, Trans. ASM **45**, 200 (1953).

16. L. R. Fleischer and J. M. Tobin, J. Crys. Growth **8**, 243 (1971).

17. K. T. Aust, in *The Art and Science of Growing Crystals*, J. R. Gilman, Ed., Wiley, New York, 1963, p. 452.

18. *Making, Shaping and Treating of Steel*, U.S. Steel Corp., City, 1957, p. 386.

19. C. Garrett and T. Massalski, *Structure of Metals*, 3rd ed. McGraw-Hill, New York, 1966.

20. C. Smithells, *Tungsten*, Chapman and Hall, London, 1952.

21. R. W. Cahn, *Physical Metallurgy*, North-Holland, Amsterdam, 1965, Chapter 19.

22. H. G. Sell and G. W. King, Res. Develop. **23,** 21 (July 1972).

23. D. M. Moon and R. C. Koo, Met. Trans. **2,** 2115 (1971).

24. A. J. Opinsky, Trans. AIME **239,** 919 (1967).

25. D. Webster, Met. Prog. **92,** 93 (Jan. 1968).

26. K. Farrell, A. Schaffhauser, and J. Houston, Met. Trans. **1,** 2899 (1970).

PROBLEMS

10.1 The purpose of this problem is to show yourself that the $P \Delta V$ term is much less than the ΔE term for plastic deformation. Clarebrough et al.[3] give the following data:

Metal	Stored Energy (cal/mole)	Fractional Change in Macroscopic Density
Ag(75% compression)	15.6	2.46×10^{-4}
Cu(70% compression)	9.4	1.93×10^{-4}
Ni(70% compression)	11.8	0.66×10^{-4}

From these data and density data calculate what fraction the $P \Delta V$ term is of the stored energy term ΔE.

10.2

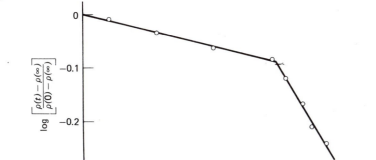

Recovery of resistivity in quenched Al [W. DeSorbo and D. Turnbull, Phys. Rev. **115,** 560 (1959)].

Aluminum specimens were heated to 284°C and quenched to 0°C. The resistivity was then measured as a function of time, $\rho(t)$. The resistivity at zero time is $\rho(0)$ and after complete recovery is $\rho(\infty)$. After a time of 95 minutes, the temperature was changed sharply to 22°C, and the data plotted as shown above. It is known that quenching a metal from high temperatures produces an excess of vacancies over the equilibrium value. At the lower temperature the excess vacancies migrate to sinks, with an accompanying decrease in resistivity. Hence, the resistivity undergoes a recovery process due solely to vacancy migration, and the kinetics of this process follow Eq. 10.8. From this equation and the above data determine the activation energy for vacancy migration in aluminum.

10.3 From a practical point of view one is usually interested in obtaining a certain grain size in a metal, and hence it is desirable to be able to relate the grain size to the parameters $\dot{N}$ and G. Assume that the Johnson–Mehl equation applies and that a sample is experimentally "completely transformed" when it is 95% transformed according to this equation. If d equals the average center-to-center distance between grains at completion of the transformation, then one finds $d = $ (const)$(G/\dot{N})^{1/4}$ (that is, Eq. 10.37).

Your problem is to calculate the value of the constant in this expression for the cases:

(a) where grains are cubes, and

(b) where grains are spheres arranged in an fcc type of close packing.

Hint: First calculate the density of grains using the Johnson–Mehl equation and then determine d.

10.4 Suppose we define the recrystallization temperature as that temperature at which we obtain 95% transformation in 1 hour. You know that this temperature will be a function of the growth rate G and the nucleation rate $\dot{N}$. Further, you know the temperature dependence of G and $\dot{N}$ both follow an Arrhenius equation:

$$G = G_0 e^{-Q_g/kT}, \qquad \dot{N} = N_0 e^{-Q_n/kT}$$

(a) From Eq. 10.36 derive an equation giving the above-defined recrystallization temperature as a function of G_0, N_0, Q_g, and Q_n.

(b) Explain how you would expect the following variables to affect G_0, N_0, Q_g, and Q_n: (1) prior strain, (2) prior grain size, and (3) purity of the metal.

(c) From your expression derived in (a) and your answer to (b) explain how you would expect each of the three variables of part (b) to affect the recrystallization temperature.

10.5 It is shown in Chapter 14 that the flow stress σ of metals generally varies with grain diameter d according to the relation $\sigma = \sigma_0 + k/\sqrt{d}$, where σ_0 and k are constants.

Suppose that you wanted to maximize the strength of an alloy by controlling the grain size, and that you intend to control the grain size by means of recrystallization.

(a) From the above equation and equations developed in this chapter, show how you can express the yield strength as a function of the parameters that govern the recrystallization process, G and $\dot{N}$. State the assumptions involved here.

(b) Assuming you have measured these parameters, at different temperatures, discuss how you would then determine at what temperature and for how long you should anneal to obtain a given strength. Why might you want to perform the anneal at fairly low temperatures?

(c) Describe how you could measure the parameters involved.

(d) You have some control over $\dot{N}$ and G. Since your object is to optimize strength, in what sense would you want to influence $\dot{N}$ and G. How could you do this in practice?

(e) In a recrystallization and grain growth process, two driving forces for growth are involved. Explain physically why the two forces arise and which of the two you are chiefly concerned with in this problem.

CHAPTER 11
PRECIPITATION FROM SOLID SOLUTIONS

When a solid solution is cooled so that it passes into a two-phase region on the phase diagram, it becomes supersaturated with respect to a new phase. A new phase then forms by a solid-state precipitation reaction. Three such solid-state transformations are indicated on Fig. 8.1, and according to the classification of Table 8.2 this is the most complex type of phase transformation. The new phase forms because by so doing it lowers the free energy of the system. Consequently, it is necessary to understand how the free energy varies as one moves across the phase boundaries of the phase diagram. A brief review of solution thermodynamics will be presented.

11.1 REVIEW OF FREE-ENERGY COMPOSITION DIAGRAMS

In general, the free energy per atom is different when the atom is dissolved in solution from what it is in the pure state. The partial molal free energy $\bar{G}_i$ is formally defined as

$$\bar{G}_i = \left[\frac{\partial G}{\partial n_i}\right]_{T,P,n_j \neq i} \tag{11.1}$$

where G is the free energy of the system and n_i is a number of i atoms. It is not apparent from this definition but it is shown in thermodynamic textbooks that $\bar{G}_i$ is exactly the free energy per atom (or mole) of i when dissolved into solution. As noted on p. 151, $\bar{G}_i$ is also referred to as the chemical potential, which is generally given the symbol μ_i.

In describing alloys one is interested in an expression for the free energy per mole of solution. It is common to define a free energy of mixing, ΔG_m, as

$$\Delta G_m = \frac{\text{Free energy change on mixing}}{\text{Mole solution}} \tag{11.2}$$
$$= G_s - G_p$$

363

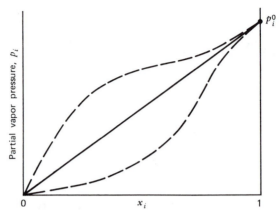

Figure 11.1   Variation of partial pressure of an element with its mole fraction, X, dissolved into solution.

where G_s is the free energy per mole of solution and G_p is the free energy per mole of the unmixed components. These two terms may be written as

$$G_s = x_1 \bar{G}_1 + x_2 \bar{G}_2 + \cdots \tag{11.3}$$

$$G_p = x_1 G_1^0 + x_2 G_2^0 + \cdots \tag{11.4}$$

where x_i is mole fraction i and G_i^0 is free energy per mole of pure i at 1 atmosphere pressure. Combining Eqs. 11.2, 11.3, and 11.4 we obtain

$$\Delta G_m = \sum_i x_i [\bar{G}_i - G_i^0] \tag{11.5}$$

The following relationship is derived in textbooks on thermodynamics[1,2]:

$$d\bar{G}_i = RT \, d \ln f_i \qquad (\text{const } T) \tag{11.6}$$

where f_i is the fugacity of component i. To quite a good approximation f_i is the partial vapor pressure of component i over the solution, p_i. For an ideal solution the vapor pressure varies linearly with the mole fraction as shown in Fig. 11.1. Hence, the fugacity is a function of the mole fraction. To use Eq. 11.6 we must integrate it and we choose the lower limit of integration as the pure state,

$$\int_{G_i^0}^{\bar{G}_i} d\bar{G}_i = RT \int_{f_i^0}^{f_i} d \ln f_i \tag{11.7}$$

This lower limit is called the standard state and it is conventionally taken as the pure component i at 1 atmosphere total pressure. Therefore we obtain

$$\bar{G}_i - G_i^0 = RT \ln \frac{f_i}{f_i^0} \approx RT \ln \frac{p_i}{p_i^0} \tag{11.8}$$

where p_i^0 is the partial vapor pressure of pure i at 1 atmosphere total pressure. The activity of component i, a_i, is defined as

$$a_i = \frac{f_i}{f_i^0} \approx \frac{p_i}{p_i^0} \tag{11.9}$$

Hence, we have the final result,

$$\bar{G}_i - G_i^0 = RT \ln a_i \quad (\text{const } T) \tag{11.10}$$

Combining this with Eq. 11.5 we obtain

$$\Delta G_m = RT \sum_i x_i \ln a_i \tag{11.11}$$

An ideal solution follows the linear solid line on Fig. 11.1, $p_i = x_i p_i^0$. Therefore, for an ideal solution of a binary alloy we have from Eqs. 11.9 and 11.11

$$\Delta G_m = RT[x_1 \ln x_1 + x_2 \ln x_2] = \Delta H_m - T \Delta S_m \tag{11.12}$$

If the i atoms tend to bond more strongly with the solvent atoms than with each other, their vapor pressure will be decreased as shown by the lower dashed curve of Fig. 11.1. Such a system will often display a compound in its phase diagram. If the i atoms tend to bond less strongly with the solvent atoms, then their vapor pressure is increased as shown by the upper dashed curve. These systems will frequently display a eutectic on their phase diagrams. In the ideal solution case there is no difference in the bonding energies of the i atoms to the solvent atoms and to each other. Therefore, $\Delta H_m = 0$ for an ideal solution and we have from Eq. 11.12

$$\Delta S_m(\text{ideal soln}) = -R[x_1 \ln x_1 + x_2 \ln x_2] \tag{11.13}$$

We are interested in the free energy per mole of alloy, which is the same as G_s of Eq. 11.3. By combining Eqs. 11.2 and 11.4 we may write

$$G_s = G_1^0 + (G_2^0 - G_1^0)x_2 + \Delta G_m \tag{11.14}$$

A plot of ΔG_m versus mole fraction 2, x_2, is shown on Fig. 11.2(a) for the case of an ideal solution, $\Delta G_m = -T \Delta S_m$ (see Eq. 11.12). The log function of Eqs. 11.12 and 11.13 is symmetric about $x_2 = 0.5$ as shown. Note that ΔG_m is everywhere negative, indicating a decrease in free energy upon mixing. The plot of G_s versus composition for the ideal solution is shown in Fig. 11.2(b). Notice that the first two terms of Eq. 11.14 describe the straight line connecting G_1^0 with G_2^0, and this line gives the free energy per mole of the unmixed alloy. Therefore, G_s is obtained by simply adding ΔG_m to this straight line as shown. It should be clear that the free energy of the solution, G_s, is everywhere less than the free energy of the unmixed alloy. This means that in ideal systems the pure components prefer to form

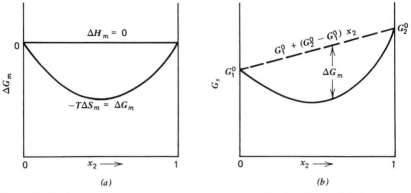

Figure 11.2   Free-energy composition diagrams for ideal solutions (a) free energy of mixing, (b) free energy of the solution.

solutions. Diagrams such as Fig. 11.2(a) and 11.2(b) are generally referred to as free-energy composition diagrams.

A very useful property of the G_s curves such as Fig. 11.2(b) is that one may determine the chemical potentials of the components of the alloy from a simple graphical construction. Refer to the figure of Problem 11.1. At composition x_2 construct a tangent to the G_s curve. The intercept of the tangent line at $x_2 = 1$ (point B) gives the chemical potential of component 2, μ_2, in the solution of composition x_2. Similarly, the chemical potential of component 1, μ_1, is given by point A.

We now consider nonideal systems. In these cases there is generally a small vibrational entropy contribution, which we will neglect. Therefore, in the absence of any ordering reactions, we may take the entropy term to be the same as Eq. 11.13. For alloys that evolve heat upon mixing (exothermic) the ΔH_m is negative as shown in Fig. 11.3(a). In alloys that require

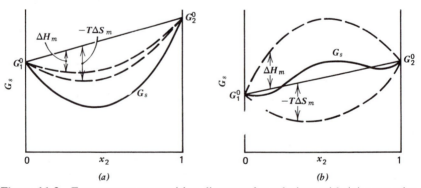

Figure 11.3   Free-energy composition diagrams for solutions with (a) a negative ΔH_m and (b) a positive ΔH_m.

heat for mixing (endothermic) the ΔH_m is positive as shown in Fig. 11.3(b). Notice that in these alloys the free energy is increased over the unmixed state for some compositions. Hence, solutions of these compositions will not form.

A. THE TANGENT RULE

Suppose an fcc single-phase alloy is cooled below the solvus line and a bcc phase precipitates from the fcc solid solution so that, in effect, a single solution has decomposed into two separate solutions. Such a reaction is illustrated schematically in Fig. 11.4(a) in which the single phase α' separates into two phases, β and α, where α has the same crystal structure as α' but a different composition. We take n_α and n_β as the total number of moles in solutions α and β, respectively. We take x_2^0 as the mole fraction of component 2 in the original solution (state A) and x_2^α, x_2^β as the mole fractions in the separated solutions (state B). Suppose the free energies of the α phase, G_s^α, and the β phase, G_s^β, are given as on Fig. 11.4(b). The compositions of the three solutions are located on the horizontal axis and the free energy of these solutions are labeled on the vertical axis. We now write expressions for the free energy per mole of the two states utilizing the curves on Fig. 11.4(b).

$$\text{State } A: \quad G_s^A = G_s^{\alpha'} \tag{11.15}$$

$$\text{State } B: \quad G_s^B = \frac{n_\alpha}{n_\alpha + n_\beta} G_s^\alpha + \frac{n_\beta}{n_\alpha + n_\beta} G_s^\beta \tag{11.16}$$

It is possible to show that Eq. 11.16 is the equation of the straight line shown dashed on Fig. 11.4(b), and the *average* free energy per mole of the mixture of the two phases, α and β, is located on this straight line at the point of their *average composition*, x_2^0 (see Problem 11.2). This value, G_s^B, is

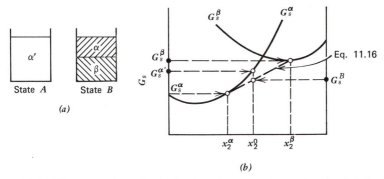

Figure 11.4 The separation of a single phase into two phases showing (a) the two states considered and (b) the corresponding free energies.

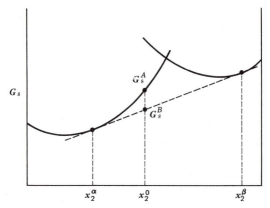

Figure 11.5   Construction showing the tangent rule.

shown on Fig. 11.4(b), and it may be seen that the free energy is lower in state B than in state A since $G_s^B < G_s^A$. Therefore, we conclude that if there are no kinetic barriers the α solution of composition x_2^0 will spontaneously decompose into two solutions of compositions x_2^α and x_2^β. It should be clear from Fig. 11.4(b) that if x_2^β were to increase slightly the dashed straight line would drop and the value of G_s^B would be lowered even further. The lowest possible value of G_s^B is obtained when the straight line lies tangent to the two curves as shown in Fig. 11.5. The compositions x_2^α and x_2^β at the tangent points must be the compositions of the two phases giving the minimum possible free energy of state B. Consequently, these compositions are equilibrium compositions and they correspond to the compositions on the equilibrium phase diagram.

This last point will be illustrated for a eutectic phase diagram. Figure 11.6 illustrates a simple eutectic phase diagram with two isotherms, T_1 and

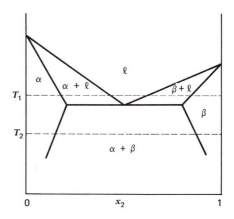

Figure 11.6   Simple eutectic diagram with two temperatures defined.

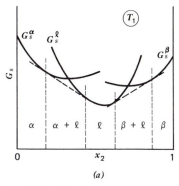
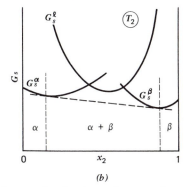

(a)   (b)

Figure 11.7   The free-energy composition diagram for the three phases at the two temperatures defined in Fig. 11.6.

T_2, indicated. First consider temperature T_1. Three phases may exist at this temperature, and so we are interested in three free-energy composition diagrams, one for each phase. A possible free-energy composition plot for temperature T_1 is shown in Fig. 11.7(a). In the two-phase regions a common tangent line gives the lowest free energy. It should be clear that these free-energy composition curves shift with temperature. At very high temperatures the curve for the liquid, G_s^ℓ lies below the other two curves. As the temperature is lowered the liquid curve rises and the solid curves drop until at the eutectic temperature all three curves touch a common tangent line. At temperature T_2 just below the eutectic temperature the G_s^ℓ curve is too high to touch the common tangent and the situation is as shown in Fig. 11.7(b).

B. SPINODAL POINTS

In Fig. 11.8(a) the curves of Fig. 11.7(b) are redrawn, where for simplicity the α and β curves are joined to form one continuous curve. (This implies the same crystal structure for α and β.) The two points on this curve at which the second derivative $d^2 G_s/dx^2$ is zero are called *spinodal points*. At compositions between the spinodal points the second derivative is negative and at compositions outside the spinodal points the second derivative is positive. The locus of the spinodal points on a eutectic phase diagram is indicated in Fig. 11.8(b).

Consider a composition within the spinodal region such as x_A on Fig. 11.8(a). If a single-phase solid solution of this composition were to decompose into two solutions of higher and lower compositions, respectively, by some statistical fluctuation, then the free energy of the system would be lowered as shown on Fig. 11.8(a). If such a fluctuation of compositions were to occur for a solid solution outside the spinodal points, such as at x_B, the free energy of the systems would be raised. Hence, small

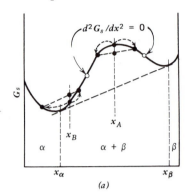 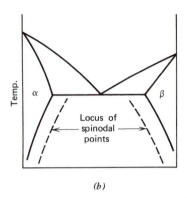

(a)   (b)

Figure 11.8   Constructions illustrating the location of the spinodal points on (a) the free-energy composition diagram and (b) the phase diagram (α and β of same crystal structure).

composition fluctuations are favored at compositions inside the spinodals but not outside the spinodals. Both solutions x_A and x_B lower the free energy of the system if they decompose into two solutions having the equilibrium compositions x_α and x_β. However, it may not be easy for the compositions to jump all the way to x_α and x_β because of diffusion problems. Such a kinetic problem would be less severe for compositions inside the spinodal since the free energy is reduced for small composition fluctuations. We will return to these ideas later in this chapter.

C.   EFFECT OF CURVATURE UPON PHASE DIAGRAM

Free-energy composition diagrams are very useful as a guide to understanding composition-temperature conditions for systems not at equilibrium. Many of the interesting and useful properties of alloys are achieved by forcing the system away from equilibrium, often by quenching. We will deal here with small departures from equilibrium due to interface curvature and show how free-energy composition diagrams provide a technique for qualitatively understanding the effects so produced upon the equilibrium phase diagram.

In order to discuss this problem it is quite helpful to establish some nomenclature that will also be useful in later discussions. Consider the simple eutectoid phase diagram of Fig. 11.9. Four possible precipitation reactions may occur upon cooling (other than the eutectoid reaction). The α phase will precipitate from hypoeutectoid γ phase, and the β phase will precipitate from hypereutectoid γ phase. Also, the α phase will precipitate β phase upon cooling and the β phase will precipitate α phase upon cooling. We will arbitrarily measure composition in terms of mole fraction

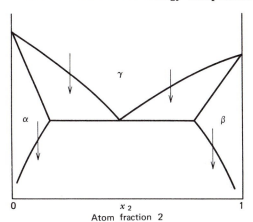

Figure 11.9   Eutectoid phase diagram illustrating possible precipitation reactions.

component 2, x_2, and observe the following nomenclature:

1.   The parent phase is called the *matrix*; the composition of the parent phase is represented by x_2^m.
2.   The phase growing out of the parent phase on cooling is called the *precipitate*; the composition of precipitate is x_2^p.
3.   *Positive precipitate* is defined as a precipitate phase with $x_2^p > x_2^m$.
4.   *Negative precipitate* is defined as a precipitate phase with $x_2^p < x_2^m$.

For example, β is a positive and α a negative precipitate. Notice that whether a given precipitate is positive or negative is arbitrary since it depends on whether one measures composition with mole fraction 2, x_2, or with mole fraction 1, x_1.

When a spherical precipitate forms in a matrix the precipitate–matrix interface is concave toward the precipitate. Following the arguments of Chapter 7 (p. 204), in order to obtain mechanical equilibrium the pressure within the precipitate will be increased by the amount $\Delta P = 2\gamma/r$. We will take the pressure of the parent phase to remain at the ambient pressure right up to the interface and the pressure within the growing precipitate phase to be increased throughout by ΔP. This assumption is only strictly true for liquids, but it provides us with a very good approximation. The solution free energy of the precipitate phase, G_s^p, will be increased by the amount $\bar{V}_s \, \Delta P$ where $\bar{V}_s$ is the atomic volume of the solid-solution precipitate phase. Hence, the G_s^p curve will be raised by $\bar{V}_s \, \Delta P$ from $G_s^p(\text{flat})$ to $G_s^p(r)$ on the free-energy composition diagram as shown for both the α and β precipitate phases on Fig. 11.10. Consider first the α phase, which is a negative precipitate on our diagram. Applying the common tangent rule it

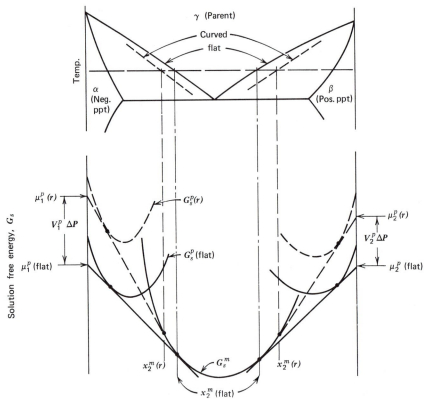

Figure 11.10 Construction illustrating effect of curvature on free-energy composition diagram and on phase diagram.

is clear that raising $G_s^p(\text{flat})$ to $G_s^p(r)$ will lower the composition of the matrix from $x_2^m(\text{flat})$ to $x_2^m(r)$. Now going back up to the phase diagram you can see that the effect of curvature will be to lower the phase boundary temperatures. For the positive precipitate, β, it is illustrated that curvature increases the composition of the matrix phase but, again, lowers the phase boundary line on the phase diagram.

Li and Oriani[3] have shown that the effect of the curvature upon the equilibrium composition is given as

$$x_2^m(r) = x_2^m \left[\frac{\dfrac{x_1^m}{x_1^p} \exp\left(\dfrac{\bar{V}_1^p \, \Delta P}{kT}\right) - 1}{x_2^p \left[\dfrac{x_1^m}{x_1^p} \exp\dfrac{\Delta P}{kT}(\bar{V}_1^p - \bar{V}_2^p) - \dfrac{x_2^m}{x_2^p}\right]} \right] \tag{11.17}$$

Since ΔP is small the following assumptions are quite good:

$$\text{(a)} \quad \exp\left(\frac{\bar{V}_1^p \Delta P}{kT}\right) = 1 + \frac{\bar{V}_1^p 2\gamma}{kTr}$$

$$\text{(b)} \quad \exp\left[\frac{\Delta P}{kT}(\bar{V}_1^p - \bar{V}_2^p)\right] = 1$$

Hence, we obtain from Eq. 11.17 after some considerable algebra

$$x_2^m(r) = x_2^m\left[1 + \frac{(1 - x_2^m)}{(x_2^p - x_2^m)} \frac{\bar{V}_1^p 2\gamma}{kTr}\right] \tag{11.18}$$

This equation shows that curvature increases the concentration in the matrix phase for a positive precipitate and decreases the matrix concentration for a negative precipitate. For the case where the precipitate phase is essentially pure one obtains a further simplification for a positive precipitate since $x_2^p \to 1$. It has become common in discussing curvature effects to present the following equation:

$$x_2^m(r) = x_2^m\left[1 + \frac{\bar{V}^p 2\gamma}{kTr}\right] \tag{11.19}$$

This equation only strictly applies for pure precipitates. For the case where the precipitate is a chemical compound, Eq. 11.17 is altered as shown by Li and Oriani.[3]

11.2 THE PRECIPITATION TRANSFORMATION

The basic requirement for the precipitation reaction is a lower solubility at lower temperatures. The lines on the phase diagram that give the maximum solubility in a solid-solution phase, as illustrated in Fig. 11.11, are

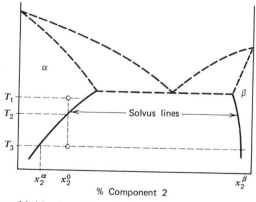

Figure 11.11 Location of solvus lines on a phase diagram.

called the solvus lines. An alloy of original composition x_2^0 heated to temperature T_1 will become a homogeneous solid solution upon annealing. If one now cools the alloy to temperature T_3 one expects the β phase to precipitate from the solid solution according to the reaction

$$\alpha(x_2^0) \rightarrow \alpha(x_2^\alpha) + \beta(x_2^\beta) \qquad (11.20)$$

This is the equilibrium reaction and it indicates that after the β phase has formed the remaining α matrix will have reduced its composition from x_2^0 to x_2^α. This equilibrium reaction only occurs if one does not cool too far below temperature T_2. If one quenches the alloy to sufficiently low temperatures the precipitation of any type of a second phase may often be entirely suppressed so that a metastable solid solution is obtained. If one quenches to moderately low temperatures it is frequently found that precipitation of a new phase occurs; however, the new phase is not β but a new metastable phase, which, of course, is not represented on the equilibrium phase diagram. It is often these types of reactions that lead to the large strengthening effects termed precipitation hardening.

There are two modes by which the equilibrium precipitation reaction, Eq. 11.20, is observed to occur.

I.   Continuous: This mode, illustrated in Fig. 11.12 is very similar to the recrystallization reaction discussed in the previous chapter. The new β phase nucleates as discrete particles, which then grow into the α matrix.

II.   Discontinuous: This mode is often called cellular precipitation. As illustrated in Fig. 11.12 the α matrix of composition x_2^0 transforms into a duplex structure consisting of plates of α phase of composition x_2^α alternating with plates of the new β phase. Transformation occurs at the moving phase boundary, which is usually observed to be a high-angle boundary between $\alpha(x_2^0)$ and $\alpha(x_2^\alpha)$. This type of precipitation reaction will

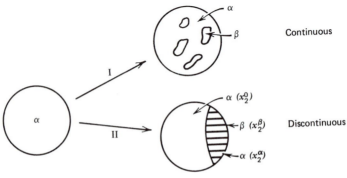

Figure 11.12   Schematic illustration of the two types of precipitation reactions.

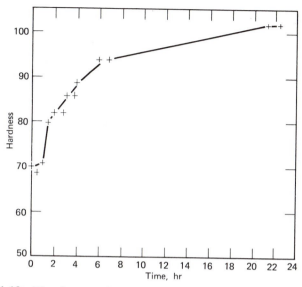

Figure 11.13 The first age-hardening curve published by A. Wilm in 1911. (Reprinted with permission from J. W. Martin, *Precipitation Hardening*, Copyright 1968, Pergamon Press Ltd.)

be discussed further in the next chapter; the remainder of this chapter will deal with the continuous precipitation, which is the more important precipitation mode since it is responsible for strengthening effects.

A. HISTORICAL

The major commercial interest in the precipitation reaction stems from its usefulness as a strengthening mechanism. It is this mechanism that has made possible the use of lightweight aluminum alloys in aircraft and is being used ever more widely in new alloys. It was discovered by accident in 1906 by Wilm.[4] In the course of experiments investigating the solid-solution strengthening of magnesium additions to an aluminum–copper–manganese alloy he found by accident that aging a quenched alloy produced a considerable increase in hardness. His initial results, first published in 1911, are shown in Fig. 11.13. Wilm could observe no microstructural changes during the strengthening period, and the source of the increased strength remained a mystery. Wilm's alloy was first produced commercially in 1909 under the trade name Duralumin. The composition of the commercial alloy varied within the following limits,[5] 3–4.5 wt% Cu, 0.4–1.0 wt% Mg, 0–0.7 wt% Mn, plus the following impurities, 0.4–1 wt% Fe, 0.3–0.6 wt% Si, balance Al. In a classical paper published in

1920,[5] P. D. Merica correctly deduced that the source of the strengthening resulted from formation of submicroscopic precipitates produced at temperatures below a solvus line. The development of other age-hardenable alloys followed this work.[6] However, it was not until the development of the electron microscope in the 1950s that the nature of the precipitation formation became physically clear.

To briefly illustrate the effectiveness of age hardening the following data are presented for the aluminum alloy 2024, a modern version of Wilm's alloy, which has a nominal composition of 4.4% Cu, 1.5% Mg, and 0.6% Mn, balance Al.

$$\text{Yield Strength (no aging)} = 20{,}000 \text{ psi } (138 \text{ MN/m}^2)$$

$$\text{Y.S. (room-temp. aging)} = 40{,}000 \text{ psi } (276 \text{ MN/m}^2)$$

$$\text{Y.S. (optimum aging treatment)} = 60{,}000 \text{ psi } (414 \text{ MN/m}^2)$$

B.   CRYSTALLOGRAPHIC DESCRIPTION OF PRECIPITATES

It frequently happens that the precipitate phase forms with a characteristic shape such as plates or needles. In these cases the appearance of the precipitate particles on the micrograph has a characteristic regularity similar to that shown on Fig. 11.14. In 1820 Carl von Schreibers published the micrograph of an iron–nickel meteorite shown in Fig. 11.14, which was prepared by Aloys von Widmanstatten using nitric acid to etch the polished surface.[7] As a result it has now become customary to term any structure having the appearance of second-phase plates or needles lying along preferred crystallographic planes of the matrix, a *Widmanstatten structure*.

If the characteristic shape of a precipitate is plate-like, then it is generally true that the precipitate will form along a specific set of $\{hkl\}$ planes in the matrix. It is customary to refer to these planes as the *habit planes* for the precipitate. Similarly, a needle precipitate will lie along the *habit directions* of the matrix.

In general, such habit forms result because they allow the precipitate–matrix interface to be a low-energy interface. It should be expected from the discussion of Chapter 7 that such a low-energy interface would require a specific crystallographic orientation between the matrix and the precipitate. As an example, when α brass (fcc) precipitates from β brass (bcc) it is found that (a) the (110) plane of β is parallel to the (111) plane of α, and (b) the [111] direction of β is parallel to the [110] direction of α. This relationship is generally given as

$$(110)_\beta \parallel (111)_\alpha$$
$$[111]_\beta \parallel [110]_\alpha$$

Figure **11.14** The Elbogen iron meteorite. A direct typographical imprint from the etched surface made by Schreibers and Von Widmanstatten in 1813. (0.59×) (Used with permission of C. S. Smith).

It is interesting to find that the close-packed plane of the β(bcc) lies parallel to the close-packed plane of α(fcc) and the close-packed direction of the β is parallel to the close-packed direction of the α. Such a result is not infrequent and it has come to be called the Kurdjumov–Sachs relationship. Such epitaxial matching at the interface may be interpreted as an effort by the system to lower the interfacial energy by optimizing the atomic matching across the interface.

C. PRECIPITATION SEQUENCE

As mentioned above, if one quenches to temperatures considerably below the solvus line no precipitate forms, so that one retains the high-temperature phase as a metastable phase at the quench temperature. If the metastable phase is allowed to age at temperatures sufficiently high, the equilibrium precipitate will eventually form. However, it is usually found

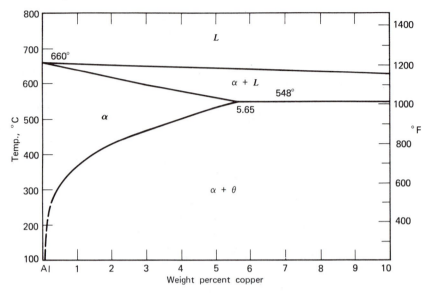

Figure 11.15 Aluminum–copper phase diagram. (Reproduced by permission, from *Metals Handbook*, American Society for Metals, 1948).

that one or two metastable precipitates will form before the equilibrium precipitate appears. The precipitation sequence that occurs in Al–Cu alloys will be discussed as an illustrative example; its phase diagram is shown in Fig. 11.15. Suppose an alloy of 4.5% Cu is annealed to produce a single-phase solid solution at 550°C and then quenched to room temperature. If this alloy is now annealed at temperatures sufficiently below the solvus temperature the following sequence of precipitates will form:

$$GP \text{ zones} \rightarrow \theta'' \rightarrow \theta' \rightarrow \theta$$

where three distinct and identifiable precipitates form and dissolve prior to the formation of the equilibrium θ precipitate.

GP Zones. Electron microscope studies[8,9] have shown that these zones have the shape of disks with a diameter of about 80 Å and a thickness of only 3–6 Å. Their average composition is around 90% Cu so that they really appear to be disks of predominately Cu atoms on the Al lattice. These zones appear to form uniformly throughout the Al matrix (homogeneously) with a density of around 10^{18} per cm^3 (see Ref. 10 for a recent study).

This type of very small zones was first detected by means of x-ray techniques in independent studies of Gunier and Preston in the 1930s.

Consequently, they are now called Gunier–Preston zones or, more simply, GP zones. These zones are frequently observed to form quite rapidly upon quenching.

θ″ Precipitate. Upon heating, the GP zones dissolve and a second more truly precipitate phase, θ″, forms. This phase is tetragonal with $a = b = 4.0$ Å and $c = 7.8$ Å and it is in plate form with a thickness of about 20 Å and a diameter of about 300 Å, having $\{100\}_{\theta''} \parallel \{100\}_{matrix}$. This phase appears to nucleate fairly uniformly in the matrix with a coherent interface between precipitate and matrix. As indicated in Table 7.1, a coherent interface produces a high lattice strain in the region of the interface in order to produce the one-to-one atom matching at the interface. This is illustrated in Fig. 11.16 in which the strained region around an edge view of a θ″ precipitate is shown by the cross hatching. The strained region due to the interface coherency shows up on an electron micrograph as a dark region under the proper conditions. This is illustrated in the electron micrographs of both GP zones and θ″ precipitates shown on pp. 657 and 658 of Ref. 9.

θ′ Precipitate. The θ′ precipitate is the first precipitate that may be observed under a light microscope and consequently it must have sizes on the order of 1000 Å. It is tetragonal with $a = 4.04$ Å and $c = 5.8$ Å and is oriented as $\{100\}_{\theta'} \parallel \{100\}_{matrix}$. The θ′ precipitate forms heterogeneously on helical dislocations and cell walls.[8] Its interface is semicoherent with the Al matrix. See Fig. 2.14 for an electron micrograph of the θ′ precipitate.

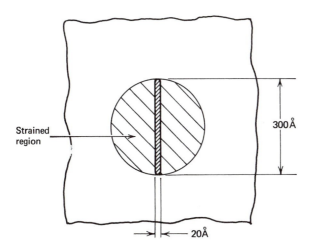

Figure 11.16 Edge view of a θ″ precipitate particle shown schematically.

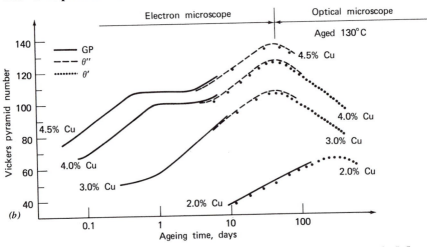

Figure 11.17   Age hardening curves for Al–Cu alloys. [From Silock et al., J. Inst. Metals **82**, 24 (1953–54).]

θ Precipitate.   The equilibrium precipitate has a tetragonal structure with $a = 6.06$ Å and $c = 4.87$ Å. It nucleates heterogeneously at grain boundaries and forms an incoherent interface with the Al matrix. It is not known for sure whether these last three precipitates form directly from the Al matrix or from the preceding precipitate.[9] However, experimental evidence indicates that it is more probable that they nucleate directly from the Al matrix.

The strengthening that accompanies the ageing of quenched Cu–Al alloys is shown in Fig. 11.17. Notice that the main strengthening occurs prior to the formation of the θ′ precipitate. This result indicates that the main strengthening must result from the interaction of dislocations with the very small coherent GP zones and θ″ precipitates. It also illustrates why early workers could not detect any microstructural changes with light microscopy prior to the optimum strengthening time.

The type of precipitation sequence described above for the Al–Cu system is often found to occur in metals that may be strengthened by precipitation reactions. Table 11.1 lists the sequence of precipitates observed in a number of precipitation-hardening alloys. Notice that the GP zones and intermediate phases do not always appear in precipitation-hardening systems. This is discussed further in Section 13.4C. They are generally observed when the size difference between the solute and solvent atoms is less than about 12% and when the crystal structure of the equilibrium precipitate is relatively simple.[8] Notice, also, that spherical GP

Table 11.1   Precipitation Sequences Observed in Several Alloys

Base Metal	Alloy	Sequence of Precipitates	Equilibrium Precipitate
Aluminium	Al–Ag	Zones (spheres) → γ′ (plates)	→ γ (Ag₂Al)
	Al–Cu	Zones (disks) → θ″ (disks) → θ′	→ θ (CuAl₂)
	Al–Zn–Mg	Zones (spheres) → M′ (plates)	→ (MgZn₂)
	Al–Mg–Si	Zones (rods) → β′	β (Mg₂Si)
	Al–Mg–Cu	Zones (rods or spheres) → S′ →	S (Al₂CuMg)
Copper	Cu–Be	Zones (disks) → γ′ ──────→	γ (CuBe)
	Cu–Co	Zones (spheres) ──────────→	β
Iron	Fe–C	ε-Carbide (disks) ──────────→	Fe₃C (laths)
	Fe–N	α″ (disks) ──────────────→	Fe₄N
Nickel	Ni–Cr–Ti–Al	γ′ (cubes) ──────────────→	γ (Ni₃Ti, Al)

Reprinted with permission from J. W. Martin, *Precipitation Hardening*, copyright 1968, Pergamon Press, Ltd.

zones form in some systems. These are only observed when the atomic misfit is less than around 3%.[8]

The general sequence of the precipitation process that occurs upon annealing of the quenched solid-solution alloys is given in Table 11.2 for alloys that form GP zones. The sequence presented in Table 11.2 does not hold exactly in all systems with zones but is presented as a general guide. For instance, in Al–Cu two intermediate precipitates form, θ″ and θ′, and in Cu–Co no intermediate precipitate forms; see Table 11.1. The puzzling question naturally arises as to why precipitation occurs in such a complex fashion. The answer to this question comes from an understanding of the nucleation process in the solid state.

Table 11.2   Precipitation Sequence Often Observed in
Age-Hardenable Alloys

Stage	Precipitate	Crystal Structure	Coherency	Nucleation
1	GP Zones	Same as matrix	Fully coherent	Uniform ~10^{18}/cm³
2	Intermediate	Diff. from matrix	Fully or partial	Heterogeneous
3	Equilibrium	Diff. from matrix	Non-coherent	Heterogeneous

11.3 NUCLEATION IN THE SOLID STATE

The problem of nucleation was discussed in Chapter 8 where it was shown that nucleation is inhibited by the positive energy required to generate the interface between the nucleus and the matrix. In solid–solid phase transformations we must also consider another factor that inhibits nucleation. Suppose that one were to remove a spherical region of α phase as shown in step I of Fig. 11.18. This α-phase sphere is now transformed to the new β phase as in step II. In order to reinsert this β sphere into the α matrix as in step III it must have the same volume as the original α sphere. Otherwise one must either compress or expand the β sphere to make it fit the hole left by the α sphere. Hence we may conclude that if $V_\beta \neq V_\alpha$, where V is volume per atom, a strain energy will arise upon formation of a β precipitate. This strain energy will inhibit nucleation and so it must be considered in any nucleation theory.

A. CLASSICAL TREATMENT

We proceed as in Chapter 8 except that now a term is included for strain energy. It also is a bit more convenient to consider the free energy per atom of the nucleus rather than the free energy per volume of the nucleus. The free energy associated with the formation of a nucleus of n atoms, ΔG, may be written as

$$\Delta G = n\,\Delta G_B + \eta n^{2/3}\gamma + nE_s \qquad (11.21)$$

where

$n = $ No. of atoms in nucleus

$\Delta G_B = \dfrac{(G_{nucleus} - G_{matrix})}{n}$

$\quad$ = bulk free energy change per atom in nucleus

$\eta = $ shape factor such that $\eta n^{2/3} = $ surface area

$\gamma = $ surface tension $\approx$ surface free energy

$E_s = $ strain energy per atom in nucleus

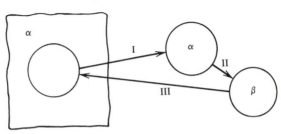

Figure 11.18 Scheme to illustrate importance of volume changes upon transformation.

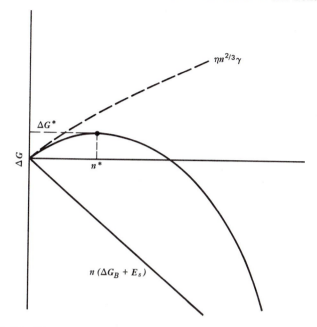

Figure 11.19   The free energy of formation as a function of the number of atoms in the nucleus.

We may regroup this equation as

$$\Delta G = n(\Delta G_B + E_s) + \eta n^{2/3} \gamma \qquad (11.22)$$

The term ΔG_B will be negative below the transformation temperature similar to the plot of Fig. 8.3, whereas E_s and γ are both positive. Hence, if $|\Delta G_B| > E_s$, then the first term is negative. A plot of the two right-hand terms will be similar to Fig. 8.4 and will appear as in Fig. 11.19. Similar to the arguments of Chapter 8 it is argued that a nucleus of n^* atoms is a critical-sized nucleus because its free energy is lowered upon addition of more atoms. The free-energy change to form the critical-sized nucleus, ΔG^*, is found by differentiating Eq. 11.22 assuming constant ΔG_B, E_s, η, and γ.

$$\Delta G^* = \frac{4}{27} \frac{\eta^3 \gamma^3}{(\Delta G_B + E_s)^2} \qquad (11.23)$$

A large strain energy E_s *reduces* the denominator and makes ΔG^* large, which means nucleation is more difficult because the critical-sized nucleus has a higher energy of formation. A small strain energy increases the denominator, lowers ΔG^*, which indicates that nucleation is easier. These same results may be seen graphically by considering the effect of E_s upon the lower curve of Fig. 11.19.

For the case of heterogeneous nucleation one must consider the effect of the nucleation interface upon the surface energy and strain energy. For this case Eq. 11.23 becomes

$$\Delta G^* = \frac{4}{27} \frac{\eta^3 \gamma^3}{(\Delta G_B + E)^2} \left[\frac{2 - 3S + S^3}{4} \right] \qquad (11.24)$$

where the factor in brackets was derived in Chapter 8 (p. 228). As shown in Table 11.2 the intermediate and the equilibrium precipitates form heterogeneously. It is often found that the intermediate precipitates form upon dislocation arrays and the equilibrium precipitates form at grain boundaries. Homogeneous nucleation is only expected to occur if γ and E are very small and if the dislocation density is low. In many cases GP zones probably form by homogeneous nucleation since they have low surface energies and misfits and in other cases they form by the spinodal decomposition process to be discussed later. There appear to be only two cases where an equilibrium precipitate forms homogeneously, Cu–Co and Ni base alloys,[12] and in these cases nucleation is homogeneous only under certain conditions. Hence, only a very small minority of solid systems form equilibrium precipitates by homogeneous nucleation.

As mentioned on p. 230 there has been much criticism of the classical theory of nucleation. Nevertheless it illustrates quite simply the major factors involved in the nucleation process. For a given degree of supercooling the major terms are functions of a number of variables,

    1. $\Delta G_B =$ function (composition, temperature)
    2. $E_s =$ function (shape, coherency)
    3. $\eta =$ function (shape)
    4. $\gamma =$ function (composition, coherency)

In general, things are extremely complicated and the theory can only offer a semiquantitative analysis.

1.   Bulk Free-Energy Change, ΔG_B.   This term is the most important in the sense that unless it is negative there is no driving force for nucleation and, hence, no nucleation. We will consider a single α-phase alloy of composition x_0 that has been quenched below a solvus line to some temperature. Figure 11.20 presents the free-energy composition diagram for this α-phase alloy and locates its free energy per mole at G_0. We now allow a small amount of precipitate to form with composition x_2 and this in turn requires the remaining α phase to lower its composition to x_1 as shown in Fig. 11.20. We will consider that the precipitate is a GP zone having the same crystal structure as the matrix and assume that its free energy G_2 is

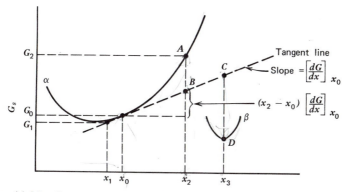

Figure 11.20 Construction for graphical determination of the free energy of the initial precipitate.

given at point A on Fig. 11.20. We now ask, what is the free energy change associated with the formation of the precipitate?

$$\Delta G' = G(\text{final}) - G(\text{initial})$$
$$= (n_1 G_1 + n_2 G_2) - (n_1 + n_2) G_0 \qquad (11.25)$$

where $\Delta G'$ is the total free energy change upon formation of the precipitate, n_2 is the number of moles of precipitate, and n_1 is the number of moles in the α matrix after precipitation. From the lever law we may write (assuming *all* of the precipitate and *all* of the matrix have uniform composition)

$$\frac{n_1}{n_2} = \frac{x_2 - x_0}{x_0 - x_1} \qquad (11.26)$$

Combining Eq. 11.25 and 11.26 we obtain for the precipitation reaction

$$\Delta G' = n_2 \left[G_2 - G_0 - (x_2 - x_0) \frac{G_0 - G_1}{x_0 - x_1} \right] \qquad (11.27)$$

At the instant of nucleation the amount of precipitate that has formed will be very small so that we may take $x_1 \approx x_0$. Under this condition Eq. 11.27 becomes

$$\Delta G_B = \frac{\Delta G'}{n_2} [\text{Initial precipitate}] = \left[G_2 - G_0 - (x_2 - x_0) \left(\frac{dG}{dx} \right)_{x_0} \right] \qquad (11.28)$$

This equation gives us the desired expression for the free energy per atom in the nucleus and it may be interpreted quite easily by graphical means. The term $G_2 - G_0$ is seen directly on the ordinate. The term $(x_2 - x_0) \times (dG/dx)_{x_0}$ is shown by the construction where it should be clear that $(dG/dx)_{x_0}$ is the slope at composition x_0. Hence, it is seen that formation of

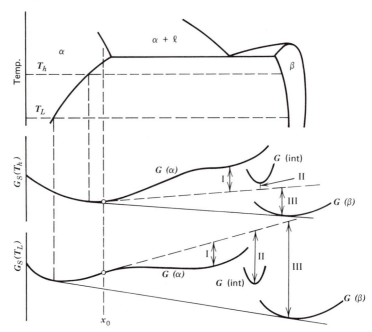

Figure 11.21 Construction illustrating the free energy of initial precipitates of (I) GP zone, (II) intermediate precipitate, and (III) the equilibrium precipitate at two different temperatures, T_h and T_L.

the precipitate of composition x_2 and free energy G_2 has a ΔG_B of $(A - B)$ on the plot. Since $(A - B)$ is positive it is illustrated graphically that this precipitate cannot form. Consider, however, the β precipitate of composition x_3 having a lower free energy per mole at point D. By arguments similar to the above you may show yourself that ΔG_B for this precipitate is $(D - C)$; furthermore $(D - C)$ is negative indicating that the precipitate may form. We conclude that *any precipitate having free energy per mole below the line tangent at x_0 will have a negative ΔG_B and its formation is thermodynamically favored.* This is a very useful result because it allows one to immediately determine ΔG_B for any potential precipitate by a simple graphical construction.

As an example consider an alloy of composition x_0 on the phase diagram at the top of Fig. 11.21. The alloy is first homogenized above the solvus temperature and then quenched to either the high temperature, T_h, or the low temperature, T_L. Prior to any precipitation at the quenched temperatures the α phase of composition x_0 is a metastable phase and we may consider a G_s curve for this phase. The free energy per mole of the metastable solid solutions must lie upon the G_s curves shown as $G(\alpha)$ at temperatures T_h and T_L on Fig. 11.21. From the above discussion we know that three types of precipitates may form, GP zones, an intermediate precipitate, or the equilibrium precipitate. The G_s curves for the equilibrium precipitate, $G(\beta)$, and the intermediate precipitate, $G(\text{int})$, are shown on Fig. 11.21. The G_s curve for the GP zones will be assumed to be the same as $G(\alpha)$ because these zones have the same crystal structure as the parent α phase. They are clusters of solute atoms on the parent crystal lattice. Constructing the tangent line at composition x_0 we may determine ΔG_B for each precipitate by inspection of Fig. 11.21. The ΔG_B for formation of the equilibrium β precipitate, $\Delta G_B(\beta)$, is shown as III and it can be seen that it is more negative at the lower temperature, T_L. For the intermediate precipitate, the $\Delta G_B(\text{int})$ is labeled II, and for the GP zone, $\Delta G_B(\text{GP})$ is labeled I on Fig. 11.21. Both of these are negative at temperature T_L and positive at temperature T_h. The relative values of these free-energy changes for the three precipitates are shown in Table 11.3 at the quench temperatures T_h T_L, and some quench temperature intermediate to these two. At the quench temperature of T_h it is not possible to form GP zones or the intermediate precipitate because their ΔG_B values are positive and, hence, there is no driving force for their formation. As the quench temperature is progressively lowered the ΔG_B first becomes negative for the intermediate precipitate and then for the GP zones. Consequently, these two precipitates may only form at lower temperatures as is shown on Fig. 11.22. This diagram is sort of a metastable phase diagram for the intermediate precipitate and the GP zones.

Referring to Table 11.3 it is seen that at temperature T_L all three precipitates have a negative free energy of formation and any of them could, in principle, form. One might expect the equilibrium precipitate always to form in preference because it has the largest driving force, ΔG_B, for formation. However, surface energy and strain energy will inhibit formation of any precipitate, and it is precisely because these factors are lower for the intermediate precipitate and GP zones that they do form before the equilibrium precipitate.

Table 11.3 Variation of Free Energy of Formation of the Three Precipitates at Three Different Temperatures

Quench Temp.	$\Delta G_B(\text{GP}) = I$	$\Delta G_B(\text{Int}) = II$	$\Delta G_B(\beta) = III$
T_h	Positive $\longrightarrow$ Positive	$\longrightarrow$ Negative	
$T(\text{Int})$	Positive $\longrightarrow$ Negative	$\longrightarrow$ More negative	
T_L	Negative $\longrightarrow$ More negative	$\longrightarrow$ Most negative	

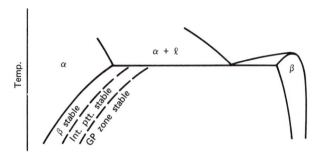

Figure **11.22** The metastable phase diagram.

2. Factors Inhibiting Nucleation. The form of the strain-energy term depends on whether the interface between the precipitate and matrix is coherent or incoherent. Russell has recently shown (Ref. 11, p. 233) that if a precipitate forms with an incoherent interface it will almost always be strain free. Hence, we will only consider the strain energy of coherent precipitates.*

Several expressions for the strain energy term E_s have been calculated from elasticity theory assuming elastic isotropy (Ref. 11, p. 233; Ref. 9, pp. 421–427). To consider the effects of precipitate shape the theoretical treatments analyze an ellipsoid of revolution which has semi-axes of R, R, and y. For $y = R$ the shape becomes a sphere, for y/R large the shape becomes a needle, and for y/R near zero the shape approximates a plate or a disk. For the case of dilational strain only (uniform expansion or contraction) the strain-energy term is independent of precipitate shape, and when matrix and precipitate have identical elastic constants the strain-energy term is found to be

$$E_s = f(\nu)\,\bar{V}E(\varepsilon_{11}^T)^2 \quad \text{and} \quad f(\nu) = \frac{1}{1-\nu} \tag{11.29}$$

where ε_{11}^T is the stress-free linear transformation strain (the strain if the transformation were unconstrained), E is Young's modulus, $\bar{V}$ is the volume per atom, and $f(\nu)$ is a function of Poisson's ratio ν. Expressions for $f(\nu)$ have been obtained for other limiting cases such as when all strain is accommodated in the matrix $[3/(1+\nu)]$ and when all strain is accommodated in the precipitate $[1.5/(1-2\nu)]$. When one allows shear strain, an

---

* The strain energy of incoherent precipitates has been treated by Nabarro (see Ref. 9, pp. 418–21). The results indicate that the strain energy depends upon the shape of the precipitate with a sphere having the maximum E_s and a flat plate the minimum E_s. The surface energy γ will be larger for the plate than the sphere because the plate has a larger surface-to-volume ratio. Hence, in this case the shape of the nucleus depends on which of these two terms dominates.

additional term is obtained that involves the shear strain, ε_{13}^T, and the shape factor y/R, and has the following form for an ellipsoid,

$$E_s = \bar{V}E\left[f_1(\nu)(\varepsilon_{11}^T)^2 + f_2(\nu)(\varepsilon_{13}^T)^2\frac{y}{R}\right] \qquad (11.30)$$

where $f_1(\nu)$ and $f_2(\nu)$ are functions of Poisson's ratio ν. For plate precipitates $(y/R \rightarrow 0)$ the shear term drops out and the strain-energy term has the simpler form of Eq. 11.29.

Consider now a plate precipitate whose strain energy is given by Eq. 11.29. Suppose that at the precipitate–matrix interface the atoms are in disregistry by a small amount along only one atomic direction. The strain then becomes identical to the disregistry δ, defined by Eq. 7.7. Taking ν as $\frac{1}{3}$ we obtain for Eq. 11.29

$$E_s = \tfrac{3}{2}\bar{V}\bar{E}\,\delta^2 \qquad (11.31)$$

The surface energy of the precipitate matrix interface is essentially equal to the surface tension of the interface, γ, as shown in Section 7.2. If the interface is incoherent its surface tension will be similar to that of high-angle boundaries and may be approximated as 500 ergs/cm^2; see p. 176. If the interface is coherent, its surface tension will be much lower, in the range of 10–30 ergs/cm^2.

Whether or not a plate precipitate will form with a coherent or an incoherent interface may be treated in the following approximate way. The two terms that inhibit nucleation are the surface-energy and the strain-energy terms. If the interface forms with an incoherent boundary the strain energy will be negligible; see the discussion on p. 180. Therefore, the energy inhibiting nucleation for the incoherent case is simply $\gamma[2\pi(At)^2 + 2\pi(At)t]$, where A, the aspect ratio, is defined as the radius to thickness of the plate, $A = r/t$. If the plate forms with a coherent interface we may neglect γ to a first-order approximation because of its low values for coherent boundaries. Therefore the energy inhibiting nucleation for the coherent interface is given by Eq. 11.31 as $[3\bar{E}\,\delta^2/2]\pi(At)^2t$. If the aspect ratio of a given precipitate is roughly independent of its size, the energy inhibiting nucleation would be proportional to t^3 for a coherent nucleus and t^2 for an incoherent nucleus. As shown schematically in Fig. 11.23 the cubed function will generally be less than squared function at low t values, so that one expects a coherent interface if the initial precipitate thickness is below t_{CR}. The critical thickness determined by equating the above two expressions gives

$$t_{CR} = \frac{4}{3}\frac{\gamma}{E\,\delta^2}\left[1 + \frac{1}{A}\right] \qquad (11.32)$$

Typical values of t_{CR} are in the range of tens to hundreds of angstroms so

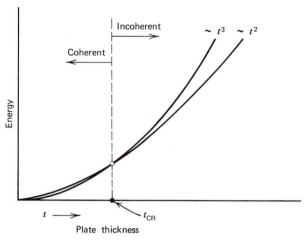

Figure 11.23   The energies associated with coherent and incoherent precipitates.

that this treatment predicts that GP zones will form with $t < t_{CR}$. Hence one expects these precipitates to initially form with a coherent interface. However, upon subsequent growth these precipitates may become incoherent when t becomes greater than t_{CR}. This treatment must necessarily be only an approximation since one would expect the precipitate–matrix boundary gradually to become incoherent by the formation of a partially coherent interface as growth procedes. Nevertheless it presents a qualitatively correct picture that explains why the precipitates generally form with a coherent interface and then form an incoherent interface at larger sizes. The preferred habit planes that give rise to the Widmanstatten appearance of many precipitates indicates that these precipitates formed originally with coherent interfaces. Because only certain habit forms will allow the precipitate–matrix interface to be coherent, the precipitate is restricted to these habit planes and it then displays a regularity as is characteristic of the Widmanstatten form. Upon growing to larger sizes the precipitates may lose their coherency.

B.   SPINODAL DECOMPOSITION

Nucleation within a solid-solution alloy requires that in certain regions of the solid solution the composition must change and these regions of new composition must be stable. In the classical theory of nucleation it is assumed that these new regions (nuclei) have dimensions of around 10 Å. Consequently, the interface between the nucleus and the solid-solution matrix will possess a distinct structural discontinuity and it will therefore possess a positive free energy and inhibit nucleation. Suppose, however, that the region of new composition was somewhat larger than these classical nuclei. The interface between the matrix and the region of new composition now becomes diffuse in nature and does not present a distinct structural discontinuity that is characterized by a positive free energy. To produce a region of new composition over a large extent would require a somewhat long-range composition fluctuation, which would tend to "unmix" the original solid solution. From the discussion on p. 369 it is apparent that such a composition fluctuation would be most likely to occur if the composition of the original solid solution were between the spinodal points of the alloy. This type of long-range composition change leading to formation of a new phase is therefore called spinodal decomposition. Such a mechanism for formation of new phases has long been the subject of debate, but recent theoretical work by M. Hillert and J. Cahn has led to critical experiments showing that such a process does occur in some cases.[12,13] It is quite probable that the GP zones in many alloys form by spinodal decomposition.

The upper part of Fig. 11.24 illustrates the composition profile associated with the classical nucleation and growth mechanism. At the distinct interface between the precipitate and matrix a composition discontinuity occurs. The composition in the precipitate nucleus rises immediately to C'_α, and as the precipitate grows the solute is transported down a composition gradient (from C_0 to C_α) toward the precipitate. In the spinodal decomposition scheme shown at the bottom of Fig. 11.24 the composition increase is initially much less and extends over a larger region. A very intriguing feature of this mechanism is that the solute must move toward the precipitate region by diffusion up the concentration gradient.

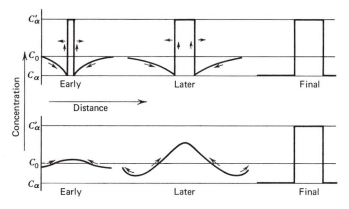

Figure 11.24 Composition variation in classical nucleation (upper curves) and in spinodal decomposition (lower curves) (from Ref. 12).

This means that the preference of the solute species to cluster together must be so strong as to allow "uphill" diffusion. As we shall now see this is in fact predicted for compositions inside of the spinodals.

In general, when one considers diffusion in concentrated alloys it is necessary to define a reference velocity. If we take a marker velocity v_m as a reference velocity, we may write the total flux of both components with respect to an observer as

$$(J_1)_T = C_1 v_m + J_1^D$$
$$(J_2)_T = C_2 v_m + J_2^D$$

$$(11.33)$$

where C_1 and C_2 are the molar volume concentrations of both components and J_1^D and J_2^D are the molar fluxes of both components due to diffusion. These equations are identical to Eq. 6.23, except here we have not yet specified the form of the diffusion terms. As before (p. 148) we assume constant molar density so that $(J_1)_T = -(J_2)_T$. Combining this equation with Eqs. 11.33 we obtain an expression for v_m that is then substituted into the lower Eq. 11.33 to obtain

$$(J_2)_T = x_1 J_2^D - x_2 J_1^D$$

$$(11.34)$$

where x_1 and x_2 are the mole fractions of the two components. The diffusion fluxes are now written using the chemical potential driving force terms (Eq. 6.34, $J_i = -C_i B_i \, \partial \mu_i / \partial Z$) and remembering that $C_i = x_i \rho$, where ρ is molar density, we obtain

$$(J_2)_{\text{TOT}} = x_1 x_2 \rho \left[B_1 \frac{d\mu_1}{dZ} - B_2 \frac{d\mu_2}{dZ} \right]$$

$$(11.35)$$

Using the Gibbs–Duhem equation, $x_1 \, d\mu_1 + x_2 \, d\mu_2 = 0$, it is possible to reduce this equation further to obtain

$$(J_2)_{\text{TOT}} = -\rho M \frac{\partial}{\partial Z} [\mu_2 - \mu_1]$$

$$(11.36)$$

where M is a sort of mutual mobility given as $x_1 x_2 [x_1 B_2 + x_2 B_1]$. If we differentiate Eq. 11.3 with respect to x_2, recognizing that $\bar{G}_i = \mu_i$ and utilizing the Gibbs–Duhem equation, we obtain

$$\mu_2 - \mu_1 = \frac{\partial G_s}{\partial x_2}$$

$$(11.37)$$

This equation is restricted to systems at *thermodynamic equilibrium*. Combination of Eqs. 11.36 and 11.37 gives

$$(J_2)_T = -\rho M \frac{\partial}{\partial Z} \left[\frac{\partial G_s}{\partial x_2} \right]$$

$$(11.38)$$

Using the chain rule we rewrite Eq. 11.38 as

$$(J_2)_T = -\rho M \left[\frac{\partial^2 G_s}{\partial x_2^2} \right] \frac{\partial x_2}{\partial Z} = -MG'' \frac{dC_2}{dZ} \qquad (11.39)$$

where we have taken $G'' = \partial^2 G_s / \partial x_2^2$. The expression generally employed for the total diffusion flux is, from Eq. 6.25

$$(J_2)_T = -\bar{D} \frac{dC_2}{dZ} \qquad (11.40)$$

Therefore, we have a relationship between the mutual diffusion coefficient $\bar{D}$ and the mutual mobility M,

$$\bar{D} = MG'' \qquad (11.41)$$

As shown on p. 369, G'' is negative inside of the spinodal compositions. This means that for compositions within the spinodals the diffusion coefficient is negative and diffusion will in fact occur "uphill," that is, from lower compositions to higher compositions.

As mentioned above, Eq. 11.37 is restricted to systems in thermodynamic equilibrium. This condition requires uniform composition, which, of course, is not the case around a composition fluctuation. A sharp composition gradient increases the energy of the lattice and this increase in energy in turn affects the chemical potential difference, $\mu_2 - \mu_1$. The energy increase may be partitioned into two forms, one due to a straining of the lattice and called a strain energy, and the other due to an effect upon the chemical bonding of the individual atoms and called the gradient energy. In a series of papers, summarized in Ref. 12, Cahn and Hilliard have shown how to account for these two forms of energy introduced by the nonequilibrium condition (see also Ref. 13).

1. Gradient Energy. In a homogeneous solid solution each atom will on the average have a certain number of like and unlike neighbors. Each atom will "feel" the presence of those atoms that lie within some interaction distance. Suppose now, however, a sharp concentration gradient exists in the solid solution as shown in Fig. 11.25. It is clear that if the curvature is sharp enough an individual atom will "feel" a different number of like and unlike atoms than it would if the composition were homogeneous. Consequently, the chemical potential of the atoms will change. The change in the chemical potential will be a function of the curvature of the composition profile, $\partial^2 x_2 / \partial Z^2$, and also a function of a parameter, K, that characterizes the interaction range of the atoms. The change in chemical potential is found to be

$$(\mu_2 - \mu_1)[\text{gradient}] = -2K \frac{\partial^2 x_2}{\partial Z^2} \qquad (11.42)$$

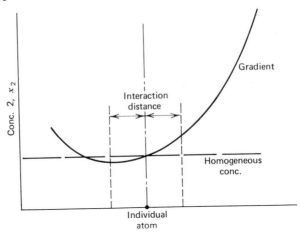

Figure 11.25 A sharp concentration change near an individual atom.

When this gradient term is added to Eq. 11.37 and substituted into Eq. 11.36 the flux equation becomes

$$(J_2)_T = -MG'' \frac{\partial C_2}{\partial Z} + M \cdot 2K \frac{\partial^3 C_2}{\partial Z^3} \qquad (11.43)$$

2. Strain Energy. Suppose we let a_0 be the lattice parameter of an unstrained solid of homogeneous composition x_2^0. If we change the composition the lattice parameter will change, and we characterize this change with the parameter η, defined as $\eta = (1/a_0)(da/dx_2)$. Consider a long rod that contains a composition gradient along its length. The lattice parameter varies along the length of the rod. If the lattice is to remain coherent along its length a coherency strain will result in order to maintain a one-to-one lattice matching along the length. The strain energy so introduced gives rise to a change in chemical potential that Cahn has shown to be of the form

$$\frac{d (\mu_2 - \mu_1)}{dZ} [\text{strain}] = 2\eta^2 Y \frac{dx_2}{dZ} \qquad (11.44)$$

where Y is a function of the elastic constants of the alloy. For elastically isotropic materials $Y = E/(1-\nu)$. In general Y has different values for different directions in the crystal lattice. In most cubic alloys Y is usually a minimum along $\langle 100 \rangle$ directions and sometimes along $\langle 111 \rangle$ directions. Consequently if the composition fluctuation localizes along these directions it will minimize its effect upon the chemical potential of the system. When both the gradient term (Eq. 11.42) and the strain term (Eq. 11.44) are added to Eq. 11.37 and substituted into Eq. 11.36 the flux term becomes

$$(J_2)_T = -MG'' \frac{\partial C_2}{\partial Z} + M \cdot 2K \frac{\partial^3 C_2}{\partial Z^3} - M \cdot 2\eta^2 Y \frac{\partial C_2}{\partial Z} \qquad (11.45)$$

Whenever diffusion occurs in a system one must necessarily have departure from thermodynamic equilibrium because the system contains a chemical potential gradient. Consequently, a strain energy and a gradient energy will always be present and, therefore, Eq. 11.45 is the proper form of Fick's first law for diffusion whether or not one is considering spinodal decomposition. However, the strain term is only applicable if coherency is maintained and the gradient term is only significant when diffusion distances become smaller than the order of a micron. Consequently, the added terms are generally negligible.

When Eq. 11.45 is substituted into the continuity equation (Eq. 6.7) we obtain the following partial differential equation governing the time–distance composition variation,

$$\frac{\partial C_2}{\partial t} = -M \left[(G'' + 2\eta^2 Y) \frac{\partial^2 C_2}{\partial Z^2} - 2K \frac{\partial^4 C_2}{\partial Z^4} \right] \qquad (11.46)$$

where we have assumed M, G'', η, Y, and K independent of composition. Now, in order to have a spinodal decomposition type of precipitation it is necessary that the solute atoms spontaneously segregate into clusters upon the parent matrix as is shown in Fig. 11.26. The solution to Eq. 11.46 describing this type of composition profile is found to be

$$C_2 - C_0 = e^{R(\lambda)t} \cos \frac{2\pi}{\lambda} \cdot Z \qquad (11.47)$$

where λ is shown on Fig. 11.26 and the time coefficient $R(\lambda)$ is given as

$$R(\lambda) = -M \frac{4\pi^2}{\lambda^2} \left[G'' + 2\eta^2 Y + \frac{8\pi^2 K}{\lambda^2} \right] \qquad (11.48)$$

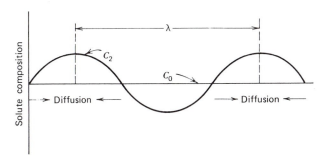

Figure 11.26 A sinusoidal composition fluctuation.

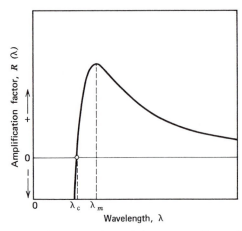

Figure 11.27 A plot of the function $R(\lambda)$ versus λ. (Reproduced by permission, from Ref. 13, American Society for Metals, 1970).

Notice that only when $R(\lambda)$ is positive will the amplitude of the clusters grow with time. Equation 11.48 shows that $R(\lambda)$ is positive only when G'' is negative, because both the strain term and gradient term are positive. Hence, a spontaneous decomposition can only occur for a system whose composition and temperature place it inside the spinodal points where G'' is negative, and where $|G''| > 2\eta^2 Y + 8\pi^2 K/\lambda^2$. The form of the function $R(\lambda)$ is shown in Fig. 11.27. The function contains a fairly sharp maximum at λ_m. Below this wavelength the gradient term begins to dominate and below wavelength λ_c decomposition is no longer possible due to the large gradient effect resulting from the shorter distances between clusters. It is theoretically possible, therefore, to form clusters of any wavelength above λ_c. However, since the time dependence of the amplitude of the composition fluctuation goes as $\exp[R(\lambda) \cdot t]$, and since $R(\lambda)$ contains a fairly sharp maximum, one expects to observe clusters mainly of spacings near to λ_m.

Experiments have verified that precipitation does in fact occur by spinodal decomposition in a few metal and also a few glass systems.[13] The value of λ_m is found to be around 50 Å in Al–Zn and around 100 Å in Al–Ag. Thus it is apparent that the gradient energy term becomes appreciable only at distances below 100–200 Å and may be neglected for diffusion distances on the order of microns or larger.

The locus of the spinodal points for an alloy were defined on Fig. 11.8 by the temperatures and compositions where $G'' = 0$. It has now become customary to refer to the *coherent spinodal* as the locus of temperatures and compositions defined by the function $G'' + 2\eta^2 Y = 0$. It is apparent from Eq. 11.48 that unless the cluster wavelength can adjust to make the gradient energy term $8\pi^2 K/\lambda^2$ less than these two terms $|G''+2\eta^2 Y|$, the spontaneous growth of clusters is not possible. Hence, the possibility of spontaneous spinodal decomposition at a given composition of an alloy is better estimated by the coherent spinodal temperature than by the spinodal temperature. For example, the Au–Ni system exhibits a large miscibility gap in its solid solutions and is a classic system for study of spinodal decomposition. However, the strain term, $2\eta^2 Y$, is large in this system and the coherent spinodal points occur at temperatures on the order of 400–800°C below the spinodal points.

The time dependence of the spontaneous decomposition may be estimated by simply determining the time it would take an atom to diffuse a distance of $\lambda_m/2$. Utilizing Eq. 6.49 we find

$$t \approx \frac{(\tfrac{1}{2}\lambda_m)^2}{6|D|} = \frac{\lambda_m^2}{24|D|} \tag{11.49}$$

Taking λ_m as 100 Å we have a decomposition time of $10^{-14}/|D|$. Hence, it is apparent that one will only be able to suppress the spinodal by rapid quenching in solids where the diffusion coefficient is smaller than around 10^{-14} cm^2/sec. The rate of precipitation via spinodal decomposition is discussed further in Refs. 12 and 13, where it is shown how one may derive a simple time-temperature-transformation diagram for this precipitation process. Conclusive evidence for the operation of the spontaneous spinodal decomposition mechanism exists for only a few alloy systems. However, it seems quite likely that in many cases the GP zones form by this process.[14]

C. SUMMARY

The precipitation sequence is summarized in general terms in Table 11.4. The term ΔG_B is the driving force for the precipitation reaction and it must be negative in order to cause precipitation. This free-energy change is a function of how far the alloy is quenched below the solvus temperature. At high temperatures only the equilibrium precipitate forms, because it is the only precipitate having a negative ΔG_B. At low temperature the driving force ΔG_B is largest for the equilibrium precipitates. However, the two factors, strain energy E_s and surface energy γ, tend to inhibit nucleation of the equilibrium precipitate. From the discussions in Chapter 8 [see Fig. 8.8(b) and Eq. 8.10] it is apparent that the time required to reach a given stage of precipitation (i.e., to precipitate N nuclei) is proportional to $\exp[\Delta G^*/kT]$. Because of its large surface energy, γ, the equilibrium precipitate has a relatively high ΔG^* value (see Eq. 11.23) and its formation is frequently sluggish compared to the intermediate precipitate and/or the GP zone. At a given quench temperature that precipitate having the lowest ΔG^* would be expected to form first. The ΔG^* for the

Table 11.4 Summary of the Major Factors Involved in the Precipitation Sequence

| Precipitate | Driving Force, ΔG_B | | Factors Inhibiting Nucleation | | Mode of Formation |
	Low Temp. Inside Spinodals	High Temp. Just Below Solvus	Strain Energy, E_s	Surface Energy, γ	
GP zone	Small negative	Positive	Low (Frequently along certain $\langle hkl \rangle$ dir.)	Low	Spinod. decomp. or homog. nucl. and growth
Intermediate	Intermediate negative	Positive	Med–high (can be lowered by precipitation on dislocations)	Low–medium	Heterogeneous nucl. and growth
Equilibrium	Large negative	Negative	Low	High [can be lowered by precipitation on surfaces (grain bdries)]	Heterogeneous Nucl. and growth

intermediate precipitate may be lowered by reducing the strain energy or by heterogeneous nucleation on dislocations or interfaces (see Eq. 11.24). The precipitate that forms will be the one most effective at reducing ΔG^* either by reducing strain energy, E_s, or by reducing surface energy, γ, and in both cases these energy contributions may be reduced by heterogeneous nucleation on interfaces. Since these intermediate precipitates generally nucleate heterogeneously on dislocations, it is apparent that their rate of formation will be influenced by the dislocation density in the sample. The GP zones form either by spontaneous spinodal decomposition or by homogeneous nucleation with a very low ΔG^* and, therefore, form quite rapidly at the low temperature. Consequently, they precede the appearance of intermediate precipitates even though they have a lower driving force.

11.4 KINETICS OF PRECIPITATION REACTIONS

A. INITIAL FORMATION

Several investigators have examined the kinetics of the formation of GP zones (see Ref. 8, p. 172 or Ref. 15, p. 110). Since these zones form by a diffusional process one can estimate a value for the diffusion coefficient required to cause the solute atoms to migrate into the clusters in the times observed for cluster formation. From experiments measuring the rate of zone formation in Al–2 at. % Cu quenched from 520°C (solution temperature) to room temperature, 27°C, Fine[15] has shown that the average Cu atom migrates 4×10^{-7} cm in 3 hours. Hence, from Eq. 6.49 we estimate D as

$$D = \frac{\overline{R_n^2}}{6t} = \frac{(4 \times 10^{-7})^2}{36 \times 3600} = 2.8 \times 10^{-18} \text{ cm}^2/\text{sec}$$

The value for the diffusion coefficient of Cu in Al has been measured by conventional techniques and found to be 2.3×10^{-25} cm^2/sec at 27°C. Hence, it is apparent that if the GP clusters form by diffusion of Cu atoms, the diffusion coefficient must be larger than that measured in conventional isothermal diffusion measurements by a factor of $2.8 \times 10^{-18}/2.3 \times 10^{-25} = 1.2 \times 10^7$.

Since Cu is a substitutional solute in Al one expects it to migrate by a vacancy mechanism. Consequently Eq. 6.56 should apply,

$$D = (\text{const}) \exp\left(-\frac{\Delta E + \Delta E_v}{kT}\right) \tag{11.50}$$

where ΔE is the activation energy for Cu migration and ΔE_v is the energy per vacancy. But from Chapter 5 we know that the equilibrium number of vacancies, n_v, is a strong function of temperature, given in Eq. 5.9 as $n_v = (\text{const}) \exp[-\Delta E_v/kT]$. Therefore, we can write Eq. 11.50 as

$$D = (\text{const}) \cdot n_v \cdot e^{-\Delta E/kT} \tag{11.51}$$

This equation suggests the cause of the above descrepancy. The GP zones form immediately following a quench from the high solution temperature. After a rapid quench from 520°C the number of vacancies present in the alloy will be much higher than the equilibrium number at the 27°C quench temperature and, consequently, the diffusion coefficient will be higher than expected. Assuming no vacancy decay after quenching, the excess number of vacancies may be taken as

$$\frac{n_v(520°C)}{n_v(27°C)} = \exp\left[\frac{\Delta E_v}{k}\left(\frac{793 - 300}{793 \times 300}\right)\right]$$

From the measured value of $\Delta E_v = 23,000$ cal/mole this ratio is found to be about 10^{10}. Since there will be some immediate vacancy decay upon quenching it is reasonable to expect the diffusion coefficient to be higher than the equilibrium 27°C value by the required amount of 1.2×10^7. Other experimental work[8] gives similar results. Hence, it is fairly well established that the initial rates of zone formation are controlled by vacancy concentration. Higher vacancy concentrations and increased vacancy lifetime both give rise to a larger average zone size at the termination of the initial growth period. One expects a rapid growth rate only initially because the vacancies decay exponentially with time after the quench and cause D to revert to its low value. However, after the initial growth rate period the zones continue to grow at a rate higher than predicted by the above type of analysis. The reader is referred to p. 174 of Ref. 8 for a discussion of two possible causes for this higher secondary growth rate.

B.   PARTICLE COARSENING

As we will see in Section 11.5 the strength of precipitation-hardenable alloys is directly related to the particle size of the coherent precipitates that are formed. Therefore, after the initial formation of the coherent precipitates it is very desirable to be able to inhibit further growth. This is extremely important in alloys that are used at high temperatures, such as in the hot stages of jet engines, because at these temperatures particle growth is difficult to avoid. The problem of particle coarsening was treated in 1900 by W. Ostwald and is often referred to as Ostwald ripening. The modern theory of particle coarsening was developed by Lifshiftz and Slyozov[16] and by C. Wagner.[17] G. W. Greenwood[18] has reviewed the theory and also has given a simplified treatment that will be presented here.

After the initial period of growth the precipitate phase closely approaches the volume fraction that one would predict from the phase diagram using the lever law. After this time, growth does not stop, but proceeds by a process in which the larger particles grow at the expense of the smaller ones in an effort by the system to reduce the surface potential γA (similar in these respects to normal grain growth). Consequently, the volume fraction precipitate phase remains essentially constant so that one may write

$$\sum_{\substack{\text{All particle} \\ \text{sizes}}} \left(\frac{\text{Rate atom loss}}{\text{Particle}} \right) = 0 \qquad (11.52)$$

If we take V as the volume of a particle and $\bar{V}$ as the volume per atom in the particle, then the rate of atom loss from a particle is $(dV/dt)(1/\bar{V})$. For a sphere $dV/dR = 4\pi R^2$, so that Eq. 11.52 becomes

$$\sum_i \frac{4\pi R_i^2}{\bar{V}} \frac{dR_i}{dt} = 0 \qquad (11.53)$$

The rate of atom loss from a particle is controlled by either the rate of transfer across the particle–matrix interface or by diffusion away from the particle into the matrix. The Lifshiftz–Wagner theory has treated both these cases, and experimental results[18,19] indicate that the growth of coherent precipitates is diffusion controlled. It is assumed that the particle is a single component (i.e., pure element). The rate of atom loss by diffusion from a pure spherical partical of radius R may be written as

$$\text{Rate} = 4\pi R^2 (-D)\left(\frac{dC}{dr}\right)_s \qquad (11.54)$$

where D is the diffusion coefficient in the matrix and $(dC/dr)_s$ is the radial concentration gradient of the precipitate atoms in the matrix at the particle–matrix surface. The concentration at this surface is increased above the equilibrium value because of the surface curvature. Since it is assumed that the particle is a pure component the surface concentration C_s may be taken from Eq. 11.19 as

$$C_s = C\left[1 + \frac{2\bar{V}\gamma}{kT \cdot R}\right] \qquad (11.55)$$

where C is the equilibrium concentration if the interface were flat and we have changed the fractional concentrations of Eq. 11.19 to volume concentrations by simply multiplying both sides by molar density. It is apparent from Eq. 11.55 that the concentration at the surface of a small particle will be increased more than at the surface of a large particle, so that diffusion will proceed from small to large particles. The concentration rise at the surface is quite small so that one can assume that the particles advance at near steady-state conditions. It is shown in Problem 11.5 that under steady-state conditions the radial concentration gradient at the surface of a sphere of radius R becomes

$$\left(\frac{dC}{dr}\right)_s = \frac{C_s - C_0}{R} \qquad (11.56)$$

where C_0 is the concentration far from the surface. Combining Eq. 11.56 with Eq. 11.54 and equating to the rate of atom loss from an individual particle (i.e., one term of Eq. 11.53) we obtain

$$\frac{dR_i}{dt} = \frac{-\bar{V}D(C_s - C_0)}{R_i} \qquad (11.57)$$

If Eq. 11.57 is substituted into Eq. 11.53 one may obtain

$$(C_0 - C)\sum R_i = nC\frac{2\bar{V}\gamma}{kT} \qquad (11.58)$$

where n is the number of particles that one sums over. The average particle size $\bar{R}$ is $\sum R_i/n$, so that with a little algebra one obtains

$$C_0 - C_s = \frac{2\gamma\bar{V}}{kT}C\left[\frac{1}{\bar{R}} - \frac{1}{R}\right] \qquad (11.59)$$

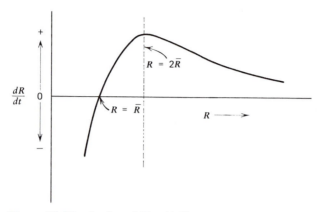

Figure 11.28   A plot of Eq. 11.60.

Substituting into Eq. 11.57 the following differential equation results:

$$\frac{dR}{dt} = \frac{2\gamma \bar{V}^2 DC}{kT}\frac{1}{R^2}\left[\frac{R}{\bar{R}}-1\right]$$

(11.60)

When this equation is plotted one obtains Fig. 11.28. The plot shows that particles of radius $R < \bar{R}$ are dissolving at a rapidly increasing rate. Also once a particle has $R > 2\bar{R}$ its growth rate slows down relative to some smaller particles. Consequently, from this simple theory of Greenwood[18] one would not expect particles of $R > 2\bar{R}$ to persist in the system. Note also that if all particles had the same size (namely, $\bar{R}$) there would be no net growth. If one solves Eq. 11.60 for particles of maximum size, $R = R_{max} = 2\bar{R}$,

$$R_{max}^3 = (R_{max})_0^3 + 6\left[\frac{\bar{V}^2\gamma CD}{kT}\right]t$$

(11.61)

This equation would require that particles of size $2\bar{R}$ be present in the initial distribution. The mean particle size will grow at a slower rate than predicted by Eq. 11.61. This simple theory cannot, however, determine the mean size growth rate.

In the Liftshiftz–Wagner theory a statistical analysis is made that accounts for the fact that a distribution of particle sizes exists in the system. The theory predicts that a "quasi-steady-state" particle size distribution is approached independently of the original size distribution. The predicted distribution has a very limited range and indicates that particles larger than $1.5\bar{R}$ should not exist, which is smaller than the prediction of the simple theory. The predicted time dependence of the mean radius is found to be

$$\bar{R}^3 = \bar{R}_0^3 + \left[\frac{8}{9}\frac{\bar{V}^2\gamma CD}{kT}\right]t \tag{11.62}$$

where $\bar{R}_0$ is the original mean particle size at onset of coarsening. Notice that if one substitutes $\bar{R} = 31R/27$ into Eq. 11.60 and integrates, the result is Eq. 11.62 in terms of R. Hence, the above simple treatment illustrates basicly how the form of Eq. 11.62 comes about.

Experimental results[18,19] have shown that the growth of coherent particles can, in a number of cases, be described quite well by Eq. 11.62. A limiting size distribution is found, but the particle size distribution is not quite as narrow as predicted. Li and Oriani[3] have presented modifications to the above theory for cases where the precipitate is not a pure element and not spherical.

In conclusion, it can be seen that since R_0 is quite small for coherent precipitates they will coarsen their radius as the one-third power of time. The average particle radius will also increase as the one-third power of the particle–matrix surface tension, γ, and as the one-third power of the equilibrium concentration of the precipitate atoms in the matrix phase, C. Hence, the theory predicts that coherent precipitates should coarsen more slowly than incoherent precipitates, and coarsening may be reduced by reducing the solubility of the particle atoms in the matrix.

C. OTHER PRECIPITATION SEQUENCES

The precipitation sequence described in Section 11.2C is illustrated schematically at the top of Fig. 11.29. This sequence is characterized by the successive formation and dissolution of precipitates, this latter process often termed reversion.

There are, however, other sequences as indicated in Table 11.1 that occur during precipitation, and some of these are found in the commercially important nickel-base superalloys. (The following account is based on Ref. 20.) The precipitates in these systems are ordered and characterized by a low value of γ at the precipitate–matrix interface. The sequence is shown by the middle row of Fig. 11.29. In this case metastable precipitates do not precede formation of the equilibrium precipitate. Rather, the equilibrium precipitate forms initially as a random array of spherical particles with a coherent precipitate–matrix interface. As the precipitates grow two interesting things occur. (1) The shape of the particles changes first to cubes and at later stages the cubes become rods or plates. (2) The cubes tend to line up along the $\langle 100 \rangle$ matrix directions and the rods and plates display a pronounced periodicity, and consequently are termed modulated structures or sometimes tweed structures. (See Fig. 3 of

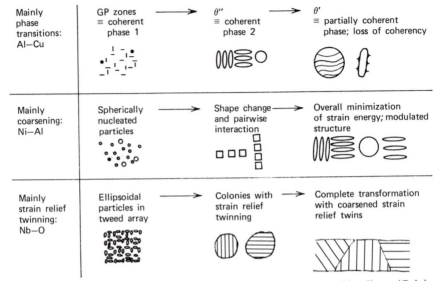

Figure 11.29   Three classes of precipitation sequences observed in alloys. (Originally published by the University of California Press; reprinted by permission of The Regents of the University of California, Ref. 20).

Ref. 21 for beautiful electron micrograph illustration.) It has been shown that these structures do not form by spinodal decomposition,[21] so that we have here an example of a modulated structure formed by a mechanism other than spinodal decomposition. Theoretical treatments (Ref. 20, p. 519) indicate that strain is minimized by formation of the periodic arrays. The theories conclude that coherent precipitation involving coherency strains and higher volume fractions of precipitate will always produce a quasiperiodic array in an effort to minimize the strain energy of the system.

A third precipitation sequence is found to occur as illustrated in the bottom row of Fig. 11.29. This precipitation sequence occurs predominantly when the precipitate is ordered and when the symmetry of the precipitate crystal differs from that of the matrix, for example, formation of a tetragonal matrix from a cubic matrix. This sequence has been observed in a number of substitutional alloys, Cu–Au, Co–Pt, Fe–Pt, and Ni–Co, but it is particularly characteristic of interstitial ordered precipitates, V–N, Nb–O, Ta–C, Ta–N, and Ta–O. Here again the initial coherent precipitate is the equilibrium phase and the sequence description of Fig. 11.29 is primarily a description of the evolution of the growth morphology of the precipitate. In these cases it is often observed that as the precipitates grow they develop a densely twinned microstructure. This twinning is thought to develop because the shape change between, for example, the tetragonal precipitate and cubic matrix, can thereby be reduced. This phenomenon is discussed at some length in connection with martensite formation (see p. 477–478).

It should be clear from the above discussion that the course of precipitation in any given system will be varied and also quite fascinating. The schemes of Fig. 11.29 present only a rough guide to a multitude of transformation sequences that may be possible.

D. OVERALL KINETICS

One can try to describe the kinetics of the precipitation process with rate equations such as the Johnson–Mehl and Avrami equations (Eqs. 10.29 and 10.30). However, a quantitative description of the entire precipitation process is extremely difficult because of the complexity and combinations of processes occurring simultaneously during the precipitation sequence. Each system is somewhat unique. However, one can find certain qualitative generalities common to all precipitation processes and Newkirk[22] offers the following five rules as a guide.

1. Rates of precipitation are greater the higher the aging temperature.

2. Precipitation is faster at a given temperature in low-melting than in high-melting alloys.

3. Damaging the matrix lattice by radiation or cold work after the solution treatment and before aging accelerates precipitation.

4. The reaction is more rapid in systems composed of widely dissimilar metals.

5. The presence of soluble or insoluble impurities usually accelerates precipitation.

11.5 PRECIPITATION HARDENING

As was emphasized in Chapter 4, the strength of a metal is controlled by the generation and mobility of dislocations. The increased strength of an age-hardened alloy is due to the interaction of the dispersed precipitate phase with dislocations. Consequently, precipitation of a coherent second phase provides the metallurgist with perhaps the most versatile strengthening mechanism at his disposal. To understand the nature of this strengthening mechanism one must examine the interaction of dislocations with a dispersed second phase. This is a very complex subject[23] and only some of the more important basic ideas can be presented here.

The interaction of the glissile dislocations with the dispersed particles will increase the critical resolved shear stress by an amount we shall term $\Delta \tau$. Theories attempt to determine $\Delta \tau$ as a function of the parameters that characterize the dislocation-particle interaction. One can break down the dislocation-particle interactions into three groups, depending on how the dislocation manages to penetrate the dispersed particles. The dislocations can (1) loop the particles, (2) cut through the particles, or (3) cross slip around the particles. Mechanism (3) will not be discussed here.

A. PARTICLE LOOPING

The basic idea of particle looping may be understood by reference to Fig. 11.30. This figure shows how a dislocation will bow out when it is held back at the points where it encounters precipitate particles. As the applied shear stress is increased the dislocation bows out sufficiently so that it begins to meet at points such as A and B at time t_2 on Fig. 11.30. Notice that the sense of the dislocation at A will be opposite to that at B. Consequently, when these dislocation segments meet they will annihilate, causing the main dislocation to separate from the looped region as shown at t_3 of Fig. 11.30. Every time a dislocation passes the precipitate particles by this looping mechanism it leaves one loop around each precipitate as shown at t_4 of Fig. 11.30. This mechanism was originally presented by E. Orowan in 1948[24] and is often referred to as the Orowan mechanism.

It was shown in Chapter 4 (Eq. 4.17) that because the dislocation possesses a line tension T, the shear stress required to bend it to a radius of curvature R is given as $T/(bR)$, where b is the Burgers vector of the dislocation. The shear stress that must be applied to force the dislocation past the particles may therefore be written as

$$\Delta\tau = \frac{T}{bR_{min}} \tag{11.63}$$

where R_{min} is the minimum average radius of curvature that the dislocation will possess in going through the various configurations shown in Fig. 11.30. Just as with the Frank–Reed generator (see p. 119) the dislocation

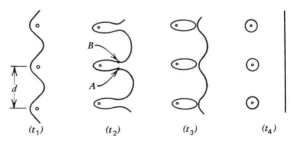

Figure 11.30   Interaction of a dislocation line with a row of precipitate particles.

achieves a minimum average radius of curvature when it becomes semicircular with a radius of half the particle spacing, that is, $R_{min} = d/2$. Substituting the expression for line tension of Eq. 4.16 without the constant term we obtain for $\Delta\tau$

$$\Delta\tau = \frac{1}{2\pi K}\frac{Gb}{d}\ln\left[\frac{d}{2r_0}\right] \qquad (11.64)$$

It was given on p. 210 that for a random distribution of particles having volume fraction f and radius r, the number of particles intercepted by 1 cm^2 of surface is $3f/(2\pi r^2)$. Therefore the average distance between particles assuming a simple cubic array would be given as

$$d = \left(\frac{2\pi}{3f}\right)^{1/2}r \qquad (11.65)$$

Combining Eqs. 11.64 and 11.65 we obtain for the looping mechanism

$$\Delta\tau = \frac{\sqrt{3}}{(2\pi)^{3/2}K}\frac{Gbf^{1/2}}{r}\ln\frac{d}{2r_0} \qquad (11.66)$$

A recent detailed analysis of this mechanism by Ashby[25] has yielded the following expression for $\Delta\tau$,

$$\Delta\tau = (\text{const})\frac{Gbf^{1/2}}{r}\ln\frac{2r}{r_0} \qquad (11.67)$$

where const is 0.093 for edge dislocations and 0.14 for screw dislocations, assuming $\nu = \frac{1}{3}$. It is apparent that the strengthening due to this looping mechanism becomes extremely large as the particle spacing d, and hence the particle radius r, becomes small. Furthermore, at large spacing and therefore large r the $\Delta\tau$ from this mechanism becomes small.

B. PARTICLE CUTTING

In the above mechanism it was assumed that the repulsive force between the dislocation and the particle was sufficiently large that the dislocation was stopped at the points of direct particle-dislocation contact. However, it is possible that the dislocation could glide right through the precipitate particle. This would cause the particle to be offset across the glide plane by one b vector as shown in Fig. 11.31. Hence, the particle is, in a sense, cut by the dislocation. There are numerous possible interaction mechanisms for this process and the more important ones will be discussed here. It is convenient to subdivide these mechanisms into two categories. If the dislocation-particle interaction distance is shorter than around $10 \cdot b$ we call it a short-range interaction, and if greater than $10 \cdot b$ a long-range interaction.

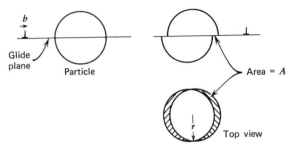

Figure 11.31 The "cutting" of a particle by glide motion of a dislocation.

1. Short-Range Interactions. The action in this case is illustrated by Fig. 11.31. Any dislocation-particle interaction mechanism causing strengthening increases the work required to cut a unit area of the glide plane occupied with particles, relative to the same area without the particles, by an amount $\Delta\tau \cdot b$ (see p. 98).

Suppose the precipitate particle is ordered. Then when the dislocation cuts the particle it will produce an antiphase boundary (APB) at the glide plane (see Problem 4.2). The energy γ_A of an APB is not negligible. For example, ordered coherent precipitates have precipitate-matrix energies of only 10–30 ergs/cm^2 compared to their APB energies of $\gamma_A \approx 100$–300 ergs/cm^2. If the radius of the particle on the glide plane, r, is large compared to b, we may take the area of the cut particle as πr^2 and write $\Delta\tau b = n\pi r^2 \gamma_A$ where n is the number of particles per area. Utilizing $n = 3f/(2\pi r^2)$ from above we obtain

$$\Delta\tau = \frac{3f}{2b}\gamma_A \qquad (11.68)$$

In this simple approach we have neglected things such as the effect of dislocation pairs and dislocation bow out, and the more complete theoretical treatment[23] gives

$$\Delta\tau = 0.28 \frac{\gamma_A^{3/2} f^{1/3}}{\sqrt{G}\, b^2} r^{1/2} \qquad (11.69)$$

A second mechanism that can be considered here involves the area A shown cross-hatched on Fig. 11.31. This area is new surface area generated between the particle and matrix by the cutting action. It can be shown that this area is approximately $A = 2rb$, and if we take the interface energy of this area as γ_s we have

$$\Delta\tau = \frac{1}{b}\left[\frac{3f}{2\pi r^2}\right] \cdot A \cdot \gamma_s = \frac{3}{\pi} \frac{f\gamma_s}{r} \qquad (11.70)$$

A more complete treatment of this mechanism allowing dislocation bowout[26] gives

$$\Delta\tau = \frac{1.1}{\sqrt{\alpha}} \frac{\gamma_s^{3/2} f^{1/2}}{Gb^2} r^{1/2} \tag{11.71}$$

where α is a function of the dislocation line tension and equals $a \cdot \ln(d/r_0)$ where a is 0.16 for edge and 0.24 for screw dislocations.

There are a number of other possible interactions that must be considered in a complete treatment.[23] For example, if the glide plane that the dislocation follows through the particle is not coplanar with its glide plane in the matrix, some sort of jog will be formed at the particle-dislocation interface. The motion of this jog can give rise to a contribution to $\Delta\tau$. If the Peierls force in the precipitate differs from the matrix, the dislocation may experience a different frictional resistance within the particle which would contribute to $\Delta\tau$. This contribution is expected to go as $f^{1/3} r^{1/2}$. It is apparent that the overall picture of the short-range interaction is quite complex but it is expected to vary with volume fraction f and particle radius r as $f^{1/3 \text{ to } 1/2} \cdot r^{1/2}$.

2. Long-Range Interactions. As the dislocation approaches a particle its strain field interacts with the strain field produced in the matrix by the particle. This retarding force experienced by the dislocation would be governed by the Peach–Koehler equation (Eq. 4.10). The contribution to $\Delta\tau$ from this mechanism has been calculated to be[23]

$$\Delta\tau = \left[\frac{27.4 E^3 \varepsilon^3 b}{\pi T (1+\nu)^3}\right]^{1/2} f^{5/6} r^{1/2} \tag{11.72}$$

where E is Young's modulus, T is the line tension, ν is Poisson's ratio, and ε is a function of the disregistry δ (see Eq. 7.7). After a dislocation overcomes the strain field barrier it is still necessary for it to cut the particle. Hence, the long-range interaction will only be important if its contribution to $\Delta\tau$ exceeds that from the short-range interactions.

C. SUMMARY

When the looping mechanism operates, the increment to the yield stress predicted by the Orowan model goes as $\Delta\tau = \alpha f^{1/2} r^{-1}$. For a system of constant volume fraction particles this curve has the form shown by curve A on Fig. 11.32. In principle, this curve rises with decreasing particle size until the theoretical CRSS is obtained. The mechanism of particle looping sets an upper limit to the strengthening effect caused by the particles. It is apparent from the above discussion that the strengthening increment from the more complex case of particle cutting will follow an equation of the

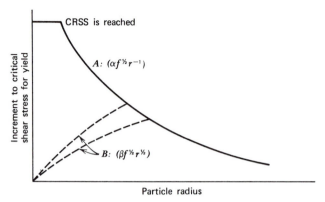

Figure 11.32   A plot of the strengthening increment equations versus particle radius.

form $\beta f^{1/2} r^{1/2}$, where the value of β and the exponent on f will depend on the controlling dislocation-particle interaction mechanisms for any individual system. Two possible curves for the particle cutting increment are shown as B on Fig. 11.32. Particle looping will only be possible if the dislocations cannot cut through the particles. Therefore as the particle radii increase from zero, one expects the yield stress increment to follow curve B until it intersects curve A. At this point looping becomes easier than cutting and the strengthening falls as the radii coarsen.

We may now summarize the main features of the strengthening that is achieved in age-hardenable alloys. Upon aging, the coherent particles (GP zones and in some cases intermediate precipitates) form and grow as their volume fraction f increases to an equilibrium value. After this time the volume fraction remains essentially constant and the particles coarsen, thus causing the mean particle radius to increase. Initially the particles are too small (or alternatively too close together) for looping to occur and yielding is controlled by particle cutting mechanisms. The strength increases because both f and r are increasing. After a short time f becomes constant but the particles continue to grow in size by the coarsening mechanism. The strength continues to increase but at a slower rate. Eventually the particles become large enough (or sufficiently far apart, $d \approx r\sqrt{2\pi/3f}$) to allow particle looping and the strength falls off. In general, to age harden an alloy effectively it is necessary to be able to produce precipitates with a spacing of less than about 1 μm. Furthermore it is found that the precipitates must be coherent or partially coherent to be effective.[8] This apparently relates to the higher surface energy of the noncoherent precipitate. Such precipitates nucleate heterogeneously with a larger initial size. In addition the high γ value increases the coarsening kinetics.

One may control the maximum achievable strength by controlling the volume fraction particles f, the particle radius r, and the dislocation-particle interaction mechanism that determines β. The volume fraction particles is limited by the phase diagram for a given alloy and the particle size is controlled by the heat treatment. The dislocation-particle interaction parameter β is most strongly influenced by particle strain fields that depend on δ and r, but it is also influenced by cutting mechanism parameters such as APB energy in ordered precipitates or stacking fault energy in, for example, fcc systems with extended dislocations. At higher volume fractions (e.g., f approaching 0.5) some of the above theoretical models do not apply but the overall picture is pretty much the same (Ref. 20, p. 526).

The above discussion has been limited to the effect of the precipitates upon the yield stress, but it is also found that the presence of a dispersed second phase will have an influence upon the work hardening rate of alloys. In general, the precipitate phase increases the work hardening rate. This effect will not be considered here but a detailed discussion may be found in Ref. 8 and summary discussions in Refs. 20 and 23.

D.   EXAMPLES

1.   Nickel-Base Superalloys. These alloys are essentially a Ni–Cr solid solution hardened by precipitation of Al and Ti to precipitate the γ' phase, $Ni_3(Al, Ti)$. The γ' phase is an ordered fcc structure that is coherent with the fcc Ni matrix (γ phase). The coherent boundary energy is only around 13–14 ergs/cm^2 compared to an APB energy of 110 ergs/cm^2. These alloys exhibit remarkable resistance to coarsening and this is one of the main reasons they may be used at the high temperatures encountered in the hot stages of jet engines. The resistance to coarsening is due to the very low γ of the coherent precipitant–matrix boundary. This reduces the rate of coarsening (see Eq. 11.62) but it also prevents particle coalescence because when two particles meet they would probably form an APB at their intersection. Since the APB energy is more than the sum of the energies of the two boundaries that would annihilate upon coalescense, the particles do not coalesce.[20]

As an example of these alloys the composition of Udimet 700 is listed in Table 11.5. This alloy is single phase γ above around 2100°F (1150°C) and γ' precipitates upon cooling. The table lists the composition of the γ and γ' phases and it is apparent that the Al and Ti partition to the γ' phase, although it still contains some Cr and Mo and significant amounts of Co. A recent study[28] has shown that the coarsening of the γ' phase in the alloy follows the Lifshiftz–Wagner equation. The dislocation-particle interaction mechanism primarily responsible for the age hardening in these alloys is

Table 11.5   Composition of Udimet 700 and its γ and γ' Phases

	Cr	Co	Al	Ti	Mo	C	Ni
Udimet 700	16.4	17.6	9.0	3.9	2.9	0.3	49.9
γ	24.3	23.5	5.3	1.5	3.9	—	41.5
γ'	2.7	8.0	13.9	8.0	0.9	—	66.5

Originally published by University California Press; reprinted by permission of The Regents of the University of California, Ref. 27.

the high APB energy.[27] A great deal of research has been devoted to the development of these alloys in the past 15 years and at present there are some 93 nickel-base superalloys available (see Ref. 29 for a listing of these alloys and their characteristic stress-rupture data). This work illustrates a fascinating example of the use of many different principles of physical metallurgy in alloy design and the student is referred to Refs. 27 and 30–32 for accounts of the factors involved.

2. Major Precipitation-Hardening Alloys. The strengthening mechanism resulting from precipitation is often present to some extent even in heat-treatable alloys that achieve their major strengthening on heat

Table 11.6   Mechanical Properties of Several Age-Hardenable Alloys Showing the Effects of the Aging Treatment

Alloy Designation	Composition	Condition	Yield Strength (psi)	(MN/m²)	Tensile Strength (psi)	(MN/m²)	Elong. (%)
Aluminum-	5.6 Zn, 2.5 Mg	Annealed	15,000	103	33,000	228	16
7075	1.6 Cu, 0.25 Cr Bal. al	Aged	73,000	503	83,000	572	11
Titanium-	6 Al, 4 V, Bal.	Annealed	128,000	882	138,000	951	12
Ti-6 Al-4 V	Ti	Aged	155,000	1069	170,000	1172	8–15
Beryllium	1.9 Be, 0.25 Co,	Annealed	32,000	221	69,000	476	47
copper-172	Bal. Cu	Aged	155,000	1069	177,000	1220	7
Inconel-718	19 Cr, 19 Fe,	Annealed	110,000	758	146,000	1010	
(Ni-base	3 Mo, 0.8 Ti,	Aged	188,000	1296	200,000	1380	20
superalloy)	0.6 Al, 5.2 Cb, Bal. Ni						
Stainless steel	17 Cr, 7 Ni,	Annealed	47,000	324	129,000	889	39
17-7 PH	1.2 Al, 0.07 C, Bal. Fe	Aged	185,000	1276	200,000	1380	9
Maraging steel	0.8 Ni, 8 Co,	Annealed	94,000	648	128,000	882	18
	5 Mo, 0.4 Ti, Bal. Fe	Aged	222,000	1530	228,000	1572	11

treatment from other mechanisms. For example, additional strengthening is obtained from carbide precipitation reactions in tempered steels. Also, certain hot-rolled low alloy steels with yield strengths in the 42,000–80,000 psi (300–552 MN/m²) range achieve a component of their strength from carbide precipitates.[33] However, there are a number of heat-treatable alloys that depend chiefly on precipitation strengthening for the added strength produced by heat treatment. Some of the more dramatic examples of important commercial alloys of the three base metals Al, Cu, and Ti are shown in Table 11.6, which illustrates the beneficial effect of this strengthening mechanism upon the yield and tensile strengths and the concurrent sacrifice in ductility. An example of a Ni-base superalloy, Inconel 718, is also included. Optimum aged properties in this class of alloys often require a double or triple aging process,[34] and added strength is obtained by hot working after one of the aging treatments.[35] In addition, two steels are included in Table 11.6; these steels achieve a large fraction of their added strength upon heat treatment from precipitation reactions. In both steels the precipitate occurs after a martensitic phase transformation, and these steels will be discussed further in Chapter 15. (See Ref. 22 for discussions of different classes of major precipitation-hardening alloy systems.)

One can strengthen an alloy by mechanically mixing into the alloy a dispersion of small inert particles. In Chapter 7 we discussed the effects of the dispersion of thoria particles on grain growth. In recent years considerable research has been directed at adding dispersions of very fine oxide particles to high-temperature alloys (see Ref. 36). Techniques are available to achieve dispersions of thoria having particle sizes in the range of

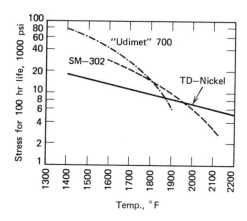

Figure 11.33 Stress rupture data as a function of temperature for TD nickel and two superalloys (from Ref. 32).

100–500 Å. The most common example of these materials is called T-D Nickel, which is nickel containing a thoria dispersion. T-D nickel as well as other thoria dispersed alloys have stress-rupture properties that are superior to superalloys at very high temperatures. This is illustrated in Fig. 11.33 where T-D nickel is compared to the wrought Ni-base superalloy, Udimet 700, as well as to a cast Co-base superalloy, SM 302. To optimize the properties of dispersion strengthened alloys it is necessary to subject them to a series of deformation and recovery treatments. Because of the thoria particles present in these alloys, they resist recrystallization at the high temperatures and therefore retain a high dislocation density. Therefore, the source of the high-temperature strength may be due in large part to the deformation structure in addition to dislocation-oxide interactions.[30]

REFERENCES

1.   L. S. Darken and R. W. Gurry, *Physical Chemistry of Metals*, McGraw-Hill, New York, 1953.

2.   J. Mackowiak, *Physical Chemistry for Metallurgists*, Allen and Unwin, London, 1966.

3.   C. Y. Li and R. A. Oriani, in *Oxide Dispersion Strengthening*, G. S. Ansell, T. D. Cooper, and F. V. Lenel, Eds. Metallurgical Society Conference, Vol. 47, Gordon and Breach, New York, 1968, p. 431.

4.   H. Y. Hunsicker and H. C. Stumpf, in *The Sorby Centennial Symposium on the History of Metallurgy*, C. S. Smith, Ed., Metallurgical Society Conference, Vol. 27, Gordon and Breach, New York, 1965, pp. 271–330.

5.   P. D. Merica, R. G. Waltenberg, and H. Scott, Trans. Met. Soc. AIME **64,** 41 (1920).

6.   Metallurgical Classics, Trans. ASM **61,** 367 (1968). Commentary by J. B. Newkirk.

7.   R. F. Mehl, in Ref. 4, pp. 245–269.

8.   A. Kelly and R. B. Nicholson, Prog. Mat. Sci. **10,** No. 3, 1963, p. 151.

9.   J. W. Christian, *The Theory of Transformations in Metals*, Pergamon, New York, 1965.

10.   V. A. Phillips, Acta Met. **21,** 219 (1973).

11.   K. C. Russell, in *Phase Transformations*, American Society of Metals, Park, Ohio, 1970, pp. 219–268.

12.   J. C. Cahn, Trans. Met. Soc. AIME **242,** 166 (1968).

13.   J. E. Hilliard, in Ref. 11, pp. 497–560.

14. J. C. Cahn, in *The Mechanism of Phase Transformations in Crystalline Solids*, Institute of Metals Monograph No. 33, Institute of Metals, London, 1969, p. 31.

15. M. E. Fine, *Phase Transformations in Condensed Systems*, MacMillan, New York, 1964.

16. J. M. Lifshiftz and V. V. Slyozov, J. Phys. Chem. Solids **19**, 35 (1961).

17. C. Wagner, Z. Elektrochem. **65**, 581 (1961).

18. G. W. Greenwood, in Ref. 14, pp. 103–110.

19. A. J. Ardell, in Ref. 14, pp. 110–116.

20. H. Warlimont, in *Electron Microscopy and Structure of Materials*, Ed. by G. Thomas, R. M. Fulrath, and R. M. Fisher, Eds., Univ. Calif. Press, Berkeley, 1972, pp. 505–533.

21. A. J. Ardell and R. B. Nicholson, Acta Met. **14**, 1295 (1966).

22. J. B. Newkirk, in *Precipitation from Solid Solution*, published by American Society of Metals, Park, Ohio, 1959, pp. 6–149.

23. H. Gleiter and E. Hornbogen, Mat. Sci. Eng. **2**, 285 (1967–68).

24. E. Orowan, in *Symposium on Internal Stresses in Metals and Alloys*, Institute of Metals, London, 1948, p. 451.

25. M. F. Ashby, in Ref. 3, pp. 143–205.

26. N. J. Olson, The Affects of Precipitation on the Mechanical Properties of the Cu-1.9 Co and Cu-2Co-6Zn Systems, Ph.D. Thesis, Iowa State Univ., 1970.

27. J. M. Oblak and B. H. Kear, in Ref. 20, pp. 566–616.

28. E. H. Van der Molen, J. M. Oblak, and O. H. Kriege, Met. Trans. **2**, 1627 (1971).

29. Metals Progress, **103**, 56 (March 1973).

30. R. W. Cahn, J. Metals **25**, 28 (Feb. 1973).

31. R. F. Decker, Strengthening Mechanisms in Nickel Base Superalloys, Steel Strengthening Mechanisms, Climax Moly. Co., Greenwich, Conn., 1969, p. 147; see also Met. Trans. **4**, 2495 (1973).

32. R. F. Decker and R. R. DeWitt, J. Metals **18**, 139 (Feb. 1965).

33. W. C. Leslie, J. Met. **23**, 31 (Dec. 1971), see p. 37.

34. J. A. Burger and D. K. Hanink, Met. Prog. **92**, 61 (July 1967).

35. B. H. Kear, J. M. Oblak, and W. A. Owezarski, J. Met. **24**, 25 (June 1972).

36. G. M. Ault and H. M. Burte, Technical Applications for Oxide Dispersion Strengthened Materials, in Ref. 3, pp. 3–60.

ADDITIONAL READING

1. J. W. Martin, *Precipitation Hardening*, Pergamon, New York, 1968. This book contains English language reprints of most of the important historical papers on the subject.

2. P. G. Shewmon, *Transformation in Metals*, Chapter 7, McGraw-Hill, New York, 1969.

PROBLEMS

11.1

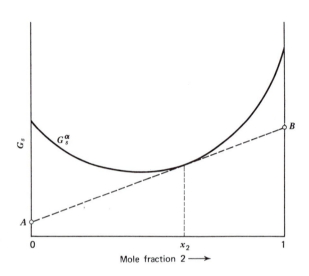

Mole fraction 2 ⟶

A tangent is constructed to the free-energy curve for the α phase, G_s^α, at composition x_2. Prove that point B is the chemical potential of type 2 atoms in the α solution, μ_2^α, and that point A gives the chemical potential of type 1 atoms in the α solution.

11.2

Given two solutions, I and II, where solution I contains n_I moles and solution II contains n_{II} moles, the free energy of mixing for the configuration of two solutions is given as

$$\Delta G_m = \frac{n_I}{n_I + n_{II}} \Delta G_m^I + \frac{n_{II}}{n_I + n_{II}} \Delta G_m^{II}$$

Your problem is to show that this equation can be transformed into the equation of the straight line shown on page 417. This means you have to express ΔG_m as a function of x_1, where x_1 is the average composition over the two solutions.

Hint: By means of a simple mass balance for component 1 you may show that $n_I/(n_I + n_{II}) = (x_1^{II} - x_1)/(x_1^{II} - x_1^I)$ and you obtain a similar expression for $n_{II}/(n_I + n_{II})$.

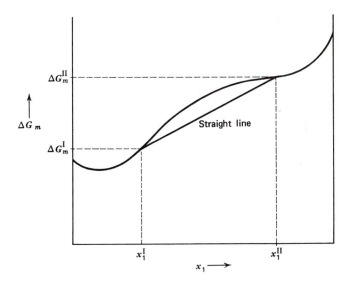

You may recognize this as the lever law. Now by substitution into the above equation you obtain ΔG_m as a function of x_1. Simplify your equation and show that it is of the form

$$y = y_0 + (x - x_0)\text{slope}$$

where $y_0 = \Delta G_m^I$, $x_0 = x_1^I$, $x = x_1$.

11.3   The following equation was used for the free energy of formation of a nucleus containing n atoms,

$$\Delta G = n\,\Delta G_B + \eta n^{2/3}\gamma + nE_s$$

where

$$\Delta G_B = \frac{(\text{free energy of nucleus} - \text{free energy matrix})}{\text{atom}}$$

$$E_s = \frac{\text{strain energy}}{\text{atom}}$$

$$\gamma = \text{surface tension}$$

$$\eta = \text{shape factor}$$

(a) Determine η for the cases where the nucleus is a cube and also a sphere, in terms of the volume per atom, V.

(b) Derive an expression for ΔG^* for the cube nucleus assuming ΔG_B, γ, and E_s constants.

11.4   (a) Sketch the FeC diagram on the top of a piece of paper using the diagram on p. 306.

(b) Below your diagram, sketch the free energy of the α, γ, and Fe_3C phases as a function of composition at a temperature of 750°C.

(c) Suppose an alloy of 0.4% carbon were heated to 900°C to form 100% γ phase and then quenched to 750°C. A very small nucleus of α phase now precipitates from this supercooled γ phase. Show by graphical construction on your free-energy composition plots of part (b) above what is the most likely composition of the *first formed* nucleus. Explain your construction fully.

11.5   The diffusion equation for spherical symmetry can be shown to be

$$\frac{\partial C}{\partial t} = D\left[\frac{\partial^2 C}{\partial r^2} + \frac{2}{r}\frac{\partial C}{\partial r}\right]$$

At steady state we can write this as

$$\frac{d}{dr}\left[r^2\frac{dC}{dr}\right] = 0$$

Inspection of this equation shows that it has solutions of the form,

$$C = A + \frac{B}{r}$$

In the problem of the coarsening of spherical precipitates a quasi-steady state is assumed so that this equation applies. From this equation show that the concentration gradient at the surface of the sphere must be

$$\left(\frac{dC}{dr}\right)_{r=R} = -\frac{C_R - C_0}{R}$$

where the surface of the sphere is taken at $r = R$, C_R is the concentration in the matrix at the surface, and C_0 is the concentration far from the surface.

11.6   (a) In discussing spinodal decomposition we said that there was a critical wavelength for decomposition below which it was predicted not to occur. Derive an equation for that wavelength as a function of the variables of the model.

(b) Derive an equation for that wavelength that the theory predicts is most likely to predominate.

(c) What are the major differences between formation of precipitates by spinodal decomposition and nucleation and growth?

11.7   For the θ'' precipitate the lattice mismatch is approximately 10%. The θ'' grows as a disk-shaped plate precipitate with a thickness of 20 Å. Assume that the lattice strain resulting from the coherent interface is given by Eq. 11.31. Calculate the diameter of the disk at which you would expect coherency to be completely lost. Take the modulus as 7×10^9 dynes/mm^2 and the energy of the incoherent interface after breakaway as 500 ergs/cm^2.

11.8   In an Al alloy that forms GP zones or solute clusters assume that (a) The energy of formation of vacancies is 20 kcal/mole (b) no vacancies are lost during quenching and (c) diffusion occurs by a vacancy mechanism. Compute the ratio of the initial rates of zone formation at room temperature for an alloy quenched from 500°C versus one quenched from 200°C.

11.9   As the bubble density increases from 100 to 400 bubbles per cm² the tensile strength of tungsten wire increases by 100%. This must mean that the bubbles inhibit the motion of dislocations. At first glance it seems strange that a bubble could inhibit the motion of a dislocation because a bubble is a void containing only a gas.

If you consider the different possible dislocation-particle interactions you will see that there is a mechanism that would produce an interaction with bubbles. Determine which mechanism this is and present an equation giving the increase in shear stress, $\Delta\tau$, produced by this mechanism. Define all terms in your equation for this specific case of bubbles.

11.10

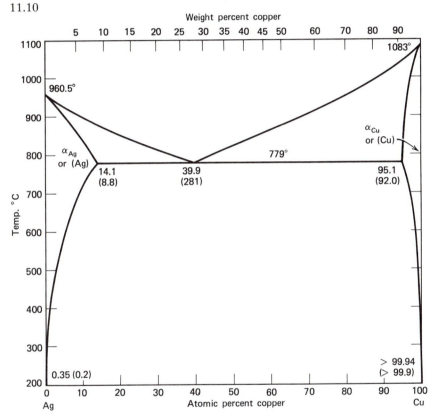

Copper–silver phase diagram. [From M. Hansen, *Constitution of Binary Alloys.* Copyright 1958 McGraw-Hill Book Co. Used with permission of McGraw-Hill Book Co.]

Silver is added to Cu in order to obtain an age-hardening reaction and thereby strengthen the Cu. Refer to the phase diagram given above. Suppose you were given the job of designing an alloy of maximum tensile properties. Outline a developmental research program designed to determine (1) alloy composition, and (2) the heat-treat cycle that would achieve this objective. Give a fairly detailed description of what variables you would control and how you would arrive at optimum values. Do not consider grain-size effects.

11.11   Suppose that as a result of Problem 11.10 you have determined that an alloy of 95 wt % Cu will give you the maximum mechanical properties. To further enhance the mechanical properties you decide to minimize the grain size by recrystallization. Your problem here then is to outline a research program that would arrive at an optimum mechanical and thermal treatment to minimize the grain size. Explain what variables you would control and how you would determine optimum values. Be careful to consider how your recrystallization treatment fits in with the precipitation hardening reaction that is also required.

CHAPTER 12

DIFFUSION-CONTROLLED GROWTH OF EQUILIBRIUM PRECIPITATES

It was pointed out in Chapter 9 that the rate of solidification is controlled by the heat flow in the system. In most solid-state transformations, however, the rate of transformation is controlled by diffusion (mass flow) rather than heat flow. In this chapter we discuss the rate of formation of equilibrium precipitates. Three common modes of precipitation are considered, single-phase precipitation, eutectoid precipitation, and discontinuous precipitation.

12.1. SINGLE-PHASE PRECIPITATES

The precipitation reaction that produces age hardening is obtained by quenching an alloy to temperatures well below its solvus-line temperature and then allowing the precipitate to grow either at the quench temperature or slightly above. As discussed in Chapter 11 this process often produces metastable precipitates. When alloys are cooled from high temperatures at slower rates it is quite common for the equilibrium precipitates to form and grow. Suppose one heats a steel of composition C_1 on Fig. 12.1 to temperature T_{SOLN} and then drops the furnace temperature to some value T_{HOLD} after which the steel is removed from the furnace and cooled to room temperature. As the austenite (γ iron) cools from T_{SOLN} and during the hold at temperature T_{HOLD}, ferrite (α iron) will precipitate and grow from the austenite. This ferrite is referred to as proeutectoid ferrite (because it precedes the formation of the eutectoid pearlite structure) and it will generally nucleate at the grain boundaries. It grows at a constant temperature, T_{HOLD}, and its growth rate is controlled by diffusion processes near the precipitate–matrix interface. The shapes of the ferrite precipitates, and of such equilibrium precipitates in general, are often found to be either plate-like or chunky. It has become fairly common to refer to the morphology of such equilibrium precipitates according to the following nomenclature.[1]

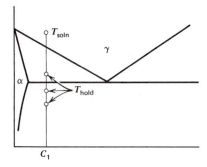

Figure 12.1   An alloy of composition C_1 quenched from temperature T_{soln} to various holding temperatures, T_{hold}.

  A.   Chunky Precipitates
  1.   *Grain-Boundary Allotriomorphs.*   In this case the precipitates form as elongated chunky blocks in the grain boundaries; see Fig. 12.2(*a*).
  2.   *Idiomorphs.*   The precipitates form as a more equiaxed chunk which can appear either at the grain boundary or within the grain; see Fig. 12.2(*b*).
  B.   Plate Precipitates
  1.   *Widmanstatten Sideplates.*   The precipitate forms as sharp needle-tipped plates growing either directly out of the boundary or from an allotriomorph; see Fig. 12.2(*c*).
  2.   *Intragranular Widmanstatten Plates.*   Plate-like precipitates form within the grains.

It is apparent that there is a striking difference between the plate morphology and the chunky morphology. This difference is related to the

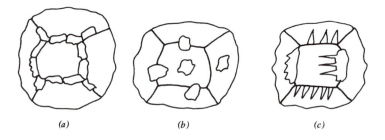

(*a*)   (*b*)   (*c*)

Figure 12.2   The shapes of various precipitates, (*a*) grain boundary allotriomorphs (GBA), (*b*) idiomorphs, and (*c*) Widmanstatten sideplates.

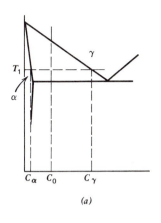

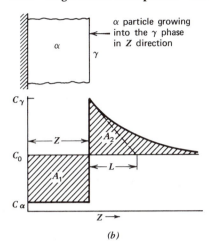

Figure 12.3 (a) Phase diagram locating ppt. composition C_0 and quench temperature T_1. (b) Composition profile along the α precipitate.

fact that there is a distinct difference between the growth front of the chunky and plate morphologies.

A. GROWTH KINETICS

1. Allotriomorphs and Idiomorphs. Consider an alloy of composition C_0 that is quenched to temperature T_1 shown in Fig. 12.3(a). A chunky α-precipitate particle nucleates and grows into the γ phase as shown in Fig. 12.3(b). At temperature T_1 this α precipitate will have composition C_α and the composition in the γ phase at the $\alpha-\gamma$ boundary must be C_γ since a local equilibrium will obtain at the growing interface. At this growing $\alpha-\gamma$ interface we may write for the solute flux with respect to the interface,

$$|\text{Flux into interface}| = VC_\gamma$$

$$|\text{Flux away from interface}| = VC_\alpha - D^\gamma \left[\frac{dC}{dZ}\right]^\gamma_{\text{INT}}$$

These two fluxes will remain balanced for all times at the interface plane and this flux balance may be written as,

$$V = \frac{-D^\gamma (dC/dZ)^\gamma_{\text{INT}}}{(C_\gamma - C_\alpha)} \tag{12.1}$$

The interface concentration gradient may be determined as a function of the parameter L from the construction shown in Fig. 12.3(b) and we obtain for the growth velocity

$$V = \frac{D^{\gamma}(C_{\gamma} - C_0)/L}{(C_{\gamma} - C_{\alpha})} \qquad (12.2)$$

Area A_1 on Fig. 12.3(b) is proportional to the mass of solute rejected from the α precipitate as it grew a distance Z. All of this solute is piled up directly in front of the interface so that the area marked A_2 on Fig. 12.3(b) must equal A_1. The further the α precipitate grows, the larger A_2 must become and consequently the bigger is L and the slower the growth velocity V, as given by Eq. 12.2. If we approximate area A_2 as $A_2 = L \cdot (C_{\gamma} - C_0)/2$ and recognize that $V = dZ/dt$, and $A_1 = Z \cdot (C_0 - C_{\alpha})$ we obtain by equating A_1 to A_2 and substituting into Eq. 12.2

$$\frac{dZ}{dt} = \frac{D^{\gamma}(C_{\gamma} - C_0)^2}{2Z(C_0 - C_{\alpha})(C_{\gamma} - C_{\alpha})} \qquad (12.3)$$

Integrating this equation we obtain

$$Z = A\sqrt{Dt} \qquad (12.4)$$

where

$$A = \frac{C_{\gamma} - C_0}{[(C_0 - C_{\alpha})(C_{\gamma} - C_{\alpha})]^{1/2}}$$

If one treats this problem in a more rigorous manner,[1,2] Eq. 12.4 is obtained with a different expression for A.

From the above discussion we may draw three pertinent conclusions regarding the growth kinetics of chunky precipitates.

1. Growth of these precipitates requires long-range diffusion. The distance L may extend out on the order of millimeters.

2. Growth will be relatively slow since it is controlled by long-range diffusion in the solid state.

3. The growth rate and, therefore, the precipitate size are time dependent. The growth rate continually decreases with time.

2. Widmanstatten Sideplates. In Chapter 9 the stability of a planar solid–liquid interface was discussed. On p. 261 it was shown that if a small bump formed on a flat interface a curvature effect would tend to stabilize the bump and a diffusion effect would destabilize it. The same ideas apply to solid-state growth of precipitates. Just as a planar solid–liquid interface becomes dendritic under certain growth conditions, the planar interface of a chunky precipitate may become unstable and form plates that can grow at much higher rates.

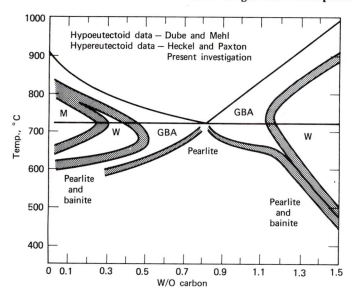

Figure 12.4 Temperature-composition regions of Fe–C alloys showing various precipitate forms present at long reaction times (from Ref. 1, p. 531).

Townsend and Kirkaldy[3] have presented a theory describing how the allotrimorph interface becomes unstable and the array of Widmanstatten side plates evolve. Their paper presents several clear photomicrographs and a good discussion of the evolution of the grain boundary allotrio-morphs (GBA) and the Widmanstatten plates (W) in steels. Figure 12.4 shows the temperature composition regions in which the various mor-phologies form in steels at late reaction times in specimens with an ASTM grain size of 0–1. [Massive ferrite (M), pearlite, and bainite will be discussed later. See Table 12.2 for ASTM grain size definition, p. 444].

Figure 12.5(a) shows an α plate growing into γ phase of composition C_0, where the composition profile shown here applies only to the centerline through the α plate. As the α plate grows, most of the solute ejected from the tip will diffuse laterally to the sides of the thin plate. Since all of the solute is not piled up ahead of the advancing tip, area A_1 does not equal area A_2. Just as above, a flux balance can be written at the tip centerline and the growth rate is again given by Eq. 12.1. Because most of the solute diffuses laterally to the sides the concentration gradient at the tip, $(dC/dZ)_{INT}^{\gamma}$, will be much steeper for a thin plate as shown in the sequence of Fig. 12.5. To a first-order approximation for plate growth we assume that L of Eq. 12.2 is linearly proportional to the radius of the plate tip, as

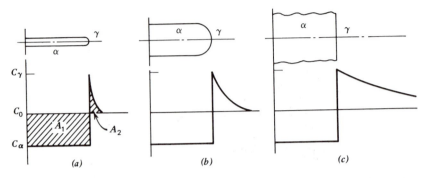

Figure 12.5 Composition profile along centerline of α precipitates having various shapes, (a) needle, (b) cylindrical, and (c) chunk.

$L = ar$, so that

$$V = \frac{D^{\gamma}(C_{\gamma}-C_0)/(ar)}{(C_{\gamma}-C_{\alpha})} \qquad (12.5)$$

where the proportionality constant a is approximately unity. The kinetics of plate growth differ from chunky growth in three main ways:

1. The radius of curvature at plate tips is frequently very small, being on the order of 100–1000 Å. Therefore $r \ll L$ and the plate growth rate will be much higher than allotrimorph growth rates.

2. Because of the small tip radius, diffusion occurs over much shorter distances with plate growth.

3. The solute profile at the centerline of the tip is essentially independent of time since the plates advance with a constant tip radius, and, hence, growth will occur at a steady-state velocity (time independent).

In Chapter 11 (pp. 370–373) it was shown that curvature can lower the lines on the equilibrium phase diagram. This means that at the sharp tip of an advancing α plate the solute concentration in the γ phase will be reduced from C_{γ} to $C_{\gamma}(r)$ as is shown on Fig. 12.6(a). This concentration change will in turn drop the concentration gradient in the γ phase at the tip as shown in Fig. 12.6(b), which means that the growth rate will slow down. We may write the growth rate from Eq. 12.5 as

$$V = \frac{D^{\gamma}}{ar} \frac{C_{\gamma}(r)-C_0}{C_{\gamma}(r)-C_{\alpha}(r)} \qquad (12.6)$$

When $C_{\gamma}(r) \to C_0$ the gradient will be zero and growth must stop and, therefore, we define the critical tip radius for this condition as r_c, $C_\gamma(r_c) = C_0$. From Eq. 11.18, dropping the solute subscript 2 and converting to volume concentrations, we may write

$$C_\gamma(r) = C_\gamma\left[1 - \frac{m}{r}\right],$$

$$C_\gamma(r_c) = C_\gamma\left[1 - \frac{m}{r_c}\right]$$

(12.7)

where

$$m = \frac{(1 - x_\gamma)2\gamma\bar{V}_1^\alpha}{(x_\gamma - x_\alpha)kT}.$$

After some algebra on Eqs. 12.7 one obtains

$$C_\gamma(r) - C_0 = (C_\gamma - C_0)\left[1 - \frac{r_c}{r}\right]$$

(12.8)

Combining Eqs. 12.6 and 12.8 we obtain

$$V = \frac{D^\gamma(C_\gamma - C_0)}{C_\gamma(r) - C_\alpha(r)} \frac{1}{ar}\left[1 - \frac{r_c}{r}\right]$$

(12.9)

and as required, the equation shows that when $r \to r_c$ growth stops.

Equation 12.9 shows that the velocity of the plate varies with tip radius. Differentiation with respect to r gives the maximum rate at $r = 2r_c$, where it is assumed that $C_\gamma(r) - C_\alpha(r) = C_\gamma - C_\alpha = \text{constant}$. This assumption should be good since both phase boundaries will drop due to curvature, see Fig. 12.6(a). It is now assumed that the tip radius will adjust to this value of $2r_c$ in order to promote growth at the fastest possible rate. Hence, this model,

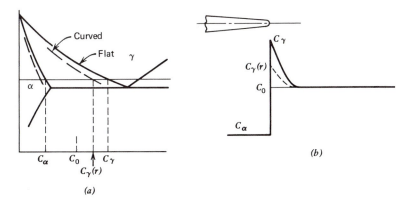

Figure 12.6 Effect of curvature on (a) phase boundaries and (b) composition profile along centerline of needle precipitates.

which is due to Zener,[4] predicts a steady-state growth velocity of

$$V = \frac{D^\gamma}{2ar_c} \frac{(C_\gamma - C_0)}{(C_\gamma - C_\alpha)} \tag{12.10}$$

Hillert[5] has presented an approximate solution for the diffusion field at the tip of plates and his analysis gives Eq. 12.10 with $a = 2(C_0 - C_\alpha)/(C_\gamma(r) - C(\alpha))$. Hence, Eq. 12.10 is known as the Zener–Hillert equation for plate growth. A more rigorously correct theory has recently been presented by Trivedi[6] but will not be discussed here. (A good comparison summary is presented in Ref. 7). In general, the agreement between theory and experiment is qualitatively correct, but there are a number of complicating factors and agreement is not as good as one would hope.[8]

Widmanstatten plate growth is found to occur quite widely and has been studied extensively. For example, it has been studied in Fe–C, Fe–P, Fe–N, Cu–Zn, Cu–Sn, Cu–Al, Al–Ag, Al–Cu, Ti–Cr, and many other systems.[8] It is usually found that an epitaxial relationship exists between precipitate and matrix and in Fe–C this is found to be the Kurdjumov–Sachs relationship given on p. 376. The sides of the plates are therefore low-angle grain boundaries between the ferrite and austenite and this accounts for their relatively slow growth into the austenite. There is currently some debate upon the mechanism by which these boundaries grow into the austenite.[9] Another interesting feature of these Widmanstatten precipitates is that at small undercoolings a needle and sometimes a blade-like morphology forms rather than a plate, whereas the plate morphology dominates at larger undercoolings. The plate–needle transition has been observed in the Fe–C and Cu–Zn systems, and a recent theoretical stability analysis has been able to account for the transition.[10] It is also interesting to note that the theories developed for Widmanstatten plate growth appear to give a reasonably good description of graphite flake growth in cast iron.[11]

12.2. EUTECTOID TRANSFORMATIONS

In Section 9.4 eutectic solidification was discussed and it was pointed out that the regular eutectic structure results from a form of coupled growth in which two solid phases grow from the liquid in a cooperative manner. Eutectoid transformations also often occur by a coupled growth process in which two solid phases, α and β, grow cooperatively from a third solid phase, γ; see Fig. 12.7. Just as in eutectic growth the microstructure of the coupled phases usually consists of alternating lamellar sheets of α and β, and there is generally some preferred epitaxial relationship between the α and β crystals at their common interfaces.

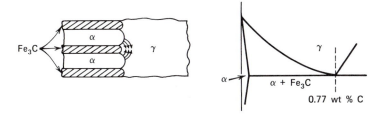

Figure 12.7 (a) Diffusion of carbon required during pearlite growth, (b) Fe–C phase diagram locating pearlite composition, 0.77 wt % C.

The most famous and commercially important eutectoid transformation occurs at 0.77 wt % C in steel. The lamellar structure consists of alternating plates of ferrite and cementite, Fe_3C, and has been given the name pearlite; see Fig. 12.7. This name apparently arose because a polished and etched pearlitic structure has the colorfulness of mother-of-pearl. The color results from the fact that the lamellar structure at the etched surface acts as a diffraction grating for light. The same fascinating appearance is found on properly etched lamellar eutectic structures. The pearlite reaction in steel has been of such predominant importance to metallurgists that it is not unusual to refer to any eutectoid transformation as a pearlite transformation.

Accordingly, we are mainly concerned here with the steel pearlite reaction because of its commercial importance and, also, because it has been studied quite extensively. Figure 12.4 shows the temperature composition region of the Fe–C phase diagram in which the lamellar pearlite structure can form. It should be noted that pearlite formation is not restricted to the eutectoid composition, 0.77% C. Also, at temperatures increasingly below the eutectoid temperature the α and Fe_3C phases may form from the γ phase by a noncoupled growth process producing a microstructure called Bainite. This process will be discussed further in Chapter 13.

A. MORPHOLOGY

It is quite generally observed that pearlite structures nucleate at the grain boundaries of the parent phase and grow into the parent phase as roughly spherically shaped grains called *nodules*; Fig. 12.8(a). Each of these nodules is composed of a number of structural units in which the lamellae are largely parallel; these structural units are generally termed pearlite colonies; see Fig. 12.8(b). It is often observed that the pearlite nodules only grow into one of the adjoining grains as is shown in Fig. 12.8(a); see p. 143 of Ref. 1 for actual micrographs.

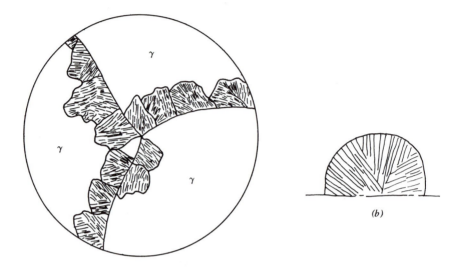

Figure 12.8 (*a*) Sketch of pearlite nodules growing from grain boundaries into austenite grains; (*b*) an individual nodule containing three colonies.

It was pointed out in Section 9.4A that the morphologies of eutectic microstructures may be classified into three categories, regular, complex-regular, and irregular. Eutectoid microstructures are most often observed to be lamellar, as is usually the case for steel pearlite. However, it is fairly common to encounter nonlamellar eutectoid microstructures in nonsteel pearlites (see Ref. 12, p. 565). These irregular structures are frequently referred to as granular pearlites and they are analogous to irregular eutectic structures.

B. NUCLEATION

Consider a hypoeutectoid steel of composition C_1 shown on Fig. 12.9. It is common to label the three pertinent phase boundaries A_3, A_{cm}, and A_1 as shown on Fig. 12.9. (Unfortunately, there is not a common general name for these types of phase boundaries as there is for the liquidus, solidus, and solvus lines.) After homogenizing the alloy at a temperature above the A_3 line we quench the alloy into a salt bath held at temperature T_1, which is between the A_1 line and T_1^*. The γ phase (austenite) is now supersaturated in solute (carbon) with respect to the α phase (ferrite)†. We therefore

---

† The γ phase becomes supersaturated with respect to α when cooled below the A_3 line, and it becomes supersaturated with respect to cementite when cooled below the A_{cm} line.

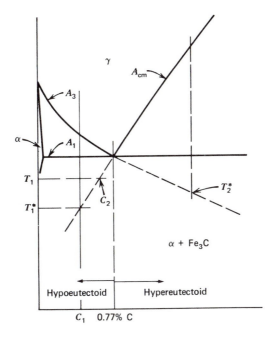

Figure 12.9   Fe–C phase diagram defining various parameters.

expect to see ferrite form as a grain-boundary allotromorph with perhaps some Widmanstatten side-plate formation. As the ferrite allotromorphs grow into the austenite grains, the composition in the austenite at the interface [C_γ on Fig. 12.3(b)] increases until it eventually reaches C_2 shown on Fig. 12.9. This austenite in contact with the GBA interface is now supersaturated with respect to cementite (Fe$_3$C), as determined by extrapolation of the A_{cm} line on Fig. 12.9. Consequently, some cementite forms and this nucleates the pearlite reaction and the remaining austenite within the interior of the grains transforms into a pearlite structure so that one obtains a microstructure such as Fig. 12.10. Notice that the actual composition of the pearlite would be a maximum at the GBA interface and decrease toward the grain centers [see Fig. 12.3(b)].

If we now changed the temperature of the salt bath to T_1^* shown on Fig. 12.9, the entire austenite grain would be supersaturated with respect to both ferrite and cementite immediately following the quench. If there is no barrier to cementite formation (there is much less chance of an effective barrier to ferrite formation since the austenite is supercooled so far below A_3), then pearlite should form immediately without formation of proeutectoid ferrite (GBA or W). Hence, it may be possible to form a pure pearlite

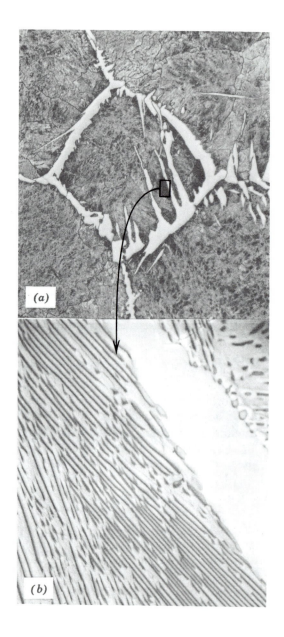

Figure 12.10 An AISI 1040 steel transformed by slow continuous cooling. (*a*) Ferrite present at the prior austenite grain boundaries as GBA with several Widmanstatten sideplates protruding into pearlite in grain interiors, 170×. (*b*) The pearlite structure near a Widmanstatten sideplate, 4300× (SEM photo), nital etch.

structure in the off-eutectoid alloy if it is quenched to temperature below T^*. This is, in fact, the case as may be seen from the data of Fig. 12.4.

It was pointed out in Section 9.4C that coupled lamellar growth of off-eutectic structures can also be obtained and the coupled growth region of Fig. 9.41 is quite similar to the pearlite region of Fig. 12.4. In the eutectic case the coupled growth region is limited by formation of dendrites due to constitutional supercooling. In the eutectoid case one might expect Widmanstatten side plates to limit the coupled region analogous to dendrite formation in eutectics because the austenite is supercooled well below A_3. This does not seem to be much of a problem in steel pearlite, as one finds the significantly wide coupled growth region shown on Fig. 12.4.‡

It is interesting to realize that the composition of the pearlite in an off-eutectoid steel is not 0.77 wt % C unless one has the equilibrium amount of proeutectoid ferrite or cementite present. Therefore, unless a steel is cooled slowly one will have a pearlite with less than 0.77%C in a hypoeutectoid steel and greater than 0.77% C in a hypereutectoid steel. The pearlite composition is a function of the relative size of the ferrite and cementite plates so that pearlite lowers its composition by simply increasing the ferrite plate width and raises its composition by increasing the cementite plate width.

In a hypoeutectoid steel the pearlite colonies will form at the interfaces between the proeutectoid ferrite and the austenite. If the proeutectoid ferrite acts as a nucleation site for the pearlite colony, then one would expect the crystallographic orientation of the ferrite to be the same in the proeutectoid ferrite and in the pearlite colony. Hillert[13] has described a careful series of experiments that determined relative crystallographic orientations and the results indicate that pearlite in steel may be nucleated by either ferrite or cementite.† In hypereutectoid steels the proeutectoid cementite nucleates pearlite, and in hypoeutectoid steels the proeutectoid ferrite nucleates the pearlite. The actual mechanism of the nucleation event is fairly complex. Hillert's study indicates that in at least some cases there is an initial competition between the two phases with first one phase

---

‡ There is a fundamental difference between off-eutectic and off-eutectoid growth. In off-eutectic growth a solute buildup is present at the reaction interface, whereas off-eutectoid growth is possible with very little if any solute buildup present in the austenite at the pearlite interface. The constitutional supercooling in the eutectic case, discussed in Chapter 9, is a function of position relative to the interface. It is generally much smaller than the supercooling present in off-eutectoid growth because the eutectic interface temperature is generally constrained to be within a few degrees of the eutectic temperature, whereas the eutectoid interface temperature may be 100 or more degrees below the eutectoid temperature.

† It had been thought for quite some time that pearlite in steel could only be nucleated by cementite.[13]

forming a film and then the other nucleating an adjoining film, until eventually crystallographic and compositional conditions are evolved that produce an alternating structure of ferrite and cementite. As this structure then grows the ferrite–cementite spacing adjusts itself by a branching process until the steady-state spacing characteristic of the reaction temperature is obtained.

C. GROWTH

It is often found that a GBA in steel will grow out from the grain boundary into one of the austenite grains but not into the other so that one obtains a flat-sided GBA as shown in Fig. 12.11(a). It is thought that the reason the GBA does not grow downward into austenite grain, γ_2, is because the lower GBA–austenite interface forms a coherent (small-angle) boundary that therefore has a low mobility. The upper GBA–austenite interface in grain γ_1 forms an incoherent (large-angle) boundary with a higher mobility and, hence, this boundary migrates upward. Hillert[13] has shown that if one quenches a growing structure, such as shown in Fig. 12.11(a) and 12.11(b), at not too fast a rate, an additional small amount of growth occurs before the remainder of the grain transforms to martensite, as shown in Fig. 12.11(c) and 12.11(d). A band of fine pearlite is found to form only at the ferrite–austenite boundaries that were incoherent prior to the ineffective quench. There is no phase formation at some of the coherent ferrite–austenite boundaries, Fig. 12.11(d), and fine Widmanstatten side plates (bainite) form at other coherent boundaries. From experiments such as these Hillert[13] has presented strong evidence for the following conclusion, which was first hypothesized in part by Smith[14]: "The ferrite and cementite constituents of pearlite can have any orientation relationships to the matrix austenite except for those which allow the formation of interfaces which are partially coherent with the matrix austenite." This results because the

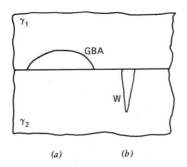

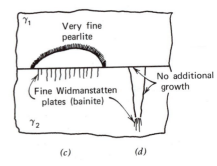

(a)   (b)   (c)   (d)

Figure 12.11 A grain boundary between two austenite grains, γ_1 and γ_2, showing the various precipitate forms discussed in the text.

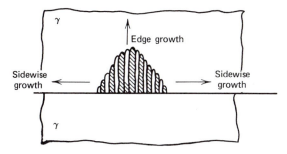

Figure 12.12   Sidewise growth and edge growth in a pearlite colony.

moving austenite–pearlite interface must be a fully incoherent boundary in order to have a high mobility. And furthermore, the reason the nodules often grow into only one of the neighboring grains, as shown in Fig. 12.8(a), is that the pearlite nuclei form a coherent boundary (low mobility) with the other grain. These conclusions have been confirmed on several pearlite colonies in a recent study[15] in which the crystallographic orientations were determined using selected area diffraction in the transmission electron microscope.

One may visualize that a pearlite colony will grow by two modes of growth as shown in Fig. 12.12. Edge growth would be accomplished by a simple diffusion process as is illustrated in Fig. 12.7, where each plate simply extends at its thin edge. Sidewise growth, however, would require, in addition to diffusion, alternate nucleation of ferrite and cementite plates. By meticulous sectioning techniques it has been found by both Forsman[16] and Hillert[13] that all of the cementite plates in a given eutectic colony are interconnected, as are all of the ferrite plates. This means that growth in the sidewise direction is accomplished by a continual branching of the plates and that the growth front of the plates must make an angle with the growing interface that is on the average closer to 90° than 0°. Although sidewise growth by repeated nucleation does not occur in most eutectoid systems, it appears that it does occur in some systems.

It was pointed out in Section 9.4D that in lamellar eutectic alloys a specific epitaxial relationship usually holds between the two phases of the lamellar eutectic. This same conclusion probably also holds for the majority of eutectoid microstructures. In steel pearlite there is an epitaxial matching at the ferrite–cementite boundaries so that these are low-energy coherent interfaces. However, it has been demonstrated by Hillert (see Ref. 1, pp. 289–292) that more than just one crystallographic matching is present in steel pearlite. Recent experimental studies[17] utilizing electron diffraction techniques have shown that the relative orientations between ferrite and cementite usually cluster around one of two distinct orientation relationships. These are:

$$(A) \quad \left\{ \begin{array}{l} (001)_{cm} \parallel (5\bar{2}\bar{1})_\alpha \\ [100]_{cm} \, 2.6° \text{ from } [13\bar{1}]_\alpha \\ [010]_{cm} \, 2.6° \text{ from } [113]_\alpha \end{array} \right\} \qquad (B) \quad \left\{ \begin{array}{l} (001)_{cm} \parallel (2\bar{1}\bar{1})_\alpha \\ [100]_{cm} \parallel [01\bar{1}]_\alpha \\ [010]_{cm} \parallel [111]_\alpha \end{array} \right\}$$

It was shown in one study[15] that relationship (B) is obtained if the pearlite nucleates upon proeutectoid cementite in the austenite grain boundaries and relationship (A) results if the pearlite nucleates on a clean austenite grain boundary.

In Section 9.4D we developed an equation for the diffusion-controlled growth of a lamellar eutectic advancing edgewise into the liquid. The interface was taken to be supercooled below the eutectic temperature by ΔT_E. The free energy therefore available, $\Delta S_f \Delta T_E$, was partitioned into that required to generate the lamellar interface and that required to drive the diffusion necessary to redistribute the atoms between the α and β plates. This same analysis may be applied to eutectoid growth so that for a eutectoid we have from Eq. 9.45

$$R = \frac{2D\Delta T_E}{(C_\gamma - C_\alpha)} \left[\frac{1}{|m_\alpha|} + \frac{1}{m_\gamma} \right] \frac{1}{S_0} \left[1 - \frac{S_{min}}{S_0} \right] \qquad (12.11)$$

where now

$C_\alpha =$ composition in the α phase at the $\gamma-\alpha$ interface

$C_\gamma =$ composition in the γ phase at the $\gamma-\alpha$ interface

$D =$ diffusion coefficient of solute in γ phase

$\Delta T_E =$ supercooling below eutectoid temperature

As mentioned in the footnote on p. 433 there is a fundamental difference in the kinetics of solid–liquid and solid–solid phase transformations. The amount of undercooling, ΔT_E, actually achieved at a moving solid–liquid interface is generally less than a few degrees in most metal alloys. Even if one initially supercools by 300–400°C the interface temperature will rise locally to within a few degrees of the liquidus as the latent heat evolves. As mentioned in Chapter 9 this result occurs because the kinetic undercooling, ΔT_K, required to achieve a high growth velocity in liquid–solid metal systems is quite small. However, this is not the case in solid–solid transformations. Therefore, if one quenches a γ phase into a salt bath at 300–400°C below the eutectoid temperature, the eutectoid reaction will occur isothermally with that ΔT_E, which is set by the salt bath temperature. Whereas the rate of solidification reactions is controlled by heat flow, the rate of most solid–solid reactions involving composition changes is controlled by diffusion.

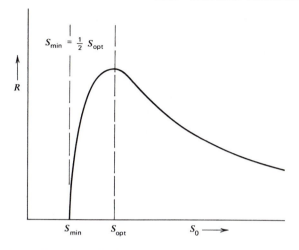

Figure 12.13 A plot of Eq. 12.11 defining the optimum and minimum spacings.

It was first proposed by Zener[4] that similar to the case of Widmanstatten side plate growth one would expect the plate spacing of the system to adjust itself to allow the eutectoid front to advance at the highest possible rate. Since the eutectoid reaction occurs isothermally, one may take all of the terms of Eq. 12.11 as essentially constant and determine the spacing, which maximizes the growth rate, S_{opt}, by differentiating R with respect to S_0. One finds $S_{opt} = 2S_{min}$ as is illustrated graphically in Fig. 12.13. Combining this result with Eqs. 9.39 and 9.41 we obtain

$$S_{opt} = \frac{4\gamma}{\Delta S_f \cdot \Delta T_E} \qquad (12.12)$$

Therefore, if the pearlite adjusts its spacing so that, $S_0 \to S_{opt} = 2S_{min}$, the steady-state rate predicted by Eqs. 12.11 and 12.12 may be written as

$$R = A \frac{D}{C_\gamma - C_\alpha} \frac{1}{S_{opt}^2} \qquad (12.13)$$

where

$$A = \frac{4\gamma}{\Delta S_f} \left(\frac{1}{|m_\alpha|} + \frac{1}{m_\gamma} \right).$$

Notice that this is the same result as that obtained for eutectic growth; see Eq. 9.48. In that case it was assumed that the plate spacing would adjust to produce a minimum undercooling at the advancing interface. It is apparent that these two optimization principles, minimum undercooling and maximum rate, are equivalent. There has been a vigorous debate in recent years over the theoretical validity of this optimization principle. There appear to be sound physical arguments that the optimum spacing for stable growth will be just slightly larger than the S_{opt} of Zener's theory (by a factor of 1 to 1.5) but at the present time there is little agreement on a theoretically correct value.[18,19]

The analysis of Section 9.4B that led to Eq. 12.11 was first presented by Zener in 1946.[4] Since then two theories[5,18] have been presented that calculate the concentration in the γ phase in a more rigorous manner, and a slightly simplified form of Hillert's results give Eq. 12.13 but with

$$A = \frac{2\gamma\pi^3 f^\alpha(1-f^\alpha)}{\Delta S_f \sum_1^\infty [\sin^2(n\pi f^\alpha)/n^3]} \left[\frac{1}{|m_\alpha|} + \frac{1}{m_\gamma} \right] \qquad (12.14)$$

where m_α and m_γ are the slopes of the A_3 and A_{cm} lines on the phase diagram and f^α is the volume fraction ferrite phase.

In the above analyses for eutectoid growth it was assumed that the solute redistributed between the α and β plates by diffusion in the γ phase as shown on Fig. 12.7. When the atoms are substitutional and diffusion occurs by a vacancy mechanism one would suspect that volume diffusion in the γ phase could be quite slow. Since the γ–eutectoid interface is a noncoherent boundary, it may well be possible that the solute redistribution could occur by grain-boundary diffusion along this interface. It was first shown by Turnbull[20] that Zener's rate equation for this process would become

$$R = KD_B \cdot \frac{\lambda}{S_0} \cdot \frac{1}{S_0} \left[1 - \frac{S_{min}}{S_0} \right] \qquad (12.15)$$

where D has been replaced by the grain-boundary diffusion coefficient, D_B, and λ is the effective grain-boundary diffusion width, and K is a constant that is to be derived in Problem 12.3. The additional term, λ/S_0, may be thought of as a reduction in the area available for lateral diffusion between the plates. Optimizing the rates as before one now obtains

$$R = A \frac{D_B}{C_\gamma - C_\alpha} \frac{1}{S_{opt}^3} \qquad (12.16)$$

Hillert[21] has presented a more complete treatment of this problem but his constant, A, for Eq. 12.16 differs by only a factor of 1.5 from Turnbull's value that you are to derive in problem 12.3.

Extensive experimental studies have been carried out on pearlite growth in steels using an isothermal growth procedure. These studies are conducted by quenching from the γ region to a specific undercooling. After an elapsed time the samples are then quenched to room temperature and the true pearlite spacing is determined by quantitative metallography. The results of such studies may be compared directly to Eq. 12.12. A plot of

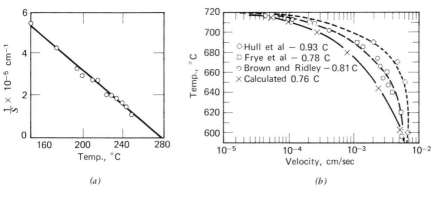

Figure 12.14 (a) Reciprocal spacing versus temperature in Al–Zn eutectoid alloys. (b) Comparison of theoretical and experimental data on velocity versus quench temperature in isothermal growth of Fe–C alloys (from Ref. 19).

$1/S_{opt}$ versus T should give a straight line, and some results on the Al–Zn eutectoid are shown in Fig. 12.14(a). Such linear results are generally obtained in conformance with the theory and one may estimate the energy of the (ferrite-cementite) interface γ from the slope. Analyzing experimental data on steel pearlite, Kirkaldy[19] determined $\gamma = 600$ ergs/cm² for the (ferrite-cementite) interface energy. This result is in reasonable agreement with experimentally determined values.

It is also possible to determine the velocity of the pearlite nodules from the isothermal growth experiments described above. The technique generally involves measurement of the largest sized nodules, knowledge of their nucleation times, and the assumption that they grew with constant velocity until termination by the final quench. Data on plain carbon steels as summarized by Kirkaldy[19] are presented in Fig. 12.14(b).

By combining Eqs. 12.12 and 12.13, and taking $D = D_0 \exp\left[-Q/RT\right]$ we find for the predicted temperature dependence of the growth rate

$$R = Be^{-Q/RT}\Delta T_E^2 \qquad (12.17)$$

where $B = AD_0\Delta S_f^2/[16 \cdot \gamma^2 \cdot (C_\gamma - C_\alpha)]$. Taking the constant B from Hillert's analysis,[5] Kirkaldy[19] has fitted Eq. 12.17 to the experimental data as shown on Fig. 12.14(b). He has chosen physical constants to make the fit as good as possible, but nevertheless it is apparent that reasonable agreement exists between theory and experiment for steel. It is interesting to examine the predicted temperature dependence of the growth rate given by Eq. 12.17. At temperatures close to the eutectoid temperature the rate is

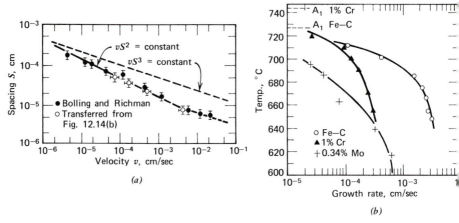

Figure 12.15 (a) Comparison of spacing-versus-velocity data on forced velocity pearlite to isothermal pearlite data [from Fig. 12.12(b)], and to theories. (b) Effect of Cr and Mo impurities upon isothermal growth velocities of pearlite (from Ref. 19).

limited by ΔT_E, which is a reflection of the fact that at these high transformation temperatures there is little free energy available ($\Delta G_B = \Delta S_f \Delta T_E$) for the reaction. At low temperatures the rate will again be limited because $\exp(-Q/RT)$ will decrease faster than ΔT_E^2 increases; in other words, diffusion limits growth at low temperatures and one therefore expects a maximum growth rate at intermediate temperatures. As we will see in the next section this maximum determines the nose of the T-T-T diagram.

Bolling and Richman[22] have recently reported experiments in which steel pearlite was grown by a technique similar to that used in controlled-solidification studies. Pearlite was forced to grow at a constant velocity down a $\frac{1}{4}$ inch (6.4 mm) steel rod by passing a hot zone down the rod at constant rate. To ensure that the pearlite–austenite growth front followed the hot zone without nucleation ahead of it, temperature gradients on the order of 2500°C/cm were used. Their results have been replotted by Kirkaldy[19] and compared to the isothermal results of Fig. 12.14(b) as shown on the plot of Fig. 12.15(a). It is seen that good agreement is obtained between the forced velocity and the isothermal results. Furthermore, the results appear to follow the volume diffusion model ($RS_{opt}^2 = $ const) down to spacings of about 700 Å, and below this size they correlate well with the grain-boundary-controlled model of Eq. 12.16. Based on these facts it has been concluded[19] that grain-boundary diffusion becomes an important mechanism for carbon redistribution at temperatures below 650°C in steel pearlite growth.

Cahn and Hagel[23] have reviewed data on a number of nonferrous pearlites and they conclude that grain-boundary diffusion is probably the significant mode of solute redistribution in these systems. These systems involve substitutional solutes whereas the steel pearlite involves an interstitial solute. Since diffusion coefficients are generally much higher for interstitial solutes than for substitutional solutes, it seems reasonable to expect grain-boundary diffusion to play a more dominant role in controlling growth rates of eutectoid reactions which only involve substitutional solid solutions.

The presence of substitutional solutes in pearlitic steels generally produces a strong reduction in the growth rate of pearlite as is shown on Fig. 12.15(b). This effect is extremely important from a commercial point of view because it allows one to delay the pearlite reaction by addition of alloying elements and thereby increase the hardenability of steels (see Refs. 25 or 26 and the next section for a further discussion of hardenability). Alloying elements that raise the eutectoid temperature, such as Cr and Mo, generally have a larger depression effect upon the pearlite growth rate than alloy additions that lower the eutectoid temperature, such as Mn and Ni. Consideration of the ternary-phase diagrams[19] indicates that Cr and Mo will partition between the ferrite and cementite plates whereas Mn and Ni will not, that is, the Mn and Ni concentrations will remain unchanged in the ferrite, cementite, and austenite phases. Hence, the strong retarding effect of Cr and Mo results because diffusion of these substitutional solutes controls the reaction. In nonpartitioning cases, such as with Mn and Ni, the reaction remains controlled by carbon diffusion. However, the pearlite growth rate is still reduced, and this effect apparently results[19] from the Mn and Ni reducing the carbon supersaturation at a given growth temperature and, also, from some sort of an interaction effect of these solutes with C that retard the diffusion rates of C in iron.

D. KINETICS OF FORMATION

Pearlite forms by a nucleation and growth process and consequently the volume transformed into pearlite will have the time dependence described by the Avrami equation presented in Chapter 10,

$$X = 1 - e^{-kt^n} \tag{10.30}$$

Pearlite nucleates almost exclusively on the interfaces between grains. Except at the higher formation temperatures there is always enough nucleation to form at least a few pearlite nodules per grain, and under these conditions, termed saturation, Cahn[23] has shown that the constant k of the Avrami equation is independent of the nucleation rate $\dot{N}$. The form of his analytic results depends on whether nucleation occurs on a grain

Table 12.1.   Values of the Constants in
the Avrami Equation for Various Sites of
Nucleation (from Cahn[23])

Nucleation Site	k	n
Grain boundary	$2AG$	1
Grain edge	πLG^2	2
Grain corner	$4\pi\eta G^3/3$	3

corner, a grain edge, or a grain boundary, and the resulting k and n values
of the Avrami equation are listed in Table 12.1. In this table G is the
growth rate, A is the grain-boundary area, L is the grain edge length, and η
is the number of grain corners per unit volume. Under these saturation
conditions transformation will be complete as soon as the pearlite nodules
migrate half way across the grains. Therefore, letting d equal the average
grain diameter and G the average growth rate we may estimate the time for
complete volume transformation, t_f, quite simply as

$$t_f \approx \frac{0.5d}{G} \tag{12.18}$$

This result as well as the results in Table 12.1 shows clearly that the rate of
pearlite formation is dependent upon the grain size of the austenite from
which the pearlite forms. Except for the highest transformation temperatures, site saturation is expected to occur in pearlite formation.[23] Therefore, since $\dot{N}$ is not involved, as shown in Table 12.1, the two important
parameters controlling the volume rate of pearlite formation are growth
rate G (given by R in Eqs. 12.13 and 12.16) and the grain size.

Cahn[23] has also considered the case at high transformation temperatures
where site saturation might not occur. Assuming that $\dot{N}$ increases with time
as at^m, he finds that the time exponent n in the Avrami equation becomes
$4+m$, and k remains a function of grain boundary area for boundary
nucleation.

1.   T-T-T Diagrams.   It is customary in describing the rate of pearlite
formation to present a time–temperature–transformation (T-T-T) curve as
is shown in Fig. 12.16 for a plain carbon eutectoid steel. These curves are
generated by quenching to some reaction temperature salt pot and after a
suitable reaction time quenching again to room temperature to arrest the
transformation reaction. For example, in quenching to 1200°F the pearlite
would not begin to form until 5 seconds after the quench, it would be 50%
complete after 17 seconds, and would be completed after 40 seconds. The

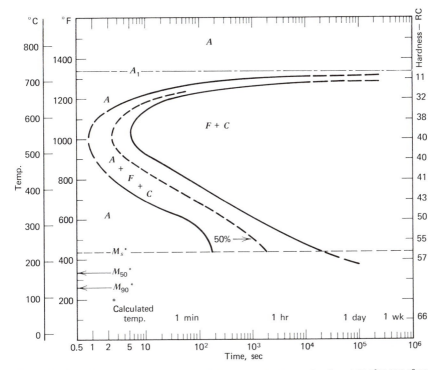

Figure 12.16   T-T-T diagram for AISI 1080 steel. Austenized at 1650°F (900°C). Grain size: 6. (From Ref. 24, copyright 1951, United States Steel Corporation.)

steel of Fig. 12.16 had an ASTM grain size of 6. From Table 12.2 we estimate the grain diameter as 44 μm so that from Eq. 12.18 we may estimate the growth rate as 6.3× 10^{-5} cm/sec. From Fig. 12.15(*b*) we see that the rate is about a factor of 5 slower than the growth rate in high-purity Fe–C steel at 1200°F (650°C), but it corresponds more to the pearlite growth rate of the impure steels. The steel used for Fig. 12.16 was a commercial 1080 steel containing 0.76 Mn and perhaps other residuals, and the slower growth rate is due to this relatively small impurity level.

It should be pointed out that below around 500°C on this T-T-T diagram pearlite does not form. A very fine Widmanstatten side-plate structure called Bainite forms down to the temperature labeled M_s, and below this temperature only martensite forms. These two structures will be discussed in the next two chapters.The important point to see here is that if one wants to obtain a structure that is fully martensite (no pearlite or bainite) it would be necessary to have the entire structure quenched from 1650°F (900°C) to below 1050°F (560°C) (the "nose" of the T-T-T curve) in less

Table 12.2.   ASTM Grain Size Numbers

ASTM No.	Grains/mm²	Approximate Grain Diameter (μm)
−1	4	500
0	8	354
1	16	250
2	32	177
3	64	125
4	128	88.4
5	256	62.5
6	512	44.2
7	1,024	31.2
8	2,048	22.1
9	4,096	15.6
10	8,200	11.0
11	16,400	7.81
12	32,800	5.52

Usual Range { 1–9

than 1 second. As will be discussed in the next chapter, steels that contain all martensite can be hardened more fully than those containing pearlite plus martensite. Therefore, to increase the hardenability of a steel one desires to shift the nose of this T-T-T curve to longer times. The nose of the T-T-T curve will be controlled by the temperature dependence of the Avrami equation. The relationship of the volume fraction curve described by the Avrami equation to the T-T-T diagram is shown in Fig. 12.17. Since Cahn's results showed the Avrami equation to be independent of $\dot{N}$ for pearlite, except at high transformation temperatures, the time-temperature position of the nose of the T-T-T curve will be controlled by two main things, (1) the grain size and (2) the pearlite growth equation, either Eq. 12.13 or 12.16. As was just shown, addition of alloying elements to steels decreases the pearlite growth rate. It should be clear, then, that by alloying additions one may shift the nose of the T-T-T curves to the right and increase the hardenability of a steel (e.g., see Fig. 13.32). The influence of various alloying elements upon the hardenability of steels is discussed at length in Refs. 25 and 26 and a more concise treatment presented in Ref. 27.

12.3.   DISCONTINUOUS PRECIPITATION

In connection with Fig. 11.12 on p. 374 it was shown that solid-state precipitation reactions may occur by either a continuous or a discontinuous mode. Discontinuous precipitation is sometimes referred to by the alternative name cellular precipitation. As shown in Fig. 11.12 the growth mode

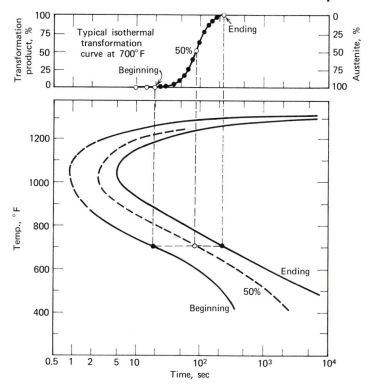

Figure 12.17   Comparison of an isothermal volume fraction transformation curve to the T-T-T diagram at the corresponding temperature. (From Ref. 24, copyright 1951, United States Steel Corporation.)

involves coupled growth of two phases (i.e., a duplex structure) and, consequently, this reaction bears a strong similarity to the eutectoid reaction. Precipitation by the discontinuous mode has generally been looked upon as something to be avoided because it interferes with the continuous precipitation reactions that are useful for precipitation strengthening. However, it has recently been shown[28] that the discontinuous precipitation reaction may be controlled so as to produce aligned composite materials, and the plate structures may be smaller by a factor of 10–100 than those produced by controlled solidification of eutectic alloys. Very small plate sizes are beneficial to the potential mechanical and electromagnetic properties of aligned composite materials.

A.   CHARACTERISTICS

We will discuss the characteristics of discontinuous precipitation reactions using the nomenclature defined on the phase diagram of Fig. 12.18(a),

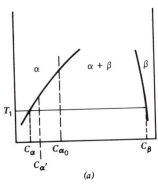

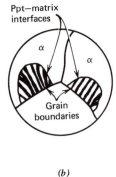

Figure 12.18 (*a*) Phase diagram showing α and β solvus lines. (*b*) Discontinuous precipitation showing sketch of microstructure of two precipitate cells.

where for later convenience we use volume concentrations C rather than fractional concentrations X. Upon cooling a solid solution of α phase having composition C_{α_0} to temperature T_1, a small cellular precipitate forms at the grain boundary and grows into the α matrix as shown schematically in Fig. 12.18(*b*). This duplex cellular precipitate is composed of alternate plates of α and β phases. The composition of the β plates is C_β and the composition of the α plates is $C_{\alpha'}$. This composition, $C_{\alpha'}$, is generally slightly greater than the equilibrium composition C_α. The reaction may be written

$$\alpha(\text{Comp } C_{\alpha_0}) \rightarrow \alpha(\text{Comp } C_{\alpha'}) + \beta(\text{Comp } C_\beta)$$

Direct in situ observation of cellular precipitation has recently been made in the high-voltage (1 MeV) electron microscope.[29] The student is referred to Fig. 7 of Ref. 29 for cine-film exposures showing the evolution of the precipitation process in an Al–Zn alloy.

Discontinuous precipitation is not a rare occurrence, but is observed in many alloy systems. For example, Bohm[30] presented data on 14 Cu base alloys, Cu–Mg, –Ti, –Be, –Sb, –Sn, –In, –Cd, –Ag, –Si, –P, –Cr, –Fe, –Mn, –Co, which showed that discontinuous precipitation occurred in the first eight of these systems. Figure 12.19 presents data on two of these systems showing the temperature range below which discontinuous precipitation occurs. Conditions favoring the formation of discontinuous precipitation have been examined by Hornbogen[30] and will be discussed later.

The name discontinuous refers to the fact that the composition of the α phase changes discontinuously at the precipitate–matrix boundary. This is illustrated by the data of Fig. 12.20(*a*), which gives measurements of the

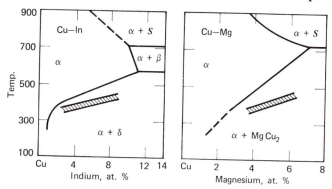

Figure 12.19 The temperature range below which discontinuous precipitation was observed in two Cu–base alloys (from Ref. 30).

α-phase lattice parameter at the precipitate–matrix boundary in an Fe–20 at. % Mo alloy. Since Mo is a substitutional solute in Fe, the composition is proportional to the lattice parameter, and hence Fig. 12.20(a) clearly shows that a sharp change in composition occurs in the α-phase across the precipitate–matrix interface. Discontinuous precipitation almost always produces an equilibrium precipitate phase rather than a metastable phase. However, as previously mentioned, the composition of the α plates in the

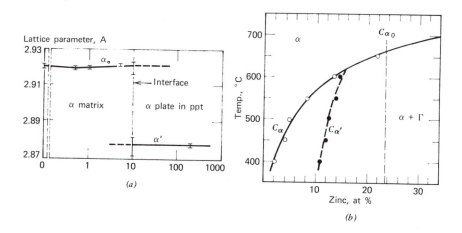

Figure 12.20 (a) Lattice parameter measurements on α phase across ppt–matrix boundary in Fe–20 at. % Mo alloy aged at 600°C (from Ref. 30). (b) Nonequilibrium segregation during discontinuous precipitation in Fe–23.5 at. % Zn alloys (from Ref. 31).

precipitate cell, $C_{\alpha'}$, is generally higher than the equilibrium composition of the α-phase, C_α; see Fig. 12.18(a). This is shown for an Fe–23.5 at. % Zn alloy by the data of Fig. 12.20(b). For example, the α plates of the precipitate formed at 400°C had a fractional composition of around 11% Zn compared to an equilibrium α-phase composition of 2% Zn. Complete precipitation has not occurred.

It is generally observed that the interface between the precipitate cell and the parent matrix (i.e., the precipitate–matrix interface) is an incoherent (high-angle) boundary. Both the α plates and β plates of the precipitate cell form an incoherent interface with the parent α phase. This means that in addition to the composition discontinuity at the α–α surfaces of the precipitate–matrix interface, there is a discontinuity in the crystallographic orientation. (There is, of course, also a discontinuity in both composition and crystallographic orientation at the β–α surfaces of the precipitate–matrix interface.) The precipitate cells are most frequently observed to nucleate at grain boundaries; and, in addition, they almost always grow into only one of the adjacent grains, as shown in Fig. 12.18(b). This behavior is similar to that observed for pearlite, and it also may be explained by the hypothesis of Smith quoted on p. 434. Apparently, the nucleus for the precipitate cell forms a coherent boundary with one of the two adjoining grains. Consequently, the cell may only grow into the other grain since its growth requires an incoherent boundary. In many cases the cells are observed to grow by edgewise extension of the plates and the lateral motion of the cells (sidewise growth of Fig. 12.12) is accomplished by branching of the plates.[29] (However, cases have been observed where sidewise growth is accomplished by repeated nucleation; see Ref. 32). The details of the nucleation mechanism for discontinuous precipitation have recently been studied. However, this topic will not be discussed here and the student is referred to Ref. 33 and 34.

B.   GROWTH

A big difference in continuous and discontinuous precipitation relates to the length of the diffusion paths involved. Figure 12.5 illustrates that, except at the tip of a plate or needle precipitate, the diffusion field extends over a relatively long range in continuous precipitation. However, as illustrated by the sketch of Fig. 12.7, the diffusion field for discontinuous precipitation extends over short distances, on the order of the plate spacing, which is generally less than 1 μm.

As might be expected the theories for discontinuous growth are similar to those for eutectoid growth. However, for discontinuous growth the solute of interest is virtually always a substitutional solute in the parent matrix. Consequently, it is generally found that growth is controlled by

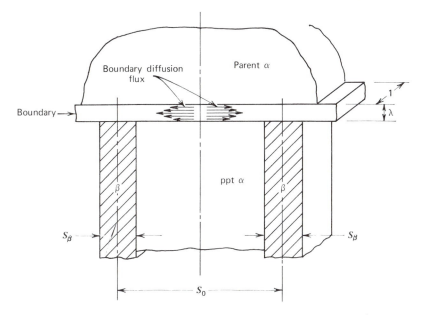

Figure 12.21 Model for discontinuous precipitation.

grain-boundary diffusion rather than by volume diffusion as is the case for pearlite growth in steel. The original theory was developed by Turnbull[3] and we will present a modified form of this theory based primarily on the presentation of Aaronson and Liu.[36] A comprehensive theory has been given by Cahn[37] and a more direct derivation of this theory is presented by Tu and Turnbull.[38] Finally, these theories have been reviewed and their shortcomings pointed out by Hillert.[39]

These theories attribute the solute redistribution to migration along the precipitate–matrix grain boundary. They imagine that the precipitate–matrix boundary is a plane of finite thickness, λ, as shown in Fig. 12.21. As growth proceeds the matrix of compositions C_{α_0} enters this boundary and the precipitate plates of compositions $C_{\alpha'}$ and C_β leave the boundary. Within the boundary the solute is pumped from in front of the α plate to in front of the β plate by grain-boundary diffusion. The flux of solute atoms ingested by a β plate may be written as

$$J(\text{ingest}) = \frac{1}{S_\beta \cdot 1} \frac{dm}{dt} = R(C_\beta - C_{\alpha_0}) \qquad (12.19)$$

where R is the rate of interface advance into the parent α phase and the surface area of the β–parent interface is taken as $S_\beta \cdot 1$, and we assume the average boundary composition in front of the β plate is C_{α_0}.

The diffusion flux along the boundary that feeds the β plate may be written as

$$J(\text{Diff}) = \frac{1}{2(\lambda \cdot 1)} \frac{dm}{dt} = D_B \frac{C_{\alpha_0} - C_\alpha}{S_0/2} \tag{12.20}$$

where D_B is the grain-boundary diffusion coefficient. The boundary area through which diffusion feeds each β plate is $2(\lambda \cdot 1)$, since each β plate receives solute from two directions in the boundary. The diffusion in the boundary proceeds from the center line of the α plates to the edge of the β plates, which are assumed to be very thin relative to the α plates so the diffusion distance is approximately $S_0/2$. The boundary composition at the α centerline is taken as C_{α_0}. The boundary composition at the β–α precipitate interface is taken as the equilibrium value, C_α, since one expects a local equilibrium to be obtained at the β–α interface. Hence, the concentration gradient driving diffusion along the boundary is given as $(C_{\alpha_0} - C_\alpha)/(S_0/2)$.

At steady state the rate of solute ingestion into a β plate, dm/dt, must equal the rate of solute supplied to the β plate by diffusion along the boundary. From Eqs. 12.19 and 12.20 we therefore obtain the expression

$$R = \frac{4D_B\lambda}{S_\beta S_0} \frac{C_{\alpha_0} - C_\alpha}{C_\beta - C_{\alpha_0}} \tag{12.21}$$

We now make a simple mass balance utilizing the fact that the average composition of the α and β plates, $C_\beta \cdot S_\beta/S_0 + C_{\alpha'}(1 - S_\beta/S_0)$, must be the same as the matrix composition, C_{α_0}, and we find

$$\frac{S_\beta}{S_0} = \frac{C_{\alpha_0} - C_{\alpha'}}{C_\beta - C_{\alpha'}} \tag{12.22}$$

Combining Eqs. 12.22 and 12.21 we obtain

$$R = \frac{4D_B\lambda}{QS_0^2}\left[\frac{C_\beta - C_{\alpha'}}{C_\beta - C_{\alpha_0}}\right] \tag{12.23}$$

where $Q = (C_{\alpha_0} - C_{\alpha'})/(C_{\alpha_0} - C_\alpha)$. Examination of this expression for Q with reference to Fig. 12.18(a) shows that it is the fraction of the initial supersaturation, $C_{\alpha_0} - C_\alpha$, that is actually removed from the α phase into the β phase. This quantity can be measured, as is evident from the data of Fig. 12.20(b). The bracket term in Eq. 12.23 will generally be close to unity since $C_\beta \gg C_{\alpha'}$ and $C_\beta \gg C_{\alpha_0}$. Therefore our final result becomes

$$R = \frac{4D_B\lambda}{QS_0^2} \tag{12.24}$$

The analysis similar to that above that requires the solute to redistribute to the β plates by volume diffusion through the parent α matrix gives the interface rate proportional to D/S_0 (see Ref. 35), where D is the volume diffusion coefficient. Hence, by comparison of rate data for discontinuous precipitation to this expression and to Eq. 12.24 it is possible to deduce whether solute redistribution occurs by volume or surface diffusion. Several studies[29,31,32,36] have shown that grain-boundary diffusion is responsible for the solute redistribution during discontinuous precipitation. The more comprehensive theories mentioned above[37–39] remove some of the assumptions of the simple theory presented here, but they do not offer much improvement on Eq. 12.24 since they introduce additional unmeasurable parameters. The expression of Eq. 12.24 already has an unmeasureable parameter, λ, which is generally taken to be 5 Å (0.5 nm).

It often happens that both discontinuous and continuous precipitation occur simultaneously in an alloy. The continuous precipitation may occur as GP zone formation, which is not detectable by optical microscopy. As would be expected, the continuous precipitation is observed to reduce the rate of discontinuous precipitation.

One would like to be able to predict when discontinuous precipitation might be expected to occur. At the present time this question can only be answered somewhat qualitatively. Hornbogen[30] concludes that the three most important factors favoring the occurrence of discontinuous precipitation are

1. A high probability of heterogeneous nucleation at grain boundaries as compared to nucleation in grain interiors.
2. A high boundary-diffusion coefficient.
3. A high driving force for precipitation.

Suppose that one were to cold work an alloy lying in the two-phase region of the phase diagram, and then produce recrystallization by a heat treatment. It is sometimes observed that the recrystallization produces a duplex microstructure of the two phases; see Refs. 40 and 41. The reaction here is very similar to discontinuous precipitation, particularly if the cold-worked alloy is initially a supersaturated single-phase matrix. Such reactions have apparently not been widely studied.

REFERENCES

1. H. I. Aaronson, in *Decomposition of Austenite by Diffusional Processes*, V. F. Zackay and H. I. Aaronson, Eds., Interscience, New York, 1962, pp. 387–546.
2. C. J. Zener, J. Appl. Phys. **20,** 950 (1949).
3. R. D. Townsend and J. S. Kirkaldy, Trans. ASM **61,** 605 (1968).

4. C. J. Zener, Trans. AIME **167**, 550 (1946).

5. M. Hillert, Jernkontorets Ann. **141**, 757 (1957).

6. R. Trivedi, Met. Trans. **1**, 921 (1970).

7. W. P. Bosze, The Effect or Crystallographic Anisotropy on the Growth Rates of Widmanstatten Precipitate, M. S. Thesis, Iowa State Univ., 1973.

8. E. P. Simonen, Edgewise Growth of Widmanstatten Precipitates, Ph.D. Thesis, Iowa State Univ., 1972.

9. H. I. Aaronson, C. Laird, and K. R. Kinsman, in *Phase Transformations*, American Society of Metals, Park, Ohio, 1970, pp. 313–396.

10. F. G. Yost and R. Trivedi, Met. Trans. **3**, 2371 (1972).

11. M. Hillert and Subba Rao V. V., in *The Solidification of Metals*, ISI Publ. 110, Iron and Steel Institute, London, 1968, pp. 204–212.

12. C. W. Spenser and D. J. Mack, in Ref. 1, pp. 549–602.

13. M. Hillert, in Ref. 1, pp. 197–236.

14. C. S. Smith, Trans ASM **45**, 533 (1953).

15. R. J. Dippenaa and R. W. K. Honeycombe, Proc. Roy. Sci. **333A**, 455 (1973).

16. O. Forsman, Jernkontorets Ann. **102**, 1 (1918).

17. W. Pitch, Acta Met. **10**, 79 (1962); K. W. Andrews and D. J. Dyson, Iron Steel **40**, 40, 93 (1967); H. G. Bowden and P. M. Kelly, Acta Met. **15**, 105 (1967); D. F. Lupton and D. H. Warrington, Acta Met. **20**, 1325 (1972).

18. K. A. Jackson and J. D. Hunt, Trans. AIME **236**, 1129 (1966).

19. M. P. Puls and J. S. Kirkaldy, Met. Trans. **3**, 2777 (1972).

20. D. Turnbull, Acta Met. **3**, 55 (1955).

21. M. Hillert, in *The Mechanism of Phase Transformations in Crystalline Solids*, Institute of Metals Monograph No. 33, Institute Metals, London, 1969, pp. 231–247.

22. G. F. Bolling and R. H. Richman, Met. Trans. **1**, 2095 (1970).

23. J. W. Cahn and W. C. Hagel, in Ref. 1, pp. 131–191.

24. *Atlas of Isothermal Transformation Diagrams*, U.S. Steel Co., Pittsburgh, 1951.

25. E. G. Bain and H. W. Paxton, *Alloying Elements in Steel*, American Society of Metals, Park, Ohio, 1966.

26. W. Crafts and J. L. Lamont, *Hardenability and Steel Selection*, Sir Isaac Pitman and Sons Ltd., London, 1949.

27. R. E. Reed-Hill, *Physical Metallurgy Principles*, Chapter 18, 2. Ed., D. Van Nostrand, New York, 1973.

28. F. M. Carpay, Acta Met. **18**, 747 (1970).

29. E. P. Butler, V. Ramaswamy, and P. R. Swann, Acta Met. **21**, 517 (1973).

30. E. Hornbogen, Met. Trans. **3**, 2717 (1972).

31. G. R. Speich, Trans. AIME **242**, 1359 (1968).

32. K. N. Tu and D. Turnbull, Acta Met. **15**, 1317 (1967).

33. H. I. Aaronson and H. B. Aaron, Met. Trans. **3**, 2743 (1972).

34. R. A. Fournelle and J. B. Clark, Met. Trans. **3**, 2757 (1972).

35. D. Turnbull, Acta Met. **3**, 55 (1955).

36. H. I. Aaronson and Y. C. Liu, Scripta Met. **2**, 1 (1968).

37. J. W. Cahn, Acta Met. **7**, 18 (1959).

38. K. N. Tu and D. Turnbull, Scripta Met. **1**, 173 (1967).

39. M. Hillert, Met. Trans. **3**, 1972 (1972).

40. J. M. Oblak and W. A. Owczarski, Trans. AIME **242**, 1563 (1968).

41. H. Kreye and E. Hornbogen, J. Mat. Sci. **5**, 89 (1970).

PROBLEMS

12.1 It was shown that the velocity of a Widmanstatten side plate could be expressed as $V = D^\gamma(C_\gamma - C_0)/[(C_\gamma - C_\alpha)2ar]$, where D is the carbon diffusion coefficient, C_γ, C_α, and C_0, are carbon concentrations in γ phase, α phase, and originally, r is the tip radius, and a is a a proportionality constant. If we apply an electric field this equation will be modified. An electric field E will produce a flux of carbon atoms, J, given by $J = CUE$, where C is carbon concentration and U is a constant (sometimes called mobility). Derive the above equation for the case where the electric field E is parallel to the growth direction of the plate. The plate is growing toward the positive electrode and carbon atoms are known to migrate toward the negative electrode in both α and γ iron.

12.2

(a) The phase diagram for Cu–In is shown below. It is found that the δ phase precipitates from the α phase by a discontinuous (sometimes called cellular) reaction that produces the lamellar structure characteristic of pearlite (and many eutectics).

(1) Outline a heat-treating process designed to produce this precipitate from an alloy of 8 at. % indium.

(2) Where in the microstructure of the parent phase would you expect the precipitate to form?

(3) At what temperature would you expect the rate of formation to be highest?

(b) An alloy of 20.15 at. % In can also be made to form in a lamellar structure.

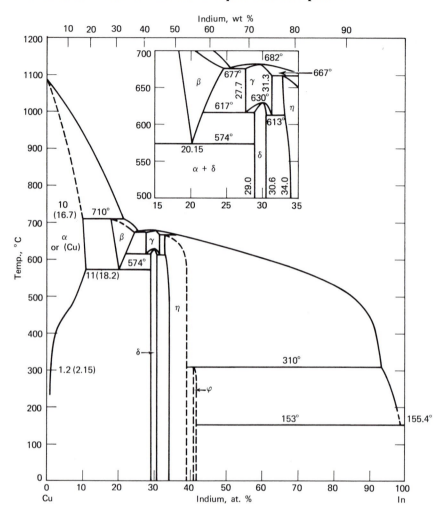

Copper–indium phase diagram. (From M. Hansen, *Constitution of Binary Alloys.* Copyright McGraw-Hill Book Co., 1958; used with permission of McGraw-Hill Book Co.)

(1)   Describe a heat-treat process designed to produce this structure in a 20.15% alloy.

(2)   Sketch a typical lamellar microstructure and show what phases would be present at around 550°C. Also indicate their compositions. Now sketch the microstructure after cooling from 550 to 0°C, showing what phases are present and indicate compositions.

(c)   In both the 8 and 20.15 at. % alloys, we obtained a lamellar structure. What are the major differences in the reactions that produced these structures?

12.3   Equation 12.15 is an expression for the rate of eutectoid growth assuming that the solute redistributes to the two product phases entirely by grain boundary diffusion along the eutectoid–parent interface. Use the model of Fig. 12.21 appropriately modified for the eutectoid reaction and derive expressions for the constant K in Eq. 12.15 and for the constant A in Eq. 12.16. Your analysis should be similar to the Zener treatment for pearlite growth that is described on p. 270.

12.4   In the derivation of Eq. 12.11, Hillert[39] has taken the diffusion distance as $S_\alpha/2$, whereas we used $S_0/2$ on p. 270. The term S_α is the width of the α plates. Show that with the $S_\alpha/2$ diffusion distance Eq. 12.11 may be written as

$$R = \frac{2D\,\Delta T_E(1-S_{min}/S_0)}{f_\alpha(1-f_\alpha)(C_\beta-C_\alpha)S_0}\left[\frac{1}{|m_\alpha|}+\frac{1}{m_\gamma}\right]$$

where for steel the β phase would be Fe_3C and the α phase would be ferrite.

CHAPTER 13
MARTENSITIC TRANSFORMATIONS

The solid-state phase transformations discussed in Chapters 10–12 as well as the solid–liquid transformation of Chapter 9 are all generally referred to as *nucleation and growth transformations*. The rate at which volume transformation occurs in these cases can generally be described by an Avrami type of kinetic equation. There exists, however, a very important class of solid-state phase transformations, called *martensitic transformations*, which do not follow the Avrami-type kinetic equation, and, consequently, these transformations are not classed as nucleation and growth transformations. Physically, however, there is a nucleation stage and a growth stage in a martensitic transformation, but the growth rate is so high that the rate of volume transformation is controlled almost entirely by the nucleation stage.

If a steel is heated into the γ region and then quenched at a sufficiently high rate a martensitic transformation occurs. In plain carbon steels with carbon contents greater than 0.6 wt % carbon, the fcc austenite transforms into a metastable body-centered-tetragonal (bct) phase. This phase grows as platelets in the austenite at such extremely high rates that the plate growth is finished within less than 0.0001 seconds of the time nucleation occurred. This bct phase in steel was first named *martensite* by Floris Osmond in 1895 in honor of the German metallographer, Adolf Martens. The formation of a new phase at such fantastically high growth rates as found in steel martensite has now been observed in a number of other alloys and in some pure metals; see Table 13.1. By analogy with the steel martensite reaction this class of rapid growth transformations is generally called martensitic transformations. The martensitic transformations in plain carbon steel proceeds from an equilibrium high-temperature phase (fcc-austenite) to a nonequilibrium low-temperature phase (bct-martensite). Since the martensite phase is metastable it only forms on very rapid cooling. In addition, a moderately low-temperature anneal will

457

Table 13.1.   Martensitic transformations in Nonferrous Metals (Ref. 12, p. 97)

Material and Composition	Structural Change	Habit Plane
Pure Ti	bcc → hcp	{8, 8, 11} or {8, 9, 12}
Ti–11% Mo	bcc → hcp	{334} and {344}
Ti–5% Mn	bcc → hcp	{334} and {344}
Pure Zr	bcc → hcp	
Zr–2½% Nb	bcc → hcp	
Zr–¾% Cr	bcc → hcp	
Pure Li	bcc → hcp (Faulted)	{144}
	bcc → fcc (Stress Induced)	
Pure Na	bcc → hcp (Faulted)	
Cu–40% Zn	bcc → fc tet (Faulted)	~{155}
Cu–11 to 13.1% Al	bcc → fcc (Faulted)	~{133}
Cu–12.9 to 14.9% Al	bcc → Orthorhombic	~{122}
Cu–Sn	bcc → fcc (Faulted) and	~{lkk}
	bcc → Orthorhombic	
Cu–Ga	bcc → fcc (Faulted) and	~{lkk}
	bcc → Orthorhombic	
Au–47.5% Cd	bcc → Orthorhombic	{133}
Au–50 at. % Mn	bcc → Orthorhombic	
Pure Co	fcc → hcp	{111}
In–18 to 20% Tl	fcc → fc tet	{011}
Mn–0 to 25% Cu	fcc → fc tet	{011}
Au–50 at. % Cu	fcc → Complex orthorhombic (order ⇌ disorder)	
U–0.4 at. % Cr	Complex tetragonal	Between $(1\bar{4}\bar{4})$
U–1.4 at. % Cr	→ complex orthorhombic	and $(1\bar{2}3)$
Pure Hg	Rhomb → bc tet	

cause the metastable bct martensite to decompose by a nucleation and growth reaction into the equilibrium structure of ferrite plus cementite. In many martensitic transformations, however, the low-temperature phase is itself an equilibrium phase rather than a metastable phase. In these cases the phase transformation occurs by the fast growth martensitic mode even with very slow cooling rates. Such is the case, for example, in the martensitic transformation of the pure metals and many of the alloys, notably Au–Cd and In–Tl, described in Table 13.1. The transformations in these systems just naturally occur martensitically and there is no need of a rapid quench to secure this fast growth mode of transformation as there is in steel. It is not unusual even in nonferrous martensites to refer to the high-temperature phase as austenite and the low-temperature phase as martensite, and this extended terminology will be used in this chapter.

Perhaps the most important aspect in martensitic reactions involves a special crystallographic relationship between the martensite phase and the parent phase which allows a fast growth mechanism to operate. The crystallography of martensitic transformations is very similar to the crystallography involved in deformation twinning. Since the twinning case is easier to understand it will be discussed first in order to provide some background and insight into the more complicated martensite case.

13.1 TWINNING

It was illustrated on pp. 13–15 that if one simply changes the stacking sequence in an fcc metal a coherent twin boundary results. These twin boundaries were briefly discussed in Chapter 7 where it was shown that coherent twin boundaries have very low energies (see pp. 180 and 187). In this chapter we will assume that all twin boundaries are coherent unless otherwise stated. Twins may form in a crystal either during deformation or during crystal growth, and accordingly twins are generally classified as either deformation twins or growth twins. Growth twins may form in a crystal whether it is growing from a vapor, a liquid, or a solid. It is quite common for fcc metals to form growth twins during recrystallization, and these twins are more often called annealling twins. It is not common for growth twins to form in metals solidifying from the melt; however, some metallic dendrites have a twin down their growth axis. Also, the graphite in cast iron and the silicon in Al–Si eutectic alloys contain many growth twins, apparently to facilitate the severe branching that occurs in these irregular eutectic structures.

For our purpose here we are only concerned with deformation twinning; growth twins have been mentioned only to clarify that twins may form by two distinctly different mechanisms. It was mentioned in Chapter 3 (p. 56) that twinning offers a mode for plastic deformation in metals. Since slip generally occurs more easily than twinning in metallic materials, twinning only becomes a significant mechanism for plastic deformation in noncubic metals, where few slip systems are available [for example, hcp metals and Sn(bct)]. However, in bcc metals deformation twinning does become significant at low temperatures. Deformation twinning is rare in fcc metals and it can only be made to occur with some difficulty at very low temperatures (Ref. 1, p. 777).

A. FORMAL CRYSTALLOGRAPHIC THEORY

Treatments of the formal crystallographic theory of deformation twinning may be found in Refs. 1–3, and this presentation is based largely after that

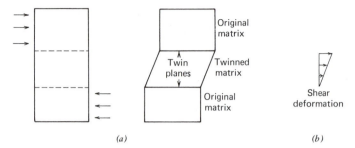

(a)   *(b)*

Figure 13.1   (*a*) Twin formation by a shear stress. (*b*) The shear deformation.

of Refs. 2 and 3. The deformation twinning process is illustrated schematically in Fig. 13.1(*a*). From a macroscopic view the applied shear stress causes a kink offset to form in the original single-crystal matrix as shown. The boundaries between the original matrix and the twin matrix are termed twin planes. The atomic arrangement in the twin matrix forms a mirror image of the atomic arrangement in the original matrix across the twin planes [see Fig. 7.13(*a*)]. There are two fundamental properties of the twinning process that form the basis for the formal description of the crystallography.

1.   The deformation is pure shear. This is illustrated in Fig. 13.1(*b*), where it may be seen that all motion occurs parallel to the shear stress involved in the process.

2.   The twin transformation must preserve the lattice structure. If one describes the crystal structure of the parent crystal with a unit cell, then the unit cell of the twin crystal must be the same as that of parent.* The only difference will be some kind of a rotation as is illustrated in Fig. 13.2.

It is apparent, therefore, that twinning may be thought of as a pure shear transformation that preserves the lengths of the unit cell vectors and, also, their mutual angles. The problem of describing the crystallography of the twinning reaction may be stated more generally as follows: For a pure shear transformation we must find three noncoplanar vectors that meet the following three requirements:

1.   They must retain their lengths upon transformation.

2.   They must retain their mutual angles upon transformation.

3.   They must be rational; that is, they must pass through atom centers in the lattice.

---

* This statement is not strictly correct in some of the more complex crystal structures where twinning restores a multiple lattice (see Ref. 2, p. 376).

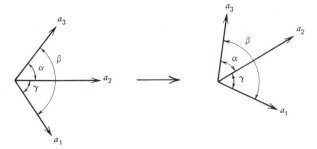

Figure 13.2   Rotation of the noncoplanar triple of vectors describing the unit cell.

The problem may be broken into two parts. First, we determine all planes that possess invariant atomic arrangements upon transformation, and second, we locate three vectors in those planes that have invariant mutual angles upon transformation.

Undistorted Planes.   Any plane upon which the atomic arrangement is the same before and after the twinning shear will be termed an undistorted plane. Consequently, our three noncoplanar vectors must be in these planes in order to satisfy requirement 1.

An edge view of the twinning plane is shown in Fig. 13.3. This twinning plane (sometimes termed *composition plane*) is the interface between the parent matrix and the twin matrix in Fig. 13.3. Since this plane belongs to

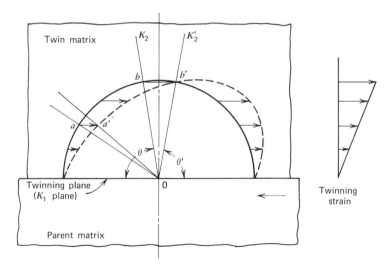

Figure 13.3   Effect of twinning strain upon an inscribed semicircle in the twin matrix.

the parent matrix as well as the twin matrix, it must be an undistorted plane. It turns out that there are only two undistorted planes involved in twinning. The twinning plane is the most obvious one and is called K_1. To locate the remaining undistorted plane imagine that the twin has not yet formed above the twinning plane of Fig. 13.3, and inscribe a semicircle on the original matrix above the twinning plane as shown. By allowing the original matrix region above the twinning plane to now undergo the pure shear strain shown on the right of Fig. 13.3, we produce a twin matrix above the twinning plane. The twinning shear strain causes each point of the circle to move to the right parallel to the twin plane by an amount proportional to its distance above the plane. Hence, the circle distorts into the elipse shown with the dashed line on Fig. 13.3. Consider the line Oa on the original matrix. After twinning this line moves to Oa', which is shorter than Oa, and therefore Oa is not an invariant length. If you now consider all points along the semicircle you will see that except for point b the twinning process changes the distance of all of these points from the origin. Therefore, direction Ob is the only direction above the twinning plane along which lengths are invariant upon twinning. We now define plane K_2 as the plane perpendicular to the page containing direction Ob. Since Ob is invariant, and since the shear does not change distances perpendicular to the page, plane K_2 is a second undistorted plane. K_2' is defined as the position of plane K_2 after the twinning shear strain operates. Two angles, θ and θ', are also shown on Fig. 13.3. Since the motion from b to b' is parallel to the K_1 plane, it is apparent from a little geometry that $\theta = \theta'$. The twinning shear strain may be easily related to the angle θ; see Problem 13.1.

Invariant Mutual Angles. We have located two planes invariant with respect to atomic arrangement, K_1 and K_2, and it is now necessary to find three noncoplanar vectors in these planes which retain their mutual angles upon shear. To do this consider the three-dimensional sketch of the K_1, K_2, and K_2' planes shown in Fig. 13.4(a). We define the common intersection of

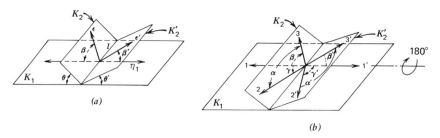

Figure 13.4 Twin of the second kind.

these three planes as I. We also define some directions that will turn out to be our required noncoplanar vectors:

$\eta_1 \equiv$ a vector in K_1 perpendicular to intersection I.
$\varepsilon \equiv$ any arbitrary vector in K_2 making angle β with η_1.

After the twinning shear, vector ε becomes ε' and angle β becomes β'. Because of the fact that $\theta = \theta'$ you may show yourself with a little geometry that $\beta = \beta'$. Hence, the angle between any given arbitrary vector ε and vector η_1 is preserved upon twinning. Therefore, we may choose our three noncoplanar vectors as η_1 plus any two noncolinear vectors in plane K_2. Three such vectors are shown on Fig. 13.4(b). Vector 1 is taken parallel to η_1 and vectors 2 and 3 are arbitrary vectors in plane K_2. In the twinned matrix vectors 2 and 3 become $2'$ and $3'$ and we take vector $1'$ parallel to vector 1 but in the opposite direction. For this triple of vectors, 1-2-3, the three mutual angles, α-β-γ, are preserved upon twinning. From the above discussion $\beta = \beta'$ and similarly $\gamma = \gamma'$. Since K_2 is invariant with respect to atomic arrangement, $\alpha = \alpha'$. Therefore, if the three vectors are rational we will have fulfilled the three requirements set out above. A rational plane is parallel to a sheet of atoms in the crystal and a rational direction is parallel to a row of atoms. Since, we have allowed vectors 2 and 3 to be arbitrary, and since they must both be rational, it follows that we must have plane K_2 rational here. Therefore, the twin described in Fig. 13.4 requires K_2 and η_1 rational and it is called a *twin of the second kind.*

Notice that if we rotate the triple of vectors, $1'$-$2'$-$3'$, $180°$ around direction η_1 (vector $1'$) we reproduce the original triple of vectors, 1-2-3, but in the negative sense. This means that for a twin of the second kind the twin matrix will look like it has been rotated $180°$ about η_1 relative to the parent matrix, and vice versa. If it bothers you that the triple of vectors reproduce in the negative sense, consider how you measure lattice orientations. The orientation relation could be measured from Laue pictures of the parent and twinned matrix. Positive and negative directions along the lattices have no meaning here, as the orientation is solely determined by atomic structure along lattice directions; that is, if you rotate a crystal having twofold symmetry $180°$ you obtain the same Laue pattern.

This brings up an important point. The above theory is purely a formal theory; it tells us nothing about how the atoms have moved. As twinning occurs the atoms of the parent matrix do not rotate a full $180°$ about some origin. Each atom actually moves only a small but equal vector relative to its neighbor, and this cooperative motion produces a different crystal structure that *looks like* a $180°$ rotation of the old crystal structure. We will discuss this further on p. 467.

If you examine Fig. 13.4(a) you will notice that the arbitrary vector ε was required to lie in plane K_2. If we repeat the above treatment with ε in plane K_1 we obtain the *twin of the first kind*. For this case we define the following directions:

$\eta_2 \equiv$ a vector in K_2 perpendicular to the intersection.

$\varepsilon \equiv$ any arbitrary vector in K_1 making angle β with η_2.

Also, η_2' is the position of η_2 in plane K_2' after twinning and β' is the angle between η_2' and vector ε. Again, by simple geometry you may show yourself that $\beta = \beta'$. Hence, the angle between η_2 and any arbitrary vector in K_1 is the same before and after the twinning shear. Consequently we chose our three noncoplanar vectors as any two arbitrary vectors in K_1 along with η_2, as illustrated for the triple of vectors, 1-2-3, in Fig. 13.5(b). In the twinned matrix vectors $2'$ and $3'$ are simply taken in the negative sense of vectors 2 and 3, and vector 1 becomes $1'$. The atomic arrangement along vector 2 will be identical to that along vector $2'$, and similar arguments hold for vectors $1 \rightarrow 1'$ and $3 \rightarrow 3'$. The mutual angles are preserved since $\alpha = \alpha'$, $\beta = \beta'$, and $\gamma = \gamma'$. By arguments similar to above, plane K_1 must be rational and direction η_2 must be rational. Hence, we have satisfied the three requirements set out above in a second way, and the twin of Fig. 13.5 is called a *twin of the first kind*. It is illustrated in Fig. 13.5(b) that for this type twin, a 180° rotation of the triple of vectors, $1'$-$2'$-$3'$, about an axis perpendicular to K_1 causes them to coincide with the original triple of vectors, 1-2-3, in the positive sense. Hence, in the first kind twin the twin matrix looks like it has been rotated 180° relative to the parent matrix about a normal to the twinning plane.

Notice that the twinning plane must be rational for the first kind twin but for a second kind twin one has an irrational twinning plane. For the twins in

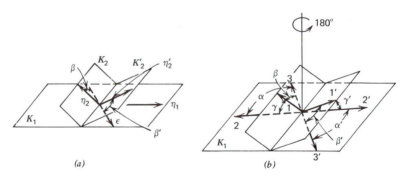

(a)   (b)

Figure 13.5   Twin of the first kind.

Table 13.2 Twinning Modes

Metal	K_1	η_1	K_2	η_2
bcc	$\{112\}$	$\langle11\bar{1}\rangle$	$\{11\bar{2}\}$	$\langle111\rangle$
fcc	$\{111\}$	$\langle11\bar{2}\rangle$	$\{11\bar{1}\}$	$\langle112\rangle$
hcp: Cd, Mg, Ti, Zn, Co	$\{10\bar{1}2\}$	$\langle10\bar{1}\bar{1}\rangle$	$\{10\bar{1}\bar{2}\}$	$\langle10\bar{1}1\rangle$
hcp: Mg	$\{10\bar{1}1\}$	$\langle10\bar{1}\bar{2}\rangle$	$\{10\bar{1}\bar{3}\}$	$\langle30\bar{3}2\rangle$
hcp: Zr, Ti	$\{11\bar{2}1\}$	$\langle11\bar{2}\bar{6}\rangle$	(0001)	$\langle11\bar{2}0\rangle$
hcp: Zr, Ti	$\{11\bar{2}2\}$	$\langle11\bar{2}3\rangle$	$\{11\bar{2}\bar{4}\}$	$\langle22\bar{4}3\rangle$
bct: Sn	$\{101\}$	$\langle\bar{1}01\rangle$	$\{\bar{3}01\}$	$\langle103\rangle$

Adapted from Ref. 1; used by permission J. W. Christian.

cubic, hexagonal, and trigonal metals K_1, K_2, η_1, and η_2 are all rational; these twins have been termed *compound twins*.[2] The twinning elements for some cubic and hexagonal metals as well as for bct tin are listed in Table 13.2.

B. ATOM MOTION DURING TWINNING

In order to obtain some insight into what sort of motion the atoms themselves undergo upon deformation twinning, it is helpful to examine in some detail the specific effects of twin formation upon atom arrangement in a particular case. We will consider the fcc twin. In Section 1.1B it was shown that an fcc twin causes a change in the stacking sequence of the {111} planes from ..ABCABC.. to ..ACBACB... The effect of the change of stacking sequence upon atom arrangement was illustrated in Fig. 1.12 by constructing an edge view of the (111) planes. In this view only those atoms on (111) planes are shown that lie in the ($\bar{1}$10) trace, as illustrated in Figs. 1.11 and 1.12.

This same geometrical scheme may be used to illustrate the atom motion that occurs as a deformation twin forms in fcc crystals. An edge view of (111) planes is shown in Fig. 13.6 with the atoms on the ($\bar{1}$10) trace illustrated by the symbol ○. The relationship of the rectangle 1-2-3-4 to the fcc unit cell is shown in Fig. 1.11. The positions of the atoms in the twin matrix above the twinning plane are shown by the symbols □. It is seen that the dashed rectangle connecting atoms in the twin matrix forms a mirror image across the twinning plane of the solid rectangle connecting atoms in the parent matrix, thus illustrating the twin relationship.

The nature of the atom motion occurring here is illustrated by considering the twinning shear shown on the left of Fig. 13.6. On the first (111) plane above the twinning plane the atoms move by the vector $\bar{t}$, which causes this plane of atoms to undergo the required transition $B \rightarrow C$. On

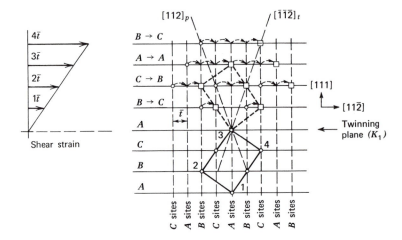

Figure 13.6   Edge view of (111) planes in fcc crystal showing atoms on the ($\bar{1}$10) trace. Original atom sites labeled ○, and atom sites after twinning shear labeled □.

each succeeding plane above the twinning plane the atoms shift by an additional vector $\bar{t}$ as is illustrated by the arrows and by the shear strain diagram of Fig. 13.6; that is, the shift is $2\bar{t}$ on the second plane up, $3\bar{t}$ on the third, and so on. Hence, each atom moves relative to its neighbor by the same vector $\bar{t}$. This is a *cooperative motion* since all atoms move through the same vector relative to their neighbor. As one moves up from the twinning plane the cooperative movements add up so that the total strain becomes large. For example, at four planes above the twinning plane the total strain is proportional to $4\bar{t}$, and at n planes above it is proportional to $n\bar{t}$. Hence, a large strain is produced by many small but cooperative atom movements.

It is quite interesting that the vector $\bar{t}$ is $[11\bar{2}]a/6$, which is the Burgers vector of the Shockley partial dislocation (see p. 113). Hence, one could produce this twin by causing Shockley partial dislocations to glide. This mechanism would require one dislocation to glide over the first plane above the twinning plane, two dislocations over the second plane up, three over the third, four over the fourth, and so on. To produce this successive increase in the number of glide dislocations as one moves up from the twinning plane a pole mechanism has been postulated. Suppose one had a screw dislocation vertically upward in the [111] direction with a b vector extending from one (111) plane to the next (that is, $b=[111]a/3$). This dislocation would distort the (111) planes into a vertical helix with the dislocation line as the helix axis. If the Shockely partial dislocation were to rotate around the vertical screw dislocation line (just as the rope of a tether ball rotates around the tether pole), the partial dislocation would automatically move up one plane from the twin plane upon each 180° rotation because of the helical distortion due to the vertical screw dislocation. There are some difficulties with this particular pole mechanism in fcc metals, but a pole mechanism has been proposed for twinning in bcc metals (see Ref. 4, pp. 31–33 and 95–101). Although conclusive evidence does not yet exist that pole mechanisms are responsible for the formation of twins in metals, it is generally held that some kind of dislocation motion is responsible for twin boundary movement.

The atom motions illustrated in Fig. 13.6 may be related to the formal crystallography approach of the last section by determining the K_1, K_2, η_1, and η_2 values in Fig. 13.6. The K_1 plane is shown directly in Fig. 13.6 as the twinning plane and it is seen to be the (111) plane. The K_2 plane may be found by first considering which atoms above the twin plane are the same distance away from the atom at 3 before and after the shear. Examination of Fig. 13.6 shows that this condition is obtained for certain atoms on the second and fourth (111) planes above the twin plane. Connecting a line through these atoms and the atom at point 3 one obtains the $[112]_p$ direction in the parent matrix and the $[\bar{1}\bar{1}2]_t$ in the twinned matrix. Hence, the K_2 plane is the plane perpendicular to the page containing the $[112]_p$ direction, which is found on Fig. 1.11(a) to be the $(\bar{1}\bar{1}1)$ plane. By inspection it should be clear that η_2 is the [112] direction and η_1 the $[11\bar{2}]$ direction. Hence, we have for the fcc twin $K_1=\{111\}$, $K_2=\{\bar{1}\bar{1}1\}$, $\eta_1=\langle11\bar{2}\rangle$, and $\eta_2=\langle112\rangle$. Notice that this is a compound twin.

The 180° relationships will now be illustrated. If the dashed rectangle connecting atoms in the twin matrix is rotated 180° about $[11\bar{2}]$ direction, it will coincide with the solid rectangle of the parent matrix. Also, rotation of the dashed rectangle 180° about an axis perpendicular to the twinning plane at the atom numbered 3 causes the twinned sites labelled □ to fall directly on parent sites labelled ○. These operations illustrate the 180° rotation relationship between twin and parent matrices in the compound twin. Notice that the atom motions causing the twinning do not rotate the entire original lattice 180°. The small cooperative atom movements upon twinning produce a new lattice orientation that *looks like* a 180° rotation of the original parent lattice.

C. TWIN FORMATION

As one increases the applied stress upon a metal that undergoes deformation twinning, a stress level is eventually reached at which twinning begins. Twin platelets nucleate within individual grains and rapidly propagate across the grains producing an appearance such as shown in Fig. 13.7(a). The flat sides of the twin plates are parallel to the K_1 planes. The twins

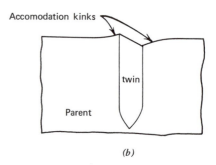

(a)   (b)

Figure 13.7   Twinning platelets formed (a) in the interior of a grain and (b) intersecting a free surface.

grow predominantly in directions parallel to K_1 planes with a smaller growth in directions perpendicular to K_1 planes, thereby producing the plate morphologies. The leading edge of the twin plates taper to a fine point so that in the region of this tip the twin–parent interface cannot be everywhere parallel to K_1 planes. Hence, in these regions the twin boundary must be partially coherent with a few mismatch dislocations present; see Fig. 7.13(b). It is common to observe the sharp tips of twin plates become squared-off upon annealing. By forming a blunt end the twin–parent interface again becomes a coherent twin boundary (the plane of the blunt end will be parallel to one of the $\{hkl\}$ set of K_1 planes). This change lowers the surface tension γ and gives rise to a net reduction of the thermodynamic potential γA.

It is apparent from the shear strain diagram of Fig. 13.6 that as the twin plate thickens in directions perpendicular to K_1 a considerable strain is required in the parent matrix if coherency is to be maintained at the twin–matrix interface. This is perhaps best illustrated by observing the cross section of the intersection of a deformation twin with a free surface as shown in Fig. 13.7(b). The free surface at the twin becomes cocked at an angle to its original orientation and the neighboring parent matrix is severely strained to maintain the coherent twin–parent interface. As shown, the strained parent region at the surface is called an accommodation kink. In metal systems it is virtually always found that the strain in the parent matrix causes the yield stress to be exceeded and, therefore, plastic flow has occurred in these regions.

If one takes a thin rod of cadmium metal and bends it back and forth a distinct clicking sound will be heard. This sound is produced by the deformation twinning that is occurring in the cadmium. The rate of growth of the twin plates is so high that sound waves are generated. It is believed that the growth rate of the twin plates in such cases approaches the speed of sound. For a more complete discussion of deformation twin formation see Section 87 of Ref. 1 and Sections 16.4–16.7 of Ref. 3.

13.2 CRYSTALLOGRAPHY OF MARTENSITIC TRANSFORMATIONS

The study of martensitic transformations has been one of the most active areas of research in physical metallurgy over the past 30 years. This is illustrated by the frequency with which review articles devoted to this subject have appeared in the literature; see Refs. 5–14. This active interest is probably due mainly to the great technological importance of martensite transformations in iron-base alloys. It has been sustained in part, however, by the success of theoretical treatments describing the crystallographic aspects of the transformation.

A. EXPERIMENTAL

In this section we discuss four experimental characteristics that are generally observed in martensitic transformations and are very significant for understanding the crystallography.

It is often found that the martensite phase appears with a distinct plate morphology within the parent matrix. Just as in deformation twinning, when these plates intersect a free surface they produce a characteristic type of surface relief that is shown schematically in Fig. 13.8. This surface relief results from a sudden shape change and it has recently been studied extensively using interference microscopy (see Ref. 13, Chapter 2). The more traditional way of studying this shape change is by placing fine scratches (called fiducial lines) on the surface prior to transformation. Careful examination of these lines following the transformation has

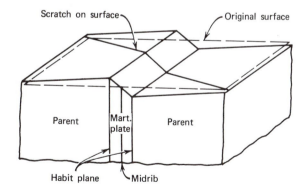

Figure 13.8 Surface relief produced by formation of a martensite plate.

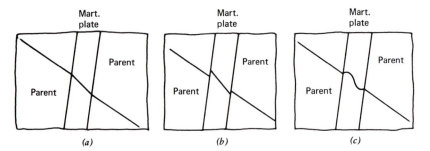

Figure 13.9 Possible distortion of a fiducial scratch. (*a*) Observed results, (*b*) loss of coherency at interfaces, and (*c*) elastic distortion in matrix. (From Ref. 1; used by permission of J. W. Christian.)

revealed considerable information about the nature of the shape change involved in martensitic transformations.

Figure 13.9(*a*) shows the type of result generally found following the transformation. There is a displacement of the parent lattice across the plate. Two important things are *not* observed and these are also illustrated in Fig. 13.9. First, it is not observed that a discontinuity results in the fiducial line at the martensite–parent interface as shown on Fig. 13.9(*b*). This observation is consistently confirmed on the optical microscope up to its limit of magnification, about 1000×. We will call this a macroscopic scale (compared to atomic dimensions) and we conclude that on the macroscopic scale the martensite–parent interface remains coherent. Hence we have our first significant characteristic:

1. On the macroscopic scale the habit plane is an invariant plane or, alternatively, a zero-distortion and zero-rotation plane (just as the K_1 plane in twinning).

Second, it is not observed that the fiducial lines in the martensite plates become nonlinear as shown in Fig. 13.9(*c*). This is true for all different scratch and surface orientations and it means that the free surface at the martensite plate remains planar. Hence, the transformation causes straight lines to transform into straight lines and planes to transform into planes. Such a transformation is called *homogeneous*. The strain that produces a homogeneous deformation with an invariant habit plane is called an *invariant plane strain*. In this type of strain the displacement of any point is a linear function of the distance of that point from the invariant plane. The simple shear that occurs in twinning and is shown in Fig. 13.10(*a*) is an example of invariant plane strain. Martensitic transformations generally involve the more complicated type of invariant plane strain shown in Fig.

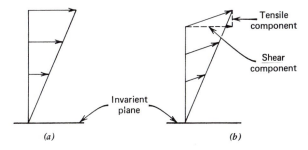

Figure 13.10 (*a*) The simple shear strain of twinning. (*b*) The invariant plane strain of martensite.

13.10(*b*), where the displacement is at a slight angle to the invariant plane. Hence, the martensitic transformations involve a simple shear combined with a uniaxial tension or compression normal to the habit plane. Examples of these strain components for various martensites are given in Table 13.3. The second significant characteristic may then be stated:

2. On a macroscopic scale the shape strain in martensite is an invariant plane strain.

It is customary to specify the habit plane of the martensite as the $\{hkl\}$ planes of the *parent phase* that are parallel to the physical plane of the

Table 13.3 Crystallographic and Deformation Data on Several Martensites. (Data Taken Mainly from Ref. 7, p. 128).

System	Structure Change	Habit Plane	Direction Displacement	Shear Strain Component	Normal Strain Compenent
Fe–C (1.35% C)	fcc → bct	~{225}	$\langle \bar{1}\bar{1}2 \rangle$	0.19	0.09
Fe–C (1.8% C)	fcc → bct	~{259}	$\langle \bar{1}\bar{1}2 \rangle$		
Fe–Ni (30% Ni)	fcc → bcc	~{9, 22, 33}	~$\langle \bar{1}56 \rangle$	0.20	0.05
Fe–Ni–C (22% Ni, 0.8% C)	fcc → bct	~{3, 10, 15}	~$\langle \bar{1}32 \rangle$	0.19	
Pure Ti	bcc → hcp	~{8, 9, 12}	~$\langle 11\bar{1} \rangle$	0.22	
Au–Cd (47.5% Cd)	bcc → orthorhombic	$\left\{\begin{array}{r} 0.70 \\ -0.69 \\ 0.21 \end{array}\right\}$	$\left\langle\begin{array}{c} 0.66 \\ 0.73 \\ 0.18 \end{array}\right\rangle$	0.05	
In–Tl (20% Tl)	fcc → fct	{011}	~$\langle 01\bar{1} \rangle$	0.02	

martensite plate. To measure this quantity one needs to know the crystallographic orientation of the parent phase relative to the martensite plates that form within it. In high-carbon steels the martensitic transformation is generally incomplete at room temperature so that retained austenite is present. Hence, the crystallographic orientation of the retained austenite is often measured by room-temperature x-ray techniques and then the orientation of the plates relative to the retained austenite is measured by examining the trace of the plates on two surfaces of the austenite at about 90° to each other. Extensive experimental investigations have established the following fact, which we list as the third significant experimental characteristic concerning the crystallography of martensitic transformations.

3. The martensite habit plane is irrational.

This fact is probably best illustrated by the experimentally observed habit planes on a standard stereographic triangle as illustrated in Fig. 13.11. For each of the three alloys it may be seen that there is a significant scatter in the data from plate to plate. In careful experimental work the habit planes can be measured with a precision of $\pm\frac{1}{2}°$, but the scatter in experimental data is significantly larger than this precision. It is customary to give as a habit plane the Miller indices of some rational plane around which the experimental data cluster, as has been done in Tables 13.2 and 13.3. The irrationality of the habit plane was first well established in a now classical paper published in 1940 by Greninger and Troiano.[15] This paper

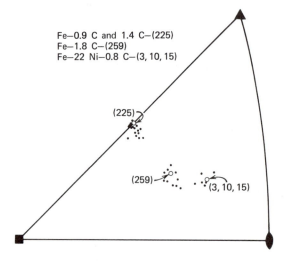

Fe–0.9 C and 1.4 C–(225)
Fe–1.8 C–(259)
Fe–22 Ni–0.8 C–(3, 10, 15)

(225)

(259)

(3, 10, 15)

Figure **13.11** Habit plane orientations of martensite plates in three steels (data on Fe–C alloys from Ref. 15 and on Fe–Ni–C alloys from Ref. 16).

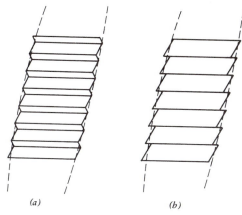

(a)   (b)

Figure **13.12** Substructure of martensite plates (*a*) internally twinned and (*b*) internally slipped (from Ref. 10).

also established that the habit plane in plain carbon steels changes from {225} to {259} when the carbon content increased from 1.4 to 1.8 wt % C. Hence, it is not uncommon in discussing steel martensite to refer to 225 (or 252) martensite and 259 (or 295) martensite. In general, addition of alloying elements to steel causes changes in the observed martensite habit (see Fe–22Ni–0.8C in Table 13.3 and pp. 191–202 of Ref. 10). The habit planes tend to lie in a band on the stereographic projection extending from the (3, 10, 15) pole to the (225) pole,[10] which are both shown on Fig. 13.11.

The final point in this section concerns the substructure of the martensite phase. Two distinct martensitic morphologies are observed in metals. In addition to the plate morphology, which we have been assuming so far in this chapter, one also observes a lath morphology. The lath martensite may be thought of as a series of adjoining parallel laths. The distinction between these two morphologies is discussed more fully in the next section. As pointed out above, when observed in an optical microscope martensite plates generally appear homogeneous, and one would assume that each plate is a single crystal.* However, examination by electron microscopy has revealed that the plate morphology has a very fine substructure. The plates are observed to be composed of stacks of very fine twins as shown schematically in Fig. 13.12(*a*). The twin spacing is on the order of 15–200 Å in steel martensite (for exccellent pictures see Ref. 14 pp. 2346 and 2378–2383). It is generally found that the lath martensite does not contain twins, but a very high dislocation density is present in the laths. It is

---

* This statement is not true for the small shape strain martensites, Au–Cd and In–Tl listed in Table 15.2. In these and some other alloys the martensite forms relatively wide twins observable by optical microscopy (see Fig. 16.37 of Ref. 3 or Fig. 2.12 of Ref. 12).

Table 13.4 Substructure of Some Iron-Base Martensites (Ref. 14, pp. 2354 and 2377).

Alloy System	Substructure of Martensite	Composition (wt %)
Fe–C	Mainly dislocated laths	% C<0.6
Fe–C	Mainly twinned plates	% C>0.6
Fe–Ni	Dislocated laths	% Ni<25
Fe–Ni	Plates twinned near midrib	% Ni>29
Fe–Cr	Dislocated laths	% Cr<10
Fe–Cr–C	Twin density decreases with increasing plate size	8% Cr–1% C

believed that this high dislocation density is due in large part to stacks of parallel regions, Fig. 13.12(b), which slip relative to each other in order to produce the lattice invariant shear described in the next section. Hence, the final significant crystallographic characteristic is:

4. On the microscopic scale martensite plates are observed to be stacks of very fine twins, and martensite laths generally have a very high dislocation density.

To further illustrate this point, data for several iron-base alloys are presented in Table 13.4. It may be seen that the type of martensitic substructure depends upon the alloy composition. The presence of a twinned substructure as shown in Fig. 13.12(a) clearly shows that the habit plane is an invariant plane only on a *macroscopic* scale as has been stated in point (1) above; and, also, the shape strain is homogeneous only on the *macroscopic* scale as stated in point (2) above.

B. THEORY

We will discuss the theoretical treatments of martensite crystallography using the high-carbon Fe–C martensite as an example. In this transformation fcc austenite becomes a bct structure. In a classic paper published in 1924, E. C. Bain proposed a scheme by which an fcc unit cell would be transformed into a bct unit cell. Figure 13.13 shows two fcc unit cells touching each other at a common (010) face. The atom at the center of this (010) face is also at the center of a tetragonal unit cell that is indicated on Fig. 13.13. This tetragonal unit cell is redrawn in Fig. 13.14(a) where it may be seen that it forms a body centered tetragonal unit cell with a c/a ratio of $1/\sqrt{2}$. If this unit cell is contracted by 18% along the $(x_3)_M$ direction and expanded by 12% along $(x_1)_M$ and $(x_2)_M$ directions, one obtains the correct unit cell for the bct martensite of Fe–C alloys. This combined expansion and contraction is often called the *Bain distortion*.

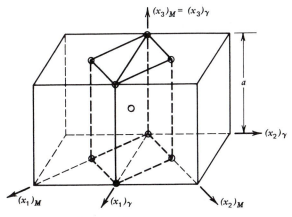

Figure 13.13 The Bain correspondence.

This proposal by Bain required a particular structural unit of the parent lattice to become the unit cell of the martensite lattice. The relationship between these two specific lattices is frequently termed the *lattice correspondence* or sometimes the Bain correspondence. A given direction in the parent matrix $[xyz]_p$ will correspond to a specific direction in the martensite matrix $[x'y'z']_M$. For example, from Figs. 13.13 and 13.14 it should be clear that the Bain correspondence requires

$$[100]_M \text{ cors} \rightarrow [1\bar{1}0]_\gamma$$
$$[010]_M \text{ cors} \rightarrow [110]_\gamma$$
$$[001]_M \text{ cors} \rightarrow [001]_\gamma$$

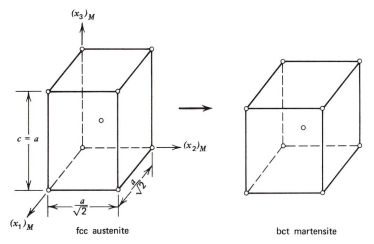

Figure 13.14 The Bain distortion.

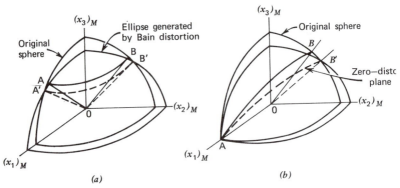

Figure 13.15 (*a*) Effect of Bain distortion upon a sphere inscribed in the austenite matrix. (*b*) Effect of a distortion with one principal strain zero, one greater than zero, and one less than zero.

Similarly, a given plane in the parent matrix $(hkl)_P$ will correspond to a specific plane in the martensite matrix, $(h'k'l')_M$, for example, from Figs. 13.13 and 13.14,

$$(112)_M \text{ cors} \rightarrow (101)_\gamma$$

From point (1) of the previous section it is required that if the Bain correspondence is to apply directly to the Fe–C martensitic transformation it must somehow produce a plane of zero distortion. To determine if there is such a plane associated with the Bain distortion we construct a sphere in the austenite matrix upon the origin of the three principle axes, $(x_1)_M$, $(x_2)_M$, and $(x_3)_M$ as shown in Fig. 13.15. If we allow the Bain distortion [18% contraction along $(x_3)_M$ and 12% expansion along $(x_1)_M$ and $(x_2)_M$], the sphere becomes an elipsoid of revolution. The only points on the ellipsoid that remain at the same distance from the origin as they were prior to the Bain distortion lie along the circle A'-B', and these points originally lay along the circle A-B. If one were to construct spheres of different diameters it becomes clear that the surface $OA'B'$ is a cone of unextended lines. Hence the only invariant vectors associated with the Bain distortion lie along a cone and, furthermore, these vectors are only invariant in length because they have rotated in direction from OAB to $OA'B'$. Hence, it is clear that there is no plane of zero distortion associated with the Bain distortion and, consequently, the Bain correspondence cannot apply directly to the martensitic transformation.

It is illustrated in Fig. 13.15(*b*) that one can obtain a plane of zero distortion following the Bain distortion if one arbitrarily collapses the lattice back to its original position along one axis, such as $(x_1)_M$. The plane OAB' is now a zero-distortion plane, but it has rotated from its original position OAB. It was pointed out by Bilby and Christian (Ref. 7, p. 149) that if an undistorted plane is to be obtained by a homogeneous deformation one of the principal strains must be zero, one must be greater than unity, and the other less than unity. Clearly, the Bain distortion by itself does not satisfy this condition since its principal strains are 1.12, 1.12, and 0.82, whereas the distortion of Fig. 13.15(b) does satisfy the condition because the principal strains are 0, 1.12, and 0.82. In general, lattice correspondences that transform one lattice into another do not satisfy these conditions on the principal strains.

Phenomenological theories of martensitic transformations were put forth in 1953 by Wechsler, Lieberman, and Reed[18] and in 1954 by Bowles and Mackenzie.[19] These theories, which are essentially equivalent, do not explain how the atoms move to bring about the transformation. Rather, they are formal analytical treatments that describe the final and initial crystallographic states of the transformation and are, therefore, able to predict certain restrictions that must apply. In this sense the theories are analogous to thermodynamics, which does not tell us how or if atoms can move from state A to B, but only whether they have the potential to move from state A to B. An excellent description of these theories may be found in Ref. 20 and only a very brief outline is presented here.

The mathematical operations of the theory may be divided into three main operations.

1. Allow a lattice (Bain) distortion to generate the new lattice.

2. Introduce an appropriate *lattice-invariant shear* to obtain a plane of zero distortion.

3. Rotate the martensite matrix so that the zero-distortion plane has its original position.

The first step produces the required lattice structure of the martensite phase. The second step provides that one principal strain be zero in order to obtain the zero-distortion plane. However, the added shear introduced here must not change the new lattice structure generated by step 1. Therefore, this shear must be a *lattice-invariant shear*. To illustrate how one can achieve a lattice-invariant shear consider the rhombus-shaped crystal shown in Fig. 13.16(a). There are two ways in which we may straighten this rhombus up into an overall rectangular shape by shear deformations without changing its crystal structure. The change is accomplished in Fig. 13.16(b) by slip along parallel planes and in Fig. 13.16(c) by generating stacks of twins. These lattice-invariant shears are thought to be present within martensite plates as shown schematically in Fig. 13.12. Hence, the formal theory requires the martensite phase to have an internal substructure of twin stacks or to be severely slipped along parallel planes.

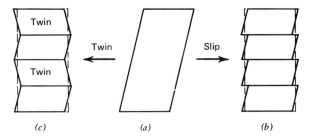

(c)   *(a)*   *(b)*

Figure 13.16   Lattice-invariant shears.

The input data for the theory consists of (a) the structure and lattice parameters of parent and martensite phases, (b) the lattice correspondence, and (c) the lattice-invariant shear. From these data the theories are able to predict (a) the habit plane, (b) the shape strain, and (c) the crystallographic orientation relationship between parent and martensite phases. The agreement between theory and experiment has been very good for martensites involving low shape strains, notably Au–Cd and In–Tl and, also, the (3 10 15) martensites in iron-base alloys (Ref. 13, p. 28; Ref. 14, p. 2327). Agreement on the (225) martensites in iron-base alloys is not good. However, recent theoretical approaches and experimental evidence indicate that a more complex substructure may be present in these (225) martensites. In conclusion it seems that the crystallographic theories present a rather good description of the martensite transformation. They are able formally to account for the two intriguing crystallographic features of martensites, the internal fine structure and the irrational habit planes displayed by martensites.

13.3   SOME CHARACTERISTICS OF MARTENSITIC TRANSFORMATIONS

A.   COOPERATIVE MOTION, INTERFACE VELOCITY, AND DIFFUSIONLESS CHARACTER

One of the major characteristics of martensitic reactions results from the fact that the structure change is produced by cooperative movement of all atoms. Just as in deformation twinning, each atom moves by the same vector relative to its neighbor. As might therefore be expected, it is observed that if the parent phase is ordered the martensite formed from the parent is also ordered. Because of this cooperative atom movement martensitic transformations are sometimes called *military transformations*. Other transformations where atoms jump across the transformation interface in a random manner are then called *civilian transformations*.[1]

Another distinguishing characteristic of martensitic transformations is the velocity at which the interface moves. As mentioned on p. 457 martensite plates generally grow to their limiting size within less than about 0.0001 seconds after they have been nucleated. The linear growth velocity has been measured to be about 10^5 cm/sec (about one-third the speed of sound) at all temperatures in the range of $-20°C$ to $-200°C$ in Fe–Ni–C martensites (Ref. 7, p. 134). Just as in deformation twinning, formation of martensite plates is often accompanied by audible clicks. When a plate forms it produces a rapid mechanical disturbance in the metal. This disturbance generates an acoustic wave that propagates to the metal surface and eventually produces the audible clicking sound at the observer's ear. The sound is not heard in plain carbon steels because of the competing sounds associated with the rapid quenching necessary to produce the martensite. However, the audible sound is very evident in, for example, the Au–Cd transformation where only slow cooling is necessary to produce the equilibrium martensite phase.

The above two features, cooperative atom motion and very rapid growth velocities produce a third characteristic of martensitic transformations. The composition of the martensite phase is identical to that of the parent phase. As an example consider a eutectoic steel, 0.77% C. If this steel is cooled slowly we expect it to consist of two phases, ferrite of composition 0.022% C and cementite of composition 6.7% C. As shown in the previous chapter these product phases change their composition from the parent phase by a diffusion process. If the eutectoid steel is rapidly quenched to form martensite, the martensite phase will have the same composition as the parent phase, 0.77% C. Therefore, no diffusion is required and, consequently, martensitic transformations are sometimes referred to as *diffusionless transformations*.

B. MORPHOLOGY

In iron-base alloys the martensite phase generally displays one of two distinct morphologies. We will term these morphologies *lath martensite* and *plate martensite*. Because the recognition of these two morphologies occurred gradually over the years there have been a number of different names used for these martensites. A partial list of the most popular alternative names is given in Table 13.5. The acceptance of the lath and plate terminology seems likely; see Ref. 14, p. 2343. In a few high alloy steels a third morphology, *sheet martensite*, is observed.

As shown in Table 13.4 Fe–C alloys form a predominantly lath martensite below 0.6 wt % C and a plate martensite above 0.6 to 1.0 wt % C. When the latch microstructure is examined in an optical microscope at high magnifications, $500 - 1000 \times$, it has a structure that is so

Table 13.5   Alternative Names for
Ferrous Martensites

Lath Martensite	Plate Martensite
Massive	Acicular
Dislocation	Twinned
Cell	Lenticular
Packet	

fine and fuzzy looking that much of it is clearly too small to be resolved in the optical microscope; see Fig. 13.17(a). The smallest clearly defined structural units in Fig. 13.17(a) are called packets, or sheaves, or blocks. These packets often appear in adjacent parallel strips, as at the right of this picture, but they will also sometimes have a more blocky appearance as at the left of the picture. Packets such as those at the left having dimensions on the order of 2×10 μm might extend out of the page from 50 to 200 μm. The packets exhibit tilts on a prepolished surface similar to that described for plate martensites on p. 469. Transmission electron microscopy has revealed that the packets themselves have a substructure (see pictures in Ref. 21). Each packet appears to consist of parallel plate-like subgrains having a small orientation difference with respect to one another. These thin parallel subgrains within the packets therefore have the shape of laths (i.e., long thin plates, like a blade of grass) and, hence, the name lath martensite. The laths range in thickness from 0.1 μm to a few microns with an average size on the order of 0.2 μm. Hence, the smaller laths are not quite resolvable in the optical microscope. The individual laths usually have an extremely high dislocation density within them, being on the order of 0.5×10^{12} dislocations per cm^2. In a few cases the individual laths are observed to have the fine internally twinned microstructure characteristic of plate martensite. See Ref. 14, p. 2345, Ref. 12, p. 74, and Ref. 36 for further discussions of lath morphology.

There are two other important ways in which lath martensite differs from plate martensite in plain carbon steels. First, the habit plane of the lath martensite is close to {111}, whereas the plate habit is either {225} or {259}; see Fig. 13.11. Secondly, the crystal structure of lath martensite is bcc compared to bct for plate martensite.

The morphology of plate martensite is usually very distinct, as shown in Fig. 13.17(b). The individual plates are not situated in adjacent parallel stacks as the laths described above, but are generally at quite distinct angles to each other. The longer plates are wider. One also generally observes a fine line running straight down the center of the plates after they

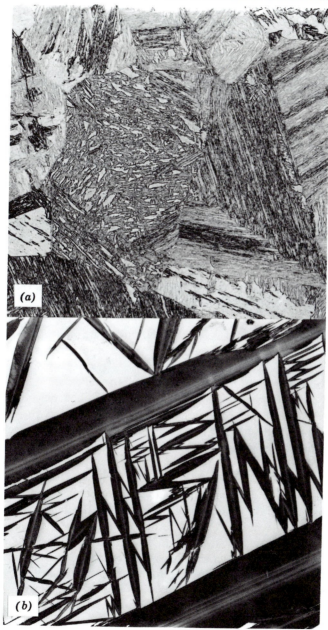

Figure 13.17 (*a*) Lath martensite in an as-quenched Fe–0.09 C steel, 500×, (*b*) plate martensite in an Fe–22.5 Ni–0.4 C alloy, 150×. (Courtesy of F. X. Kayser, A. Litwinchuk, and D. Diesburg.)

have been etched. This feature is termed the *midrib*; at present the cause of the midrib remains a mystery. There is good evidence, however, that the midrib is the first part of the plate to form and that there is a small difference in the twinned substructure at the midrib region, which, therefore, produces the different etching character.

The mode of formation of these two types of ferrous martensites is quite distinct. The formation of lath martensite is characterized by the growth of adjacent parallel laths. A group of laths sometimes grow simultaneously in a cooperative manner as an array, and sometimes they grow successively by nucleation and growth parallel to the first formed laths. In plate martensite the first formed plates grow across the entire austenite grains. Additional plates then form in the remaining austenite by nucleation and growth between the first formed plates and the grain boundaries. As the austenite is further partitioned by the martensite plate formation, the austenite regions become smaller, and the plates formed therein become smaller. Hence, in plate martensite we have nonparallel plates with relatively large size variations, whereas in lath martensites we have mainly parallel lath formation with relatively uniform lath sizes within packets (Ref. 14, pp. 2345–2352 and Ref. 12, pp. 72–80).

The sheet martensite formed in a few high alloy steels has the appearance of parallel sheets. It is always hcp and is formed parallel to {111} austenite planes. The sheets are so thin that little or no substructure is revealed in the TEM; see Ref. 21.

The morphologies of many nonferrous martensites are quite similar to the plate martensite in ferrous alloys. There are, however, a number of nonferrous alloys (for example, In–Tl, Mn–Cu) where the martensite appears as parallel bands. The bands have a substructure consisting of parallel twinned regions and both the bands and the twins are large enough to be resolved in the optical microscope. Near the martensite–parent interface, however, the twins become quite fine. Hence, this *banded martensite* is similar in appearance to the lath martensite of ferrous alloys, except that the scale is much coarser than for the lath martensite and banded martensite has a twinned rather than a dislocation substructure. The shear component of the banded martensite is generally small compared to the lath and plate martensites. Therefore, it seems apparent that a smaller shape change may be accomodated at the interface by a wider spacing of the twin stacks. For a further discussion of the morphologies of nonferrous martensites see Ref. 12, Section 5.4 and Ref. 9, p. 6.

C.   INTERFACE STRUCTURE

In the case of deformation twinning, the twin boundary interface (K_1 plane) is a plane of complete coherency between parent and twin crystals. The

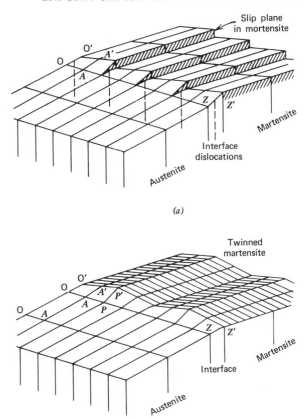

(a)

(b)

Figure 13.18   Schematic models of the martensite–austenite interface structure (from Ref. 9, p. 5).

martensite–parent interface was shown to be coherent only on a macroscopic scale. Because of the twinned or slipped nature of the martensite plates, it is apparent that the martensite–parent interface will not be completely coherent and is more appropriately described as a semicoherent boundary. To accommodate the required shear at the interface some kind of an interface dislocation structure is usually postulated. Figure 13.18 presents schematic models of the interface structure for a slipped and for a twinned martensite (Ref. 9, p. 5). In Fig. 13.18(a) the lattice invariant shear is produced by the set of parallel screw dislocations while in Fig. 13.18(b) it is produced by alternate twinning. The interface in Fig. 13.18(a) lies in the plane defined by the parallel set of interface dislocations. In both pictures the lines OZ and $O'Z'$ are macroscopic vectors on either side of the interface which, you will note, display the macroscopic coherency of the interface. The vectors OA and $O'A'$, however, are lattice vectors on either side of the interface and they display the lattice invariant shear that must be accommodated at the interface. Whatever the detailed fine structure of the interface, it is clear that it must possess the ability to move at extremely high velocities. Interface dislocation arrays would, in principle, possess this ability, and it is generally held that the martensite shape change is accomplished by the cooperative gliding of arrays of interface dislocations.

D. KINETICS OF FORMATION

The temperature at which martensite begins to form upon cooling an austenite phase is generally quite well defined and reproducible. This temperature is termed the martensite start temperature and is given the symbol M_s. In plain carbon and low alloy steels the amount of martensite formed is only a function of how far one cools below the M_s temperature, as is illustrated in Fig. 13.19(a). Eventually a temperature is reached where all austenite has transformed to martensite or where the transformation stops, and this is called the martensite finish temperature, M_f. (The temperature T_E on Fig. 13.19(a) represents the thermodynamic equilibrium temperature between the austenite and martensite phases.)

The M_s temperature is, of course, an extremely important temperature for consideration in the heat treatment of steels. Experiments have shown that addition of virtually any chemical element to steel will lower the M_s,

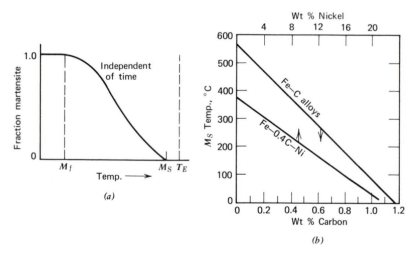

Figure 13.19   (a) The volume fraction martensite as a function of temperature. (b) Variation of M_s temperature with composition in two steels.

provided the element is in solution in the austenite; and, further, interstitial elements are most effective at lowering M_s. This is illustrated by the empirical equation of Steven and Haynes[22] for low alloy steels,

$$M_s(°C) = 561 - 474\,C - 33\,Mn - 17\,Ni - 17\,Cr - 21\,Mo \qquad (13.1)$$

where C is the wt % carbon, Mn the wt % manganese, and so on. This formula should only be regarded as an approximation since it is known that there is a small dependence of M_s on grain size (Ref. 6, p. 604). Figure 13.19(b) shows the composition dependence of M_s predicted for a plain carbon steel and for Fe–0.4% C–Ni steels.

Experiments have shown (Ref. 14, p. 2395) that the M_s temperature is constant to within ±20°C at cooling rates as high as 50,000°C/sec. And, as previously mentioned, the martensite plates (or laths) form in a matter of microseconds. Accordingly, the volume fraction martensite has been found to vary with time in low alloy steels as is shown in Fig. 13.20(a). The transformation is terminated in milliseconds and a continued isothermal hold will not produce any more martensite. Additional martensite may only be obtained by quenching to a lower temperature [see Fig. 13.19(a)]; but again the transformation occurs virtually instantaneously and is then terminated until the temperature is again lowered. Of course, once the temperature reaches M_f no more austenite remains and the transformation is complete. The martensites produced by this type of reaction have been called *athermal martensites*. This term is a bit confusing because the volume fraction martensite is certainly temperature dependent. The quantity that is truly athermal here (independent of temperature) is the rate of interface motion. As previously mentioned the interface velocity is estimated to be on the order of one-third of the speed of sound in the alloy, independent of the temperature of formation. The edgewise growth velocity of plate martensites in Fe–Ni alloys has been measured as about 2×10^5 cm/sec over the temperature range −20 to −195°C (Ref. 14, p. 2400). In Chapter 6 we

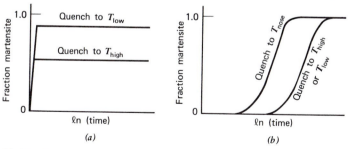

Figure 13.20 Time dependence of martensite formation in (a) athermal and (b) isothermal martensites.

showed that an interface velocity could be expressed as a mobility times a driving force. Since mobility is proportional to a diffusion coefficient, the predominant temperature dependence of velocity is the $\exp(-Q/RT)$ term from the mobility, where Q is an activation energy for the diffusive motion of atoms required for interface advancement. Hence, for athermal martensites it is apparent that the activation energy for the growth mechanism must be essentially zero, since the growth velocity is temperature independent.

Not all martensitic transformations follow the athermal kinetics described by Figs. 13.19 and 13.20(a). It is found that some alloys (notably Fe–Ni–Mn alloys) follow a kinetic equation similar to the Avrami equation, as illustrated schematically in Fig. 13.20(b). If one quenches to some temperature below the M_s and holds, the martensite phase continues to appear as time passes. In contrast to athermal martensites, a continued isothermal hold does produce additional formation of martensite. This type of martensite is termed *isothermal martensite*. The kinetics of these type of martensites displays the typical C curves similar to T-T-T diagrams such as Fig. 12.16. As shown on Fig. 13.20(b) quenching to the nose temperature of the C curves will produce the maximum rate of transformation.

In spite of the fact that the volume transformation rate of isothermal martensites is slower than for athermal martensites, the rate of interface motion in isothermal martensites has been found to be essentially as high as in athermal martensites (Ref. 1, p. 813). Also, it is generally observed in isothermal martensites that once a plate has formed and grown across the austenite grain to a limiting size in a fraction of a second, further growth of this plate does not occur. Therefore, the continued transformation that occurs in isothermal martensites upon holding at temperature must result from nucleation of new martensite plates. The initial increase in the rate of volume fraction transformed is believed to be due to an increase in the nucleation rate. This increase in the nucleation rate is thought to be due to an autocatalytic effect, whereby when one plate runs into another plate conditions are generated which produce nucleation of more plates. The final decrease in nucleation rate is thought to result from the fact that the martensite plates partition the austenite into smaller and smaller regions so that the probability of a nucleation event occurring in these regions decreases. Hence, it appears that the main difference in the kinetics of athermal and isothermal martensites must relate to their nucleation characteristics. In isothermal martensites the number of nuclei available is a function of temperature and time, whereas in athermal martensites the number of nuclei available is a function of temperature but *not* time.

In many Fe–Ni and Fe–Ni–C alloys a significantly different mode of transformation occurs than in the athermal or isothermal martensites

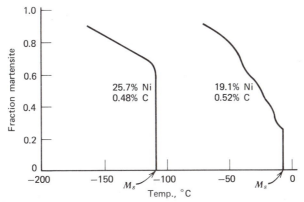

Figure 13.21 Transformation curves for burst martensites (from Ref. 14, p. 2398).

discussed above. At the martensite start temperature a large fraction of the austenite is transformed to martensite within milliseconds as shown in Fig. 13.21 for two Fe–Ni–C alloys. Accordingly these martensites are referred to as *burst martensites*. It is common in these martensites for the plate structure to display zig-zag features as shown on Fig. 13.17(*b*). It is believed that an autocatalytic nucleation occurs in order to produce the large burst of martensite formation at the M_s temperature. Each plate is limited in size by impingement upon another plate or a grain boundary; at this impingement location conditions are generated that nucleate another plate and, hence, a cooperative nucleation occurs producing the zig-zag pattern and the very large volume transformation at one temperature. Apparently, the shape change of the first plate produces a partial accommodation of a second plate on a crystallographic equivalent habit plane in the parent austenite phase. These burst martensites are found to occur only when M_s is below 0°C, and the M_s is observed to be a function of the prior austenite grain size. (See Ref. 14, p. 2398 for further details of these martensites.)

It seems clear that the differences in the three types of martensites discussed here, athermal, isothermal, and burst, are related to differences in nucleation characteristics. At the present time little is known about the factors controlling which type of nucleation characteristics a given alloy will display.

13.4 THERMODYNAMICS

Just as in the previous phase transformations that have been discussed, the martensite phase forms because it has a lower free energy. The temperature dependence of the free energy for the austenite and martensite phases must have the form shown on Fig. 13.22 (the same dependence as in Fig. 8.3). There will be some temperature at which the two phases are in thermodynamic equilibrium and this is designated T_E. The martensite start temperature M_s is located significantly lower than T_E. There is a strong nucleation barrier to the formation of martensite so that a significant supercooling must occur before enough free energy is available to form the martensite phase. The free energy required to form the martensite at the M_s temperature may be estimated from Eq. 9.3 as

$$\Delta G_{\gamma \to M}(\text{start}) = \Delta S_f [T_E - M_s] \tag{13.2}$$

where ΔS_f now refers to the entropy change for the $\gamma \to M$ transformation. Estimates of the curves of Fig. 13.22 (Ref. 14, p. 2392) show that the value of $\Delta G_{\gamma \to M}$ at the M_s temperature in iron-base alloys is on the order of 250 cal/mole (1050 J/mole).

A second interesting feature relating to the thermodynamics of martensitic transformations concerns the diffusionless characteristics of the transformation. Consider as an example the lath martensites in plain carbon steels where an fcc austenite transforms into a bcc martensite. The equilibrium phase diagram between the fcc and bcc phases is shown at the top of Fig. 13.23. The free-energy composition diagram at temperatures T_1 and T_2 are shown directly below the phase diagram. Suppose the austenite phase were quenched to temperature T_1. Now, in order to produce a

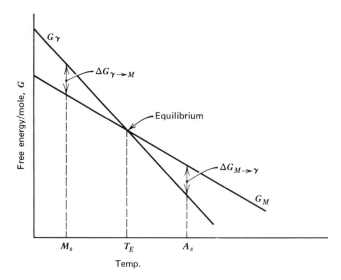

Figure 13.22   Free-energy curves for the austenitic and martensitic phases as a function of temperature.

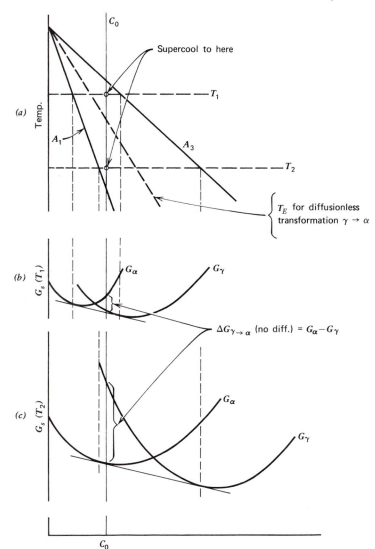

Figure 13.23   Free-energy composition curves at two temperatures illustrating the ΔG for diffusionless transformations.

diffusionless transformation from γ to α the α phase must form with the same composition as the parent γ phase. As shown on Fig. 13.23(b) the $\Delta G_{\gamma \to \alpha}$ (No diff) is a positive quantity at temperature T_1. Similar arguments at temperature T_2 [Fig. 13.23(c)] show that at this temperature the value of $\Delta G_{\gamma \to \alpha}$ (No diff) is a negative number and the reaction may proceed. The temperature at which $\Delta G_{\gamma \to \alpha}$ (No diff) becomes zero would define the equilibrium temperature T_E for the martensitic transformation from the fcc γ phase to the bcc α phase. By drawing free-energy composition diagrams at various temperatures between T_1 and T_2 you may show yourself that $\Delta G_{\gamma \to \alpha}$ (No diff) goes to zero at a temperature that must be roughly midway between the A_1 and A_3 lines at composition C_0. Hence, the equilibrium temperature T_E for the martensite transformation is roughly midway between A_1 and A_3 as shown on Fig. 13.23(a). (See Fig. 3.13 of Ref. 8 for an actual calculation on an Fe–Ni alloy.)

Note that the results here are general and not limited to martensitic transformations. In principle, any phase transformation may occur without a change in composition between parent and product phases. If the parent phase can be supercooled to more than roughly half way between the phase boundaries on the phase diagram, the driving force for the diffusionless transformation becomes negative. There are, in fact, a class of diffusionless phase transformations called massive transformations that are quite different from martensitic transformations; see p. 499.

13.5   THERMOELASTIC MARTENSITES

There are a number of nonferrous martensites that display thermoelastic characteristics, and an understanding of this phenomenon provides some insight into the nature of martensite growth in general. For discussion purposes we will class martensites into two groups, A or B, depending on. the magnitude of their shear components,

A: Large shear component, for example, Fe–C ($\gamma = 0.19$), Fe–30 Ni ($\gamma = 0.20$).
B: Small shear component, for example, Au–Cd ($\gamma = 0.05$), In–Tl ($\gamma = 0.02$).

As previously mentioned, a number of martensites can be retransformed into austenite upon heating by the fast-growth martensitic type of transformation mechanism. Transformation curves for two such martensites, Fe–30 Ni (A type) and Au–Cd (B type) are shown in Fig. 13.24. The reverse martensitic transformation back to the high-temperature phase upon heating begins at a distinct temperature called the austenite start temperature, A_s. The A_s temperature is always significantly above the M_s temperature and the transformation curves display the pronounced hysteresis shown in Fig. 13.24. As shown, the hysteresis is found to be significantly larger in A-type martensites. The width of the hysteresis loop is estimated as 420°C for Fe–Ni, compared to only 16°C for Au–Cd.

The M_s and A_s temperatures are both located on Fig. 13.24. Since the

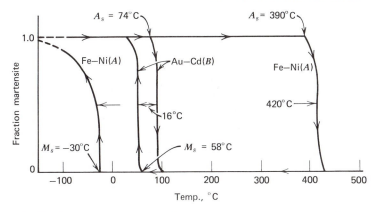

Figure 13.24 Hysteresis loops displayed in transformation curves for a type-A martensite (Fe–Ni) and a type-B martensite (Au–Cd).

austenite formation reaction is the reverse of the martensite formation reaction it may be assumed that the equilibrium temperature lies roughly midway between M_s and A_s (see Ref. 8, p. 180 for limitations of this statement). Hence, one may estimate T_E from the hysteresis loop. It is apparent from this discussion and Eq. 13.2 that we may estimate the free energy required to start the martenstic reaction as $\Delta G_{\gamma \to M}$ (start) $= \Delta S_f[\Delta T(\text{loop})]/2$, where $\Delta T(\text{loop})$ is the temperature width of the hysteresis loop. Hence, one sees that a larger free energy is required to start the type-A martensite, and this result is no doubt a consequence of the fact that a larger shear is required to form these martensites.

A significant difference occurs in the mode of formation of type-A and type-B martensites. In both cases, upon lowering the temperature to some value below M_s, plates form and grow quickly to a limiting size. (Actually in type-B martensites the martensite morphology has a band character which we are calling plates in this discussion.) Upon further lowering the temperature, additional transformation occurs by both nucleation of new plates and growth of old plates in type-B martensites. However in type-A martensites additional transformation only occurs by nucleation of new plates. The old plates, once formed, will no longer grow even at lower temperatures. Continued growth of old plates in type-B martensites occurs by a jerky motion. The growth rate of these plates remains at the very high values characteristic of martensitic transformations, but growth occurs over short distances as more free energy becomes available on lowering the temperature.

Figure 13.25 shows a lenticular martensite plate surrounded by a sphere of the same radius, r, as the lens-shaped plate. The shape change is illustrated by the several fiducial lines shown. It is apparent that a

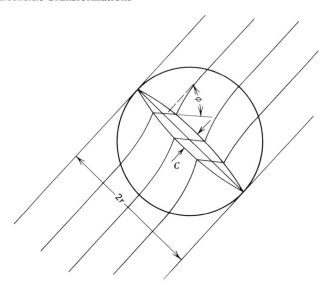

Figure 13.25   Strain around a lenticular martensite plate (after Ref. 23).

considerable strain has been introduced into the austenite phase within the spherical region surrounding the plate. It may be shown (Ref. 23, Section 2.4) that the strain energy per unit volume in the austenite region, E_s, is given approximately as

$$E_s = \frac{G\phi^2}{2}\frac{C^2}{r^2} \tag{13.3}$$

where G is the austenite shear modulus, C is the plate thickness, and ϕ is the shear angle defined on Fig. 13.25. It is apparent that as the plate thickens (C becomes larger), additional strain must be introduced into the surrounding austenite lattice. Eventually this increased strain causes the yield stress to be exceeded in the austenite. It is thought (Ref. 1, p. 815) that when sufficient plastic flow occurs the special interface arrangement between austenite and martensite which allows the fast growth mechanism is destroyed and growth stops. Additional growth will be extremely difficult and it is easier for the system to nucleate new plates than allow growth of the old plates. In type-A martensites the shear strain is large and the supercooling, $T_E - M_s$, is large so that this mechanism of permanent growth termination is thought to operate. However, in type-B martensites there is smaller shear and smaller supercooling. In these cases growth is terminated when the free energy available to drive the reaction, $\Delta G_{\gamma \to M}$, is counterbalanced by the strain energy generated in the austenite phase, Eq. 13.3. Plastic flow does not occur and a balance is achieved between elastic strain energy and the free energy made available by the lower free-energy state of the martensite phase. Hence, this martensite is called *thermoelastic*, and type-*B* martensites are therefore often thermoelastic. As the temperature is lowered more free energy becomes available to form martensite and the plates grow until sufficient strain energy is generated to achieve a new balance. The rate at which the bulk interface moves is governed by how fast the temperature is lowered. However, the local interface itself moves at very high rates in short jumps in order to produce the net motion observed for the bulk interface. With these thermoelastic martensites it is possible to perform what are called single-interface experiments. A single crystal of the high-temperature phase is slowly withdrawn from a furnace. The low-temperature phase outside the furnace is separated from the high-temperature phase inside the furnace by a single interface, which is probably located near the M_s isotherm of the system. The single interface then propagates into the single crystal, in the jerky fashion previously described, at a bulk rate controlled by how quickly the crystal is withdrawn from the furnace. The experiment is quite similar in many respects to control solidification experiments in which the interface position is also controlled by heat flow. The experiment dramatically illustrates the shape change involved in martensitic transformations because one observes a distinct angle between the single-crystal parent phase and the fine twinned martensite phase (see Ref. 13, p. 7 for a picture).

13.6   ADDITIONAL CHARACTERISTICS OF MARTENSITIC TRANSFORMATIONS

A. REVERSIBILITY

One of the more fascinating properties of martensitic transformations is the fact that in many cases they are reversible. This is dramatically illustrated in the Ti–55 wt % Ni martensitic transformation, which occurs at about 60°C (see Refs. 24 for a discussion of memory alloys). First, straight wires of this material are prepared in the high-temperature phase. The wires are then cooled to room temperature and wound into a helix. Upon heating these wires back to the high-temperature phase with a hand torch the wires immediately uncoil into their original shape, in this case a straight wire. This same dramatic effect is nicely illustrated with a thin rod of the Au–Cd alloy described in Table 13.3. With the straight rod cantilevered from a support, it is heated to produce the high-temperature phase. A small weight is now hung on the free end and as the rod cools below the transformation temperature it bends sharply under this weight. If the rod is now reheated to the high-temperature phase it will quickly straighten back up raising the weight with it. As it does so one hears the characteristic clicking associated with martensitic reactions and deformation twinning. The reversibility after plastic deformation is only so demonstrated in these two alloys if the plastic deformation is carried out on the low-temperature martensite phase. It is currently thought that this dramatic display of reversibility will only occur in thermoelastic martensites having the twinned substructure [Ref. 24(b)].

Experiments on other martensites have shown (Ref. 6, p. 620; Ref. 1, p. 814) that, upon heating, the martensite plates disappear to give the original grain structure of the parent phase; and upon cooling the position of the martensite plates and their order of formation is almost exactly the same as when they formed the first time. These results are interpreted as evidence that the same nuclei for formation of the martensite are operative in the same order each time the alloy is cycled from austenite to martensite. This conclusion is supported by the fact that if the alloy is annealed at high temperatures, well above the transformation temperature, the reversibility can be destroyed. Hence, it appears that the reason the Ti–Ni and the Au–Cd revert to their original shape upon heating is because the nuclei operative at the austenite start maintain their original orientations and cause the original high-temperature crystal structure to be obtained. Some reversibility characteristics would be expected to apply to all martensitic transformations, but they are not observed in steels because the martensite phase is thermodynamically metastable and before the reverse reaction can occur, competing reactions develop.

B. STABILIZATION

Suppose that after cooling an austenite phase to some temperature below M_s but above M_f the specimen is held at this temperature for a time interval Δt; see Fig. 13.26. It is often found that upon further lowering of the temperature additional formation of martensite does not occur until the temperature has dropped by some amount ΔT, and that even at very low temperatures one does not obtain as much transformation as would have occurred upon continuous cooling. This same type of phenomenon occurs during the reverse reaction to form austenite, see Fig. 13.26, and the phenomenon is termed *stabilization*.

This stabilization effect has been observed in both ferrous and nonferrous martensites. For example, in single-interface experiments if one stops the interface motion a significant ΔT is required to get it moving again. It has been well established that the effect does not occur unless interstitial solutes are present, and a number of theories have been proposed to explain stabilization (Ref. 11, p. 138). All of these theories attribute the stabilizing effect to some locking mechanism produced by migration of the interstitial atoms during the holding time interval Δt.

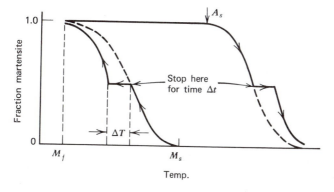

Figure 13.26   Result of stabilization effect upon transformation curve.

This stabilization effect is more than a laboratory curiosity, such as the reversibility property discussed above, because it occurs in steels. To reduce the amount of retained austenite prior to tempering in steels where the M_f temperature is below room temperature, it is necessary to refrigerate the steel after quenching to room temperature. In such cases an extended time interval between the quench and the refrigeration can produce stabilization, which will limit the amount of retained austenite that may be removed by refrigeration. This effect is important in certain tool steels as may be seen from examination of Fig. 13.27.

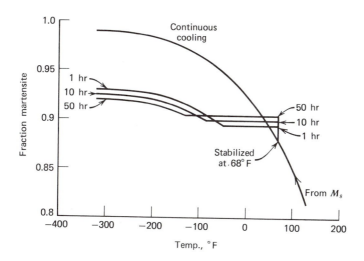

Figure 13.27   Stabilization effect in a type-W1 tool steel (adapted from Ref. 25).

C. EFFECTS OF PLASTIC DEFORMATION AND APPLIED STRESS

It is observed that plastic deformation has a significant effect upon the volume fraction martensite versus temperature relationships such as are shown on Fig. 13.24. The main experimental results may be summarized as follows.

1. If the austenite is first cooled to some temperature between M_s and M_f and plastic deformation carried out at this temperature the amount of martensite is increased by the deformation.

2. If the austenite is cooled to some temperature above M_s but below a temperature called M_D, plastic deformation will cause martensite to form. M_D is the highest temperature above M_s at which plastic deformation will cause martensite to form. In alloys displaying the reverse reaction from martensite to austenite, the reverse reaction is affected by plastic deformation in an analogous manner. The A_d temperature is defined as the lowest temperature below A_s at which plastic deformation will cause austenite to form. The relationship of the M_D and A_D temperatures to the M_s and A_s temperature is shown in Fig. 13.28 for some iron–nickel alloys. As indicated on this figure the equilibrium temperature is estimated as $\frac{1}{2}(M_d + A_d)$. In small shear martensites (type B) it is often found that M_d and A_d are essentially equal.

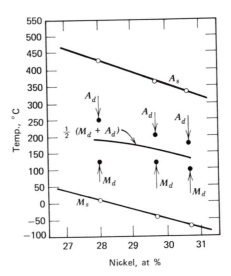

Figure 13.28 The M_S, M_D, A_S, and A_D temperatures in some Fe–Ni alloys. (Reprinted with permission from Ref. 8, copyright 1958, Pergamon Press Ltd.)

At temperatures below the M_D temperature the free energy of the martensite phase is lower than that of the austenite phase. The martensite phase is inhibited from nucleating primarily because of the large strain energy associated with the shape change, which is produced when a martensite plate forms. The function of the applied stress producing the deformation is to reduce somehow the nucleation barrier that results from the shape change and, thereby, permit martensite to form at a lower driving force than it would form in the absence of the deformation. Since there is a distinct shear strain and tensile strain associated with the formation of each martensite plate, it seems very reasonable that the plastic deformation produced by the applied stress would generate many regions in the metal where the shape change would be mechanically assisted and the nucleation barrier therefore reduced.

If plastic deformation is carried out above the M_D temperature, an effect is produced upon the M_s subsequently measured upon cooling. If a small deformation is applied one sometimes finds that M_s is raised, presumably due to defects that assist nucleation (Ref. 14, p. 38). Severe deformation above M_D will generally reduce M_s, an effect known as *mechanical stabilization*. This effect is thought to result from an increase in difficulty of plate propagation into the austenite phase due to work hardening.

If one simply applies an elastic stress to an austenite phase, an effect is observed upon the M_s temperature, but it is usually smaller than that produced by the plastic deformation described above. The applied stress field will either assist or inhibit the shape change of a given martensite plate. For the case of a pure hydrostatic stress field, the sign of the interaction depends only on the volume change of the transformation and, hence, M_s may in principle be either raised or lowered. For example, in Fe–C alloys a pressure of 40 kilobars drops M_s from 150 to 300°C as the carbon content increases from 0.2 to 1.2 wt % (Ref. 11, p. 138). When a shear stress is present in the stress field the M_s will usually be raised because there will generally be some habit planes favorably oriented for the stress to assist the transformation shear. One may estimate the energy equivalent of the work done by the applied stress when the plates form, ΔW. The effect of the stress upon the M_s temperature, ΔM_s, may then be estimated from Eq. 13.2 as $\Delta M_s = \Delta W/\Delta S_f$. Such calculations have given fairly good agreement with experimental results (Ref. 8, p. 204).

13.7 NUCLEATION OF MARTENSITE

As pointed out in the introduction to this chapter, the formation of martensite is essentially controlled by its nucleation characteristics, because the growth rate is so high that from a practical point of view it does not control martensite formation. Unfortunately our understanding of the nature of the martensite nucleation event is quite poor in spite of considerable efforts on this problem.

There seems to be general agreement that the classical theory of nucleation can not apply to formation of martensite. For example, taking the nucleus as an oblate spheroid of radius r and semithickness c one may write for the free-energy change upon formation of a nucleus[26]

$$\Delta G = \tfrac{4}{3}\pi r^2 c\, \Delta G_B + \tfrac{4}{3}\pi r c^2 A + 2\pi r^2 \gamma \tag{13.4}$$

where A is a strain energy factor defined to give Ac/r as the strain energy per unit volume of the austenite-to-martensite reaction and the other terms have their usual meanings. Proceeding in the usual manner one finds

$$c^* = \frac{-2\gamma}{\Delta G_B}$$

$$r^* = \frac{4A\gamma}{(\Delta G_B)^2} \tag{13.5}$$

$$\Delta G^* = \frac{32\pi A^2 \gamma^3}{3(\Delta G_B)^4}$$

Substituting known values in these equations the value of ΔG^* is found to be orders of magnitude too large to account for martensite formation at the conditions where it is known to form.[26]

Alternative approaches to a theory of martensite nucleation generally postulate that a distribution of subcritical nuclei exist within the austenite; they do not have to be formed by a thermally activated process as in the classical nucleation case. These subcritical nuclei are usually thought to be some sort of crystal defect. For example, a stacking fault in an fcc lattice

$$\downarrow$$

changes the stacking sequence as $...ABCABABABC....$ In the immediate region of this fault the stacking has become $ABAB$, which is the same stacking sequence as occurs in hcp metals. Hence, for martensite transformations between fcc and hcp crystals, such as in Co, a stacking fault, which could easily be produced by formation of partial dislocations, might act as a nucleus for martensite formation. In other cases certain dislocation arrangements might serve as the subcritical nucleus. The athermal and isothermal martensites may then be accounted for as follows:

Athermal:

(a) A distribution of subcritical nuclei exists.

(b) At M_s the largest (or most mobile) nuclei become critical and grow.

(c) As the temperature is lowered, smaller (or less mobile) nuclei become critical, due to the higher driving force, and grow.

(d) Holding at temperatures of M_s and below does not allow new nuclei to form.

Isothermal:

(a) A distribution of subcritical nuclei exists.

(b) At temperatures of M_s and below certain of these nuclei are critical and will grow.

(c) Holding at temperatures of M_s and below causes additional critical nuclei to form by some unknown thermally activated process.

The difficulty with this type of theory of martensite nucleation is that there is no evidence from electron microscope studies of the preexistant subcritical nuclei and, also, the theory is not quantitative. For further discussions of nucleation see Ref. 1, pp. 916–925 and Ref. 12, pp. 32–35.

13.8 SUMMARY AND COMPARISON WITH MASSIVE TRANSFORMATIONS

One of the central features of the martensitic change pointed out by Christian[1] that has not been adequately stressed so far in this chapter is that the martensite reaction occurs because of the existence of an easy growth mechanism, not requiring atomic diffusion, which produces the new phase. Hence, understanding the nature of the dislocation arrangement at the austenite–martensite interface is a very important part of understanding martensite reactions. It is because some sort of a special arrangement can exist between two particular crystal structures that the transformation between these structures can occur martensitically. The interface structure and consequent fast growth mechanism appears to be essentially similar between almost all martensites. This is true in spite of the various characteristics of the different martensites, for example, athermal, isothermal, thermoelastic, lath type, plate type.

When an alloy is quenched very rapidly through a solid-state phase transformation to a low temperature it is often possible to completely suppress the formation of the low-temperature equilibrium phase or the formation of any metastable low-temperature phases. If a low-temperature phase forms martensitically, however, it is not possible to suppress its formation even at the highest possible quench rates. It is sometimes found that a low-temperature phase may form during a fast quench in a manner we have not yet discussed. These transformations are termed *massive transformations* because the new phase forms a chunky morphology. This morphology is in sharp contrast to the plate or needle morphology of most fast growth transformations. The main features of the massive transformation are listed as follows:

1. Parent and product phases have the same composition (diffusionless).

2. The growth rate is fast but not as fast as in the martensitic case.

3. No shape change is seen on a free surface as with martensite.

4. The parent–product interface which migrates is an incoherent boundary.

Hence, the massive transformation behaves similar to the martensitic transformation in that the new phase forms rapidly without any composition change. However, the fast growth mechanism of the martensitic case is not present, and one simply has a high-angle boundary migrating very rapidly due to a large driving force. This driving force, $\Delta S_f \cdot \Delta T$, (where ΔT is the supercooling below the T_E of the diffusionless transformation shown on Fig. 13.22) is generally lower than for martensitic reactions; and in some systems a massive transformation will occur at low undercooling while a martensitic transformation occurs at higher undercoolings. Such is the case in low carbon steels; see Fig. 12.4. A massive transformation occurs in brass, and the reader is referred to Ref. 27 for a good discussion of this transformation (see also Ref. 13, pp. 433–486 for additional discussion of massive transformations).

In summary then, if one severely quenches an alloy through a solid-state phase transformation a number of possibilities exist:

1. The high-temperature phase may be retained as a metastable phase at the quench temperature.

2. Small particles of a metastable phase may form as discussed in Chapter 11 by either spinodal decomposition or homogeneous nucleation.

3. A fast growth precipitate phase such as Widmanstatten side plates or bainite (see next section) may form.

4. A massive transformation may produce a low-temperature phase.

Christian[1] has presented a comparison summary between martensitic reactions and nucleation and growth reactions. This summary provides an excellent comparison of the two types of reactions and at the same time offers a good brief review of the main characteristics of both types of reactions. The summary is presented on pp. 11–16 of Ref. 1 and the student is urged to study this presentation.

13.9 BAINITE

When one quenches an austenitic steel to temperatures just slightly below the eutectoid temperature the characteristic lamellar pearlite structure is obtained at the interior of the grains. Depending on carbon composition Widmanstatten side plates of either ferrite or cementite are often seen growing from grain-boundary allotriomorphs. These structures, pearlite, Widmanstatten side plates, and GBA were described in Chapter 12. As the quench temperature is lowered the pearlite spacing gets finer and the Widmanstatten side plates become thinner and thinner and eventually disappear, as may be deduced from Fig. 12.4. From the discussions presented in Chapter 12 on pearlite growth, one might expect the pearlite simply to become finer and finer as the quench temperature is lowered until the M_s is reached, where the austenite would then transform into the martensitic structures described in this chapter. However, at quench temperatures slightly above M_s a microstructure forms that is distinctly different from fine pearlite. Bainite is often shown metallographically as a group of very fine plate-like projections growing out extremely close to each other having an appearance resembling a feather; see Refs. 30. This feathery microstructure was identified in a report by Davenport and Bain[28] in 1930 and it has subsequently become known as *bainite* in honor of E. C. Bain. Similar structures are also observed in certain nonferrous alloys[33] and the formation of this type of structure is generally referred to as a *bainitic transformation*. The bainitic transformation is a particularly complex reaction because it involves features common to both martensitic reactions and diffusion-controlled nucleation and growth reactions, and some aspects of bainitic transformations still remain a subject of controversy.[29] The major review articles on bainite are Refs. 9 and 29–33.

A. BASIC CHARACTERISTICS

The major characteristics of the bainitic transformation may be subdivided into three categories.

1. Morphologies. Bainite has a two-phase microstructure composed of ferrite and iron carbide. There is a variation in the morphology of bainite and in the type of carbide (Fe_3C or ε carbide, $\sim Fe_{2.5}C$) depending upon the quench temperature and composition. A fairly distinct change in morphology occurs between high and low quench temperatures; see Fig. 13.29. These two morphologies are referred to as *upper bainite* (high quench temperature) or *lower bainite* (lower quench temperature). The variation of the transition temperature with carbon content for a series of steels has been given by Pickering (Ref. 31, p. 115) and is shown in Fig. 13.30. There is a small temperature range in which both forms are present, and the transition temperature here is the highest temperature at which lower bainite is observed.

a. Upper Bainite. The external shape of upper bainite is quite irregular and therefore difficult to determine from examination of a single

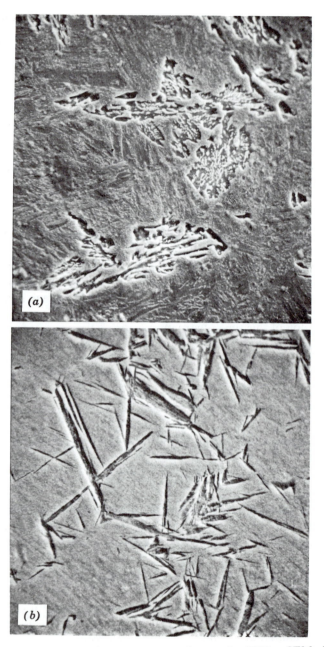

Figure 13.29 (a) Upper bainite in a martensite matrix, 2200×, SEM photo. (b) Lower bainite in a martensite matrix, 2300×, SEM photo. AISI 4340 steel, austenitized at 843°C (1550°F), quenched to T for 6 min followed by ice brine quench and nital etch. $T = 468°C$ (875°F) for (a) and 300°C (572°F) for (b).

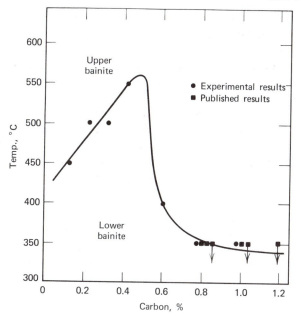

Figure 13.30 Effect of carbon content on the transition temperature from upper to lower bainite (from Ref. 31, p. 115).

surface. Two-surface analysis has shown[33] that a primary characteristic of the external morphology of upper bainite is a lath or needle-like shape, that is, one dimension longer than the other two. The cross sections perpendicular to the long axis vary widely, for example, plate, elliptical, or blocky. Because of this fact plus the fact that the polish surface intersects the long axis at random angles, a wide variety of external shapes are observed on a single surface, as illustrated in Fig. 13.29(*a*). The upper bainite of this photo is present in a martensite matrix produced by the final rapid quench. The white particles within the bainite are Fe_3C. Transmission electron microscope studies have shown the internal structure of this upper bainite to consist of ferrite laths running parallel to the long axis with the carbide particles precipitated primarily at lath boundaries. The ferrite laths may contain subboundaries arranged so that the laths themselves are composed of sublaths (sometimes called sheaves). For example, a $3 \times 7 \times 30\ \mu m$ lath might be composed of a stack of sublaths having dimensions on the order of $\frac{1}{2} \times 7 \times 30\ \mu m$. Sometimes the sublaths have carbides in their boundaries.[29] The sublaths have a very high dislocation density. Hence, the external morphology of upper bainite is lath or needle shaped and internally it is composed of laths of ferrite running parallel to its long axis with Fe₃C particles at lath boundaries; these ferrite laths may in turn be subdivided into sublaths having high dislocation densities.

It is not uncommon for some retained austenite to be present between ferrite laths. As the ferrite laths grow they reject carbon into the surrounding region and, hence, lower its M_s temperature so that chances for retained austenite are increased. The effect of lowering the quench temperature on the upper bainite morphology is to cause the laths to become finer and closer together, thus producing a closer spacing of the carbide particles between laths (see Fig. 2 on p. 400 of Ref. 33). In hypereutectoid steels the leading phase of bainite may become cementite rather than ferrite, and these structures are sometimes called inverse bainite.

b.   *Lower Bainite.* In lower bainite the ferrite forms as plates rather than laths, as indicated in Fig. 13.29(*b*) and carbides precipitate out inside of the ferrite plates. The carbide precipitates are on a very fine scale and they frequently have the shape of rods or blades. The carbide rods or blades are more or less aligned parallel to each other making an angle generally between 55° and 65° with the growth axis of the ferrite plate. See Ref. 31, pp. 18–28 and 110–114, for excellent electron microscope pictures of both upper and lower bainite. The plates themselves are found to be composed of several subplates that have distinct low-angle boundaries separating them. The subplates have a high dislocation density. The iron carbide within lower bainite is found to be ε carbide when silicon is present as an alloying addition; otherwise, the carbide may be a mixture of cementite and ε carbide or all cementite. It is generally thought that the alloying additions do not form significant amounts of carbides in bainite because of the low reaction temperature.

2.   Crystallography. One of the fundamental characteristics of upper and lower bainite is that they both exhibit a definite *surface relief*, whose appearance is quite similar to that displayed by martensite plates and deformation twins. See p. 362 of Ref. 34 for an excellent photograph illustrating this surface tilting caused by bainite plate formation. Lower bainite has been found to exhibit a single uniform surface relief across each plate. In upper bainite each lath exhibits a multiple surface relief.[33] This suggests that each relief in upper bainite is associated with the formation of the individual sublaths. The ferrite component of bainite forms with a definite crystallographic orientation relative to the austenite. This relationship is generally found to be the Kurdjumov–Sachs lattice relationship (p. 376) as is also usually found for Widmanstatten ferrite plates.

When one tempers martensite one of the first things that occurs is the precipitation of ε carbide within the martensite plates (see next chapter). It has been known for many years[30] that lower bainite has an appearance quite similar to tempered martensite. As noted above, the carbides in lower bainite occur at a specific angle to the growth direction of the bainite plates, and this indicates a specific epitaxial relationship between the carbide and the ferrite plates. This relationship has been found to be (Ref. 9, p. 131)

$$(001)_{Fe_3C} \parallel (211)\alpha$$
$$[100]_{Fe_3C} \parallel [0\bar{1}1]\alpha$$
$$[010]_{Fe_3C} \parallel [1\bar{1}1]\alpha$$

in lower bainite. The same epitaxial relation is found in tempered martensite and the similarity is sometimes interpreted to mean that the carbides in lower bainite form by precipitation from the ferrite rather than by formation at the austenite–ferrite growth interface. There is also evidence for an epitaxial relationship between the carbide of upper bainite and the parent austenite, and this result supports the belief that the upper bainite carbide precipitates from austenite trapped between the ferrite laths.

3. Kinetics. The bainite transformation follows its own C curve on a T-T-T diagram. In plain carbon steels there is a significant overlap between the bottom of the pearlite reaction curve and the top of the bainite reaction curve; see Fig. 13.31(a). Addition of certain alloying elements suppresses the rate of pearlite formation but has less effect on the rate of bainite formation. In these cases there is little overlap of the pearlite and bainite C curves, as shown on Fig. 13.31(b).

The volume fraction transformation curves for bainite display the characteristics of nucleation and growth transformations. However, they

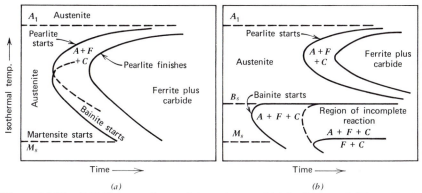

(a)

(b)

Figure 13.31 T-T-T curves for a plain carbon and high alloy steel. [From R. F. Heheman and A. R. Troiano, Met. Prog. **70**, 97 (Aug. 1956).]

have some features that are similar to martensitic transformations. In alloy steels one has to quench the austenite below a specific temperature, called the B_s temperature, before the bainite reaction will start; see Fig. 13.31(b). At temperatures just below B_s it is often not possible to completely transform the austenite to bainite even after considerable times. There is a lower temperature, however, B_f, below which complete transformation of the austenite to bainite is possible. This B_f temperature may be above or below the M_s temperature, and if the latter is true then one cannot obtain a fully bainitic steel. In analogy with the M_s and M_f of martensites, the bainite transformations in alloy steels display B_s and B_f temperatures.

The presence of the bainitic transformation accounts for the unusually complex shape of the T-T-T curves of many alloy steels. Unless the bainite and pearlite curves merge as shown in Fig. 13.31(a) the classical C shape is not obtained. As an example for a popular alloy steel the T-T-T curves for AISI 4340 steel are presented in Fig. 13.32. Notice that the curves terminate at the M_s temperature for the obvious reason that the martensite

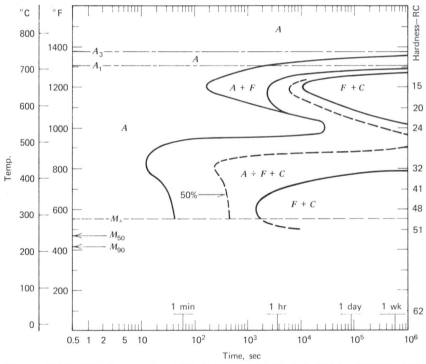

Figure 13.32   T-T-T curve for AISI 4340 steel. (0.42 C, 0.78 Mn, 1.79 Ni, 0.80 Cr, and 0.33 Mo austenitized at 800°C, grain size 7–8.) (From Ref. 35, copyright 1951 United States Steel Corp.).

reaction controls below this temperature. Although not marked on these curves, it should be clear to you which lines locate the start of the bainitic reaction, the pearlitic reaction, and the GBA or Widmanstatten plate formation of ferrite.

B. REACTION MECHANISM

It is apparent from the above discussion that the bainitic transformation displays characteristics of both martensitic and nucleation and growth transformations. For this reason the reaction mechanism for bainite formation continues to be a subject of considerable controversy, and a brief (and therefore incomplete) summary can only be presented here.

Utilizing the surface relief produced by bainite, one may determine the rate at which the plates or laths form by hot-stage optical microscopy. The observed edgewise growth velocities of bainite plates are much smaller than martensite growth velocities. One can analyze the growth rate using a model such as Fig. 13.33(a) where it is assumed that growth is controlled by diffusion of carbon from the bainite tip into the austenite parent. The bainitic reaction occurs at sufficiently low temperatures and high rates that the diffusive motion of substitutional solutes does not appear to play a role in the kinetics. As in the case of Widmanstatten plate growth, the Zener–Hillert equation (Eq. 12.10) is employed and satisfactory agreement is found between growth rate and theory; see Ref. 33, p. 420. Hence, it was held for some time that the bainitic reaction was a shear transformation (because of surface relief evidence) whose growth rate was controlled by diffusion of carbon into the austenite.

The recent electron microscope studies, which have shown bainite ferrite to be composed of sublaths or subplates, have suggested an alternative growth mechanism that does not require the continuous growth of a unique interface. It is postulated that each subplate advances into the austenite with its own tip configuration. One model is shown in Fig. 13.33(b) where each subplate forms as a ledge upon the adjacent subplate. Thickening occurs by nucleation of additional subplates and growth to the right. Direct

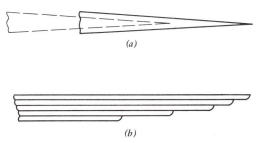

(a)

(b)

Figure 13.33 Models for growth of bainite.

observations of the growth of upper bainite plates in the thermionic emission microscope have confirmed this model in an Fe–0.66 C–3.3 Cr alloy.[29]

There are two interpretations of the mechanism for this growth process. The interpretation supported by Hehemann[29] holds that the individual subunits grow at very rapid rates to a size limited by something such as volume strain. Hence, the observed growth rates of bainite plates or laths are determined largely by nucleation of the subunits, but the mechanism for this nucleation is not known. The growth velocity of the individual subunits is therefore very high as in martensitic transformations. This theory is supported by studies showing that the crystallographic parameters of lower bainite can be satisfactorily explained in terms of the phenomenological theories of martensite formation.

An alternative interpretation is advanced by Aaronson and Kinsman.[29,32] This interpretation holds that the shear evidenced by surface relief plays no role in the growth mechanism. Growth is thought to occur by a continuous diffusion-controlled motion of the individual ledges shown in Fig. 13.33(b). The surface relief is produced because of the fact that the interfaces between subunits are low-angle, low-mobility boundaries. Hence, as each ledge reaches the surface the trailing subunit adds an increment to the surface relief because of the tight constraint to its neighboring subunit. This interpretation is supported by agreement with theories of diffusion controlled ledge motion and by thermionic emission experiments, which reveal slow continuous ledge motion rather than fast jerky motion. Also, it is interesting to note that Widmanstatten ferrite plates exhibit a surface relief. These transformations may occur at temperatures above T_E where it is thermodynamically impossible for a martensitic shear mechanism to operate. Hence, a shear mechanism is generally held to be nonimportant in their formation but the thickening kinetics may be explained in terms of this latter theory.

REFERENCES

1. J. W. Christian, *The Theory of Transformations in Metals*, Chapters 20–23, Pergamon, New York, 1965.

2. R. W. Cahn, Advan. Phys. **3**, 363 (1954).

3. R. E. Reed-Hill, *Physical Metallurgy Principles*, 2nd ed., Van Nostrand, New York, 1973.

4. *Deformation Twinning*, R. E. Reed-Hill, J. P. Hirth, and H. C. Rogers, Eds., Gordon and Breach, New York, 1964.

5. A. R. Troiano and A. B. Greninger, Met. Prog. **50**, 303 (1946).

6. M. Cohen, in *Phase Transformations in Solids*, Wiley, New York, 1951, p. 588.

7. B. A. Bilby and J. W. Christian, in *The Mechanism of Phase Transformation in Metals*, The Institute of Metals, London, 1956, pp. 121–217.

8. L. Kaufman and M. Cohen, in *Progress in Metal Physics* Vol. 7, Pergamon, New York, 1958, pp. 165–246.

9. *Physical Properties of Martensite and Bainite*, Special Report 93, The Iron and Steel Institute, 1965.

10. C. M. Wayman, in *Advances in Materials Research*, H. Herman, Ed., Interscience, New York, 1968, pp. 147–304.

11. J. W. Christian, in *The Mechanism of Phase Transformations in Crystalline Solids*, The Institute of Metals, London, 1969, pp. 129–142.

12. *Martensite, Fundamentals and Technology*, E. R. Petty, Ed., Longman, London, 1970.

13. *Phase Transformations*, Chapters 1 and 3, American Society for Metals, 1970.

14. Symposium on the Formation of Martensite in Iron Alloys, Met. Trans. **2,** 2327–2462 (1971).

15. A. B. Greninger and A. R. Troiano, Trans. AIME **140,** 307 (1940).

16. D. P. Dunne and J. S. Bowles, Acta Met. **17,** 201 (1969).

17. E. C. Bain, Trans. AIME **70,** 25 (1924).

18. M. S. Wechsler, D. S. Lieberman, and T. A. Read, Trans. AIME **197,** 1503 (1953).

19. J. S. Bowles and J. K. Mackenzie, Acta Met. **2,** 129, 138, 224 (1954).

20. C. M. Wayman, *Introduction to the Crystallography of Martensite Transformations*, Macmillan, New York, 1964.

21. *Metals Handbook*, 8th ed., Vol. 8, 194–201 (1973).

22. W. Stevens and A. C. Haynes, J. Iron Steel Inst. **183,** 349 (1956).

23. M. E. Fine, *Phase Transformations in Condensed Systems*, Macmillan, New York, 1964.

24. (a) C. M. Jackson, H. J. Wagner, and R. J. Wasilewski, 55-Nitinol—The Alloy with a Memory, NASA-SP5110 (1972); (b) K. Otsuka et al., Met. Trans. **2,** 2583 (1971); (c) H. C. Tong and C. M. Wayman, Met. Trans. **6A,** 29 (1975).

25. P. Payson, *The Metallurgy of Tool Steels*, Wiley, New York, 1962, p. 49.

26.   M. Cohen, Met. Trans. **3,** 1095 (1972).

27.   P. G. Shewmon, *Transformations in Metals,* Sec. 6.9, McGraw-Hill, New York, 1969.

28.   E. S. Davenport and E. C. Bain, Trans. AIME **90,** 117 (1930).

29.   R. F. Hehemann, K. R. Kinsman, and H. I. Aaronson, Met. Trans. **3,** 1077 (1972).

30.   R. F. Hehemann and A. R. Troiano, Met. Prog. **70,** 97 (Aug. 1956); R. F. Hehemann, *Metals Handbook,* Vol. 8, p. 194 (1973).

31.   *Transformation and Hardenability in Steels,* Climax Molybdenum Company of Michigan, Ann Arbor, 1967.

32.   H. I. Aaronson, in Ref. 11, pp. 270–281.

33.   R. F. Hehemann, in Ref. 13, pp. 397–432.

34.   G. R. Speich, in *Decomposition of Austenite by Diffusional Processes,* V. F. Zackay and H. I. Aaronson, Eds., Interscience, New York, 1962, pp. 353–366.

35.   *Atlas of Isothermal Transformation Diagrams,* United States Steel Co., 1951.

36.   C. A. Apple, R. N. Caron, and G. Kruss, Met. Trans. **5,** 593 (1974).

PROBLEMS

13.1

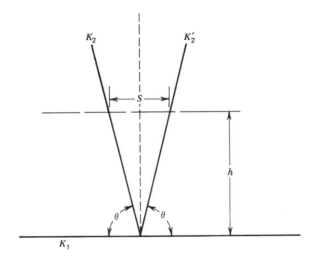

The shear strain associated with a twin is defined as

$$\varepsilon = \frac{s}{h}$$

where s and h are defined on the above diagram. Derive the relationship between this shear strain ε and the angle θ,

$$\varepsilon = 2 \tan [90 - \theta]$$

13.2   Twinning in bcc metals occurs on $\{112\}$ with $\langle 111 \rangle$ as the shear direction. The second undistorted plane, K_2, is another plane of the same type as K_1, that is, $\{11\bar{2}\}$. Prove that the twinning shear strain is 0.707. Do this problem with the aid of a stereographic projection. Show the trace of K_1, K_2, K_2' and also η_1, η_2, η_2' on your projection.

13.3   Repeat Problem 13.2 for the fcc system using the information of Table 13.2. The twinning shear strain is also 0.707.

13.4   Calculate the shear strain ε (see Problem 13.1) for Zr given that: (a) the c/a ratio is 1.589 and (b) the twinning system is

$$K_1 = \{11\bar{2}2\} \qquad \eta_1 = \langle 11\bar{2}3 \rangle$$
$$K_2 = \{11\bar{2}\bar{4}\} \qquad \eta_2 = \langle 22\bar{4}3 \rangle$$

13.5   The hardness of as-quenched steel increases to a maximum as the carbon content increases to 0.8%; see the figure in Problem 6.1.

(a)   Would you expect the shear component of the martensite reaction to increase or decrease as carbon content is increased? Explain your reasoning.

(b)   What effect would this have on the $\Delta G_{\gamma \to M}$ required to form the martensite?

(c)   Based on your answer to part (b), deduce what effect increased carbon content should have on the M_s temperature.

13.6   Suppose a lenticular plate of martensite has formed having a thickness C and a radius r. Take the volume of the plate to be $\pi r^2 C$. Take the volume of the stressed region around the plate to be $\frac{4}{3}\pi r^3 - \pi r^2 C$. It can be shown that the strain energy per unit volume in the stressed region may be approximated as $G\phi^2 C^2 / 2r^2$ (see Eq. 13.3).

Assume now that the plate begins to thicken (C increases) while its radius r remains fixed, perhaps due to intersection of its periphery with other plates. Show that due to the increased strain energy in the surrounding material as the plate thickens, it will not be able to grow to a thickness greater than C_{max}, which is determined from the equation

$$-\Delta G \pi r^2 = \frac{1}{6} G \phi^2 \pi C_{max} [8r - 9C_{max}]$$

where ΔG is the free energy change in going from austenite to martensite.

Hint: To work this problem write an expression for the total free energy of the system: plate plus stressed region. Include both volume and strain energy terms. Allow the plate to thicken at constant radius r and determine the thickness that minimizes the total free energy.

13.7   Assume that $r \gg C$ for all values of C of interest, so that we can neglect the $9C$ term in the bracket of Problem 13.6.

(a)   Determine C_{max} from the equation of Problem 13.6.

(b)   Now show that if the average yield strain of the austenite phase is greater than $3\Delta G/(4G\phi)$ you will have a thermoelastic martensite, and if it is less you will not. For this problem take the strain per unit volume as simply $G\varepsilon^2/2$, where ε is an average strain.

CHAPTER 14
SOME APPLICATIONS OF PHYSICAL METALLURGY

The structure of an alloy on the atomic scale is influenced by such things as dislocation densities, vacancy concentrations, and composition variations. The structure of an alloy on the microscale is influenced by such things as the kinds of phases present and their grain shape and size distributions. Both of these, atomic structure and microstructure, exert a significant influence upon the various physical properties of alloys. The purpose of a study of physical metallurgy is to allow one to understand the nature and the control of the atomic structure and microstructure of alloys, so that they can be manipulated to achieve the particular physical properties that one desires. In this chapter we will discuss how the structures of a few steels have been manipulated to optimize their strength and ductility. Numerous other important and interesting examples could also be considered. For example, the structure of weld joints, solder joints, and powder metallurgical parts are governed by the principles of physical metallurgy. It is generally true that whatever combination of physical properties is responsible for the use of an alloy in a particular device, the optimization of those properties for that particular device requires an understanding of the principles of physical metallurgy.

It was pointed out in Chapter 3 (pp. 71–74) that the strength of metals is considerably less than is theoretically possible due to the presence and mobility of dislocations. Using an equation for the theoretical strength of a metal, such as Eq. 3.6, Thomas et al.[1] have presented the data given in Table 14.1. It is apparent that the maximum strengths achieved to date are much less than is theoretically possible. However, the maximum strengths are very much higher than is found in pure metals. For example, the tensile strengths of annealed pure Al, Cu, and Fe are 6.8 ksi (47 MN/m^2), 28 ksi (193 MN/m^2), and 35 ksi (241 MN/m^2). Comparison to Table 14.1 shows that considerable success has been achieved at increasing the strength of these three metals. This has been done by manipulating the

513

Table 14.1.   Representative Metal Properties (From Ref. 1, p. 7)

Metal	Theoretical Tensile Yield Strength (ksi)	(MN/m²)	Maximum Observed Tensile Strength (ksi)	(MN/m²)	Ratio obs./theo.
Al(fcc)	600	(4,140)	100	(689)	0.17
Cu(fcc)	750	(5,170)	214	(1480)	0.35
Be(hcp)	2400	(16,550)	80	(552)	0.03
Mo(bcc)	2400	(16,550)	80	(552)	0.03
W(bcc)	4000	(27,580)	300	(2070)	0.075
Fe(bcc)	1800	(12,410)	500[a]	(3450)	0.3
			700[b]	(4830)	0.4
			1000[c]	(6890)	0.56

[a] Ausformed alloy.
[b] Drawn wire.
[c] Whiskers.

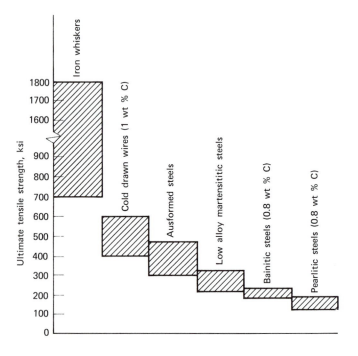

***Figure* 14.1**   Influence of several compositional and microstructural variables on the yield strength of iron (adapted from Ref. 2).

structure to achieve various strengthening mechanisms. Nevertheless, with the exception of whiskers,* the maximum strengths are still considerably less than half of the theoretical limit in these three metals. In other metals, such as Be, Mo, and W, it is apparent that proper control of structure could lead to much improved strengths, and such a possibility presents an exciting area for the application of the principles of physical metallurgy.

Every phase transformation that we have discussed in this book occurs in steels. Because of this and other fortuitous events, steel offers us perhaps the largest number of potential strengthening mechanisms in one alloy system. This is illustrated in part by the selection of steels presented in Fig. 14.1. The variation of the yield strengths of these steels is a function of their structure and composition. To understand this variation it is necessary to consider the major mechanisms contributing to the strength of an alloy.

14.1. STRENGTHENING MECHANISMS

It was pointed out in Chapter 4 that the strength of a metal is controlled by the number and motion of dislocations. The stress required to move dislocations (Peierls–Nabarro stress, p. 81) is quite low in pure metals. Consequently, to strengthen metals one must restrict the motion of dislocations by either generating internal stresses that oppose their motion, or by placing particles in their path that require them to either loop or cut the particles. There are four major strengthening mechanisms that will be discussed here. (See Refs. 4–7 and most books on dislocations, e.g., Ref. 7 of Chapter 4, for more detailed discussions of strengthening mechanisms.)

A. WORK (DISLOCATION DENSITY) HARDENING

The work-hardening mechanism has been described on p. 121 in connection with the interpretation of plastic flow in single crystals. The significant strengthening that occurs in the stage II region of the stress–strain diagram is produced by the interaction of dislocations with each other. Because of the long-range stress field associated with each dislocation, the energy of the system can be lowered by grouping of the dislocations. Consequently, deformation generates a tangled mess of dislocations, as discussed in Chapter 10, which provide effective barriers to the additional motion of dislocations. Several theories have been put forth to account for the strengthening that results from possible dislocation interactions produced by deformation. All of these theories predict that the flow stress be a function of the square root of the dislocation density ρ. The shear flow

---

* Whiskers are extremely thin filaments of metals usually grown from the vapor phase. They are believed to be essentially dislocation free and owe their high strength to the absence of dislocations.

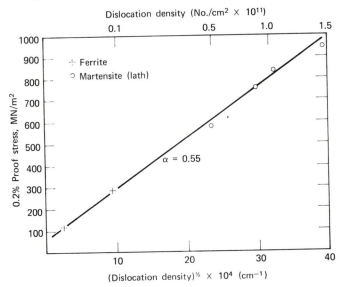

Figure 14.2   Dependence of yield stress upon dislocation density in Fe–C alloys, 0.01–0.1 wt % C, room temperature (from Ref. 8).

stress τ is predicted to vary as

$$\tau = \tau_0 + \alpha G b \sqrt{\rho} \tag{14.1}$$

where τ_0 is the original flow stress, G is the shear modulus, b is the Burgers vector, and α is a constant. Kelly[4] presents a brief summary of four current theories of dislocation interactions that all predict a value of α around 0.3. Equation 14.1 has strong support from experimental work on many different metals (e.g., see Fig. 1 of Ref. 4) that consistently show a linear dependence between τ and $\sqrt{\rho}$, but with α closer to 0.5 than the theoretical value of 0.3. This is illustrated by a specific example in Fig. 14.2, which is a plot of the 0.2% proof stress (yield stress) as a function of $\rho^{1/2}$. As indicated, the structure consists of ferrite at the lower densities and lath martensite at the higher densities. It is interesting that the high dislocation densities were produced here not by work hardening, but by rapid quenching. The variation in dislocation density was achieved in the ferrite by varying the heat treatments and in the martensite by varying carbon content. In conclusion, we have very good evidence that the strength of a metal increases with the square root of the dislocation density.

B. MICROSTRUCTURE HARDENING

The flow stress of a metal is almost universally observed to increase as the grain size decreases. The experimental data virtually always display a linear relationship between flow stress and the reciprocal of the grain diameter d to the one-half power,

$$\sigma = \sigma_0 + k/\sqrt{d} \tag{14.2}$$

where σ_0 and k are constants. This equation is termed either the Petch equation or the Hall–Petch equation, and it is illustrated for data on mild steel in Fig. 14.3. Notice that the data cover a very large range of grain sizes with the sizes below 5 μm being unusually small (see Table 12.2). Equation 14.2 is generally found to apply whether the metal is subdivided by grain boundaries or by subgrain boundaries. For example, severe cold working often produces a cellular substructure as discussed in Chapter 10; and in such cases the flow stress is observed to vary as the reciprocal one-half power of the cell diameter (see for an example Fig. 6 of Ref. 1).

The strengthening produced by the microstructure results from blockage of dislocation motion by the interfaces of the microstructure. If a source within a given grain is feeding dislocations into a grain boundary, the dislocation motion will eventually become blocked at that boundary. Further motion then requires generation of new dislocations in the neighboring grain by means of the local stress field generated by the blocked dislocations in the parent grain. The value of k in Eq. 14.2 is therefore interpreted as a measure of the difficulty required to unlock or generate dislocations in the neighboring grain. Experimental results indicate that k/G ranges from about 10^{-3} psi/in.$^{\frac{1}{2}}$ (43 MN/m$^{\frac{3}{2}}$) in pure fcc metals to 4×10^{-3} psi/in.$^{\frac{1}{2}}$ for bcc and hcp metals.[4]

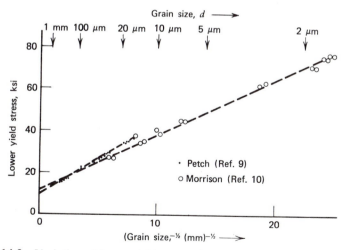

Figure 14.3 Variation of lower yield stress of mild steel with grain size (adapted from Ref. 4).

C. SOLID SOLUTION HARDENING

When a foreign atom dissolves in a solid metal it may act as an atomic sized obstacle to the motion of dislocations. Suppose that each foreign atom produces a restraining force F upon the dislocation line, and these foreign atoms are spaced at an average distance d along the dislocation line. The ratio F/d then gives the force per unit length of dislocation line that must be overcome by the applied stress τ. The increment in applied stress necessary to overcome this restraining force per length of line would be $\Delta\tau b$ (see Eq. 4.8), so we have

$$\Delta\tau = \frac{F}{bd} \tag{14.3}$$

The spacing of foreign atoms along the line, d, is a function of the atom size D, their distribution within the structure, and the fractional concentration of these foreign atoms, x_f. The value of d has been calculated[4] as $0.82b/\sqrt{x_f}$, where the atom size D and the Burgers vector b have been related. This result gives

$$\Delta\tau = \frac{F}{0.83b^2}\sqrt{x_f} \tag{14.4}$$

The value of F depends on the nature of the interaction of the dislocation with the foreign atoms. Two general interactions are usually considered, one of a chemical nature and the other an elastic nature. The difference in chemical bonding between the foreign atoms and the parent atoms is reflected in the difference in their elastic shear moduli ΔG. This difference gives rise to a dislocation-atom force F, which leads to the expression $\Delta\tau = \Delta G\sqrt{x_f}$.[4]

If the foreign atom has a different size than the parent atoms, then a misfit strain field will be produced around the foreign atom that may interact with the dislocation strain field. We will take the fractional difference in foreign atom diameter and parent atom diameter as ε. Then, for a substitutional solute or an interstitial solute with only a symmetrical strain field, analysis of the strain field–dislocation interaction[4] gives an expression for F that leads to $\Delta\tau = G\varepsilon\sqrt{x_f}/4$. Placing a carbon atom in the octahedral void (largest void) of fcc iron will produce a symmetrical strain field as may be determined from Fig. 1.8. However, placing a carbon atom in the tetrahedral void (largest void) of bcc iron will produce an unsymmetric strain field. As may be seen from Fig. 1.10 the carbon atom will expand the bcc iron matrix more in one of the principal lattice directions than in the other two. The shear component of this tetragonal strain will interact with both screw and edge dislocations thus leading to a stronger interaction than for the pure dilatational case. Calculations[4] lead to the relation

$\Delta\tau = G(\varepsilon_1 - \varepsilon_2)\sqrt{x_f}/7.5$, where ε_1 is the strain along the direction of maximum distortion and ε_2 is the strain at right angles thereto. Experiments generally show reasonable agreement with the predicted linear relationships between τ and $\sqrt{x_f}$.[6] In conclusion, we may expect the solid solution hardening to increase as the square root of the atom fraction solute in solution, and further, large tetragonal distortions produced by interstitial solutes will be the most effective at increasing strength.

D. PARTICLE HARDENING

Particles may be introduced into the crystal lattice either by precipitation or by direct mechanical implanting. These particles will interact with the dislocations causing the dislocations to either loop the particles or cut through them. Details of these interactions and consequent effects upon strength were discussed in Section 11.5.

14.2. STRENGTH AND DUCTILITY

If a glass rod is etched in dilute hydrofluoric acid for several hours and then tested in tension, it is not unusual to obtain tensile strengths of 500,000 psi (3450 MN/m^2). Similar results are obtained if the rods are tested immediately following careful hand drawing.[11] Experiments on fine glass filaments [around 0.001 inches (25 μm)] tested at liquid nitrogen temperatures commonly give tensile strengths as high as 2,000,000 psi (13,790 MN/m^2). In such a study by Morley, Andrews, and Whitney[12] the tensile strengths ranged from 1,700,000 psi (11,720 MN/m^2) to 2,300,000 psi (15,860 MN/m^2). Comparison of these tensile strengths with the data for various steels in Fig. 14.1 reveals that the tensile strengths achievable in glasses are clearly superior to those of the highest strength alloy steels. This is a very interesting observation because it is apparent to everyone that glasses are of no use in the design of machinery or devices subject to even small tensile stresses. The reason, of course, is that ordinary glass is extremely brittle. This example illustrates quite nicely that simply achieving a high tensile strength in a material does not necessarily mean the material will be useful for high-strength design applications; in addition to strength, the material must also be ductile.

If an ordinary glass object such as a window glass is tensile tested, the observed tensile strengths usually fall in the range of only 2000–20,000 psi (13.8–138 MN/m^2).[5,11] The theoretical estimates of the tensile strength of glass based on calculations similar to those presented in Section 3.5 fall in the range of 3,400,000–4,000,000 psi (23,440–27,580 MN/m^2). The tensile tests discussed above show that glass filaments tested under appropriate conditions can achieve a significant fraction of their theoretical strengths. The obvious question, then, is why are ordinary glass objects so brittle when they should be so strong? To understand the answer to this question one considers the mechanism by which a brittle fracture occurs in a material. There is good evidence indicating that these fractures occur by the formation of a small crack that propagates across the material. Propagation of the crack requires that stresses approaching the tensile strength of the material be exceeded only *locally* at the tip of the crack. It is well known from elasticity theory that local stress concentrations occur at any sharp corner on a surface. First-order calculations[5,11] show that the stress at the tip of a crack is increased by a factor of $2(a/r)^{\frac{1}{2}}$, where a is crack length and r the radius of curvature of the tip. If the crack tip is very sharp, say $25\,\text{Å}\,(2.5\,\text{nm})$ then it would only have to be 0.001 inches $(25\,\mu\text{m})$ long to locally increase the stress by a factor of 200. An applied stress of $20{,}000\,\text{psi}\,(138\,\text{MN/m}^2)$ would then produce local stresses on the order of $4{,}000{,}000\,\text{psi}\,(27{,}580\,\text{MN/m}^2)$ at such crack tips. Consequently, it is believed that the cause of such high brittleness in glasses is the presence of surface notches or cracks that nucleate mobile cracks that cause fracture. Etching or fresh drawing of the glass apparently eliminates or rounds out these defects and thus allows higher tensile loads to be applied prior to nucleation of a mobile crack. By special cooling techniques it is possible to prepare glasses such that all exposed surfaces are in a permanent state of compression. It is then more difficult to nucleate surface cracks, and such commercially prepared glasses commonly display tensile

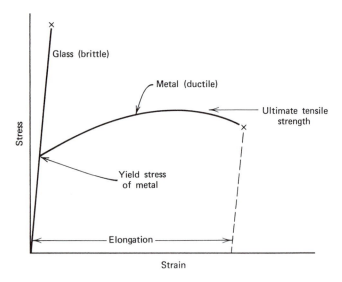

Figure **14.4**   Stress–strain curves for a glass and a metal.

strengths in the 30,000–100,000 psi (205–682 MN/m^2) range.[11] These glasses, sometimes termed tempered glass, are used in automobile windows.

As opposed to glass, metals are usually quite ductile, and it is their consequent property of strength without brittleness that accounts in large part for the widespread use of metals in our technological society. When the stress at the tip of a crack or a surface notch in a metal reaches the yield stress, dislocations begin to move. The consequent plastic flow acts to round out the crack tip and thereby relieve the stress concentration and the crack becomes immobile. It is interesting to realize that it is due to the presence and mobility of the dislocation defect in metal grains that metals have such broad use in technology. Glasses have an amorphous rather than a crystalline structure and, consequently, there is no mechanism for plastic flow such as is provided by dislocations in crystals. Figure 14.4 illustrates the difference between the mechanical properties of glasses and metals by means of a stress–strain diagram. The ductility may be gauged by the fractional amount of elongation at fracture. Application of a bulk stress of less than the metal yield stress will not generally cause the metal to fail. However, application of the same level of stress to the glass will generally cause failure in spite of its much higher inherent strength, due to the crack propagation discussed above.

The fracture that occurs in glass is generally referred to as a brittle fracture. Metals may fail by either a brittle fracture or a ductile fracture. A brittle fracture is one where very little plastic deformation occurs in the region of the crack tip. Theories of brittle fracture in metals postulate nucleation of cracks by sudden motion of many dislocations followed by crack propagation along cleavage planes. Ductile fracture in metals is thought to occur by nucleation of voids followed by either coalescence of the voids or propagation of cracks between them. The high-strength steels discussed in this chapter generally fail by ductile fracture. It is apparent that application of physical metallurgical principles to development of high-strength alloys requires a study of fracture. (See Ref. 13, Chapter 19 and Ref. 14 for discussions of fracture.) Also, it can be quite misleading to judge the resistance of a metal to brittleness by use of ductility as measured by elongation or the reduction in area at the fracture surface. The toughness of a metal is better gauged by the Charpy impact test or by notched tensile tests. However, even better techniques have recently been developed based on the study of fracture mechanics. The fracture toughness of a material is characterized by a stress intensity factor that provides the engineer with a parameter useful in design. For discussions of fracture toughness see Ref. 13, Chapter 19 and also Refs. 15 and 16.

Drill rod (also known as W-1 tool steel) is essentially an approximately 1 wt % plain carbon steel that is usually available in machine shops. If you heat a piece of drill rod to a bright red heat with a hand torch and then quench into water you will find that it is extremely brittle and may be easily broken in your fingers. Its stress–strain diagram will be similar to that shown for glass in Fig. 14.4, with the potential yield and breaking stress both probably at around 500,000–600,000 psi (3450–4140 MN/m^2). The quenched structure formed in the steel has so effectively tied up the dislocations that no plastic flow mechanism is available to effectively halt crack propagation. Consequently, it is necessary to gently heat (temper) the steel to make it useful for most applications. Depending on how you have carried out the tempering process the ductility may increase to give an elongation of around 15%, but at the same time the yield strength will have dropped to around 150,000 psi (1030 MN/m^2). Nevertheless this is a very high strength level and you will find the heat-treated portion of the drill rod now impossible to bend with your hands. This example illustrates a general difficulty encountered in attempting to produce high-strength alloys. It is usually observed that increasing the tensile strength reduces the ductility. This is illustrated for quenched and tempered steels in Fig. 14.5. The horizontal axis gives the hardness (measured on a Brinell Hardness machine), which has been increased by increasing the carbon content. In steels, hardness and strength are somewhat synonymous since linear relationships exist between the hardness measurements and tensile strength; see Ref. 17. As hardness increases the yield and tensile strengths increase but the ductility, as measured by elongation, decreases. Since strength is increased by immobilizing dislocations and ductility requires mobile dislocations, there is an inherent conflict in nature between strength

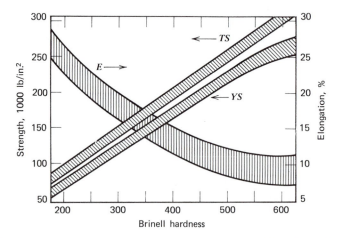

Figure 14.5 The inverse relationship between strength and ductility as hardness is increased in tempered martensitic steels (from Ref. 2).

and ductility. The challenge in alloy design is trying to introduce strengthening effects that somehow allow local plastic flow at the crack tips and do not promote void nucleation and coalescence during plastic flow.

It is interesting to point out here that there are a number of examples where an increase in strength is accompanied by an increase in ductility. The most outstanding example was discussed on p. 315, where it was noted that changing the morphology of cast irons from the flake to the nodular form increased both strength and ductility quite dramatically. The flake graphite provides ready-made locations for stress concentration and these locations are removed by nodularization. In general, if a process that increases strength also eliminates internal defects acting as stress concentrators, the process may increase both strength and ductility. Additional examples occur in the heat treatment of castings containing second-phase particles (see Table 9.7), and in the heat treatment of powder metallurgical samples that contain pores (see Ref. 18). In both cases the heat treatment is effective at removing sites of potential stress concentration.

14.3.   THE PHYSICAL METALLURGY OF SOME HIGH-STRENGTH STEELS

The physical metallurgy of steel is such a vast subject that a complete textbook would be required to cover all of the various aspects involved. It is interesting to realize that the origin of physical metallurgy as a separate subject for study stems from efforts to control the strength and ductility of steels. A good review of this development, as well as a brief summary of the modern physical metallurgy of steels, is presented in the articles of Ref. 19. The purpose of this section is to present examples illustrating some of the physical metallurgy principles involved in a few high-strength steels and in the tempering of steel. The development of high-strength steels is naturally one of the most exciting and dramatic applications for physical metallurgy and much recent research has been directed toward this end. However, to provide a proper perspective the following quote is reproduced from Ref 19a:

> "If the majority of current papers on the physical metallurgy of steel were read it would be concluded that the major interest of the user industries was greater strength. This is quite wrong, however, apart from those untypical industries concerned with space travel and defense. What is required is the extension of systematic physical metallurgy studies to other aspects such as solidification, segregation, and homogenization, to the surface and matrix effects of reheating processes and to hot deformation processes. There should also be a

greater application of the existing physical metallurgy knowledge to other properties such as fracture and wear; to corrosion and stress corrosion; and to weldability, formability, machinability, and not least reliability."

A.   HIGH-STRENGTH STEEL WIRE

One often associates the high strength of steels with the ability to form martensite. However, the strongest known metal (disregarding whisker filaments) is drawn pearlitic wire containing no martensite. Fine wires having ultimate tensile strengths (UTS) as high as 700,000 psi (4830 MN/m^2) with a 20% elongation have been reported.[20] Not only is this an extremely high strength, but it is an amazing ductility at this strength level. A major limitation upon this material is that it can only be produced by drawing wire to a reduction in area of over 90% and, therefore, is only available as fine wires; see Table 14.2. However, there are a large number of commercial uses for strong fine wire with high ductility. This point is illustrated in Table 14.3, which presents a partial list of the strength and applications of commercially produced steel wire. The wire is produced commercially by a continuous process known as patenting. A scheme for a laboratory patenting process is shown in Fig. 14.6(a). The wire is heated to the austenizing temperature by the AC voltage, after which it is quenched to the desired transformation temperature by direct immersion into a molten lead bath. Following this isothermal quench the wire is cooled to room temperature and subsequently drawn to the desired size. The purpose of this heat treatment is to achieve the finest possible pearlite spacing prior to drawing. Ideally, one would like to cool to the

Table 14–2.   Drawn Pearlitic Wire, 0.9% C, 0.4% Mn, 0.2% Si. Initially fine pearlitic wire 0.0246 inches (0.62 mm), austenized at 980°C (1800°F) and transformed at 496°C (925°F) (Ref. 20).

Wire Diameter		0.2% Proof Stress		Ultimate Tensile Strength	
(mils)	(mm)	(ksi)	(MN/m^2)	(ksi)	(MN/m^2)
24.6	0.62	170	(1170)	218	(1500)
21.4	0.54	258	(1780)	278	(1920)
16.7	0.42	255	(1760)	274	(1890)
12.9	0.33	284	(1960)	295	(2030)
7.7	0.20	377	(2600)	387	(2670)
5.7	0.15	439	(3030)	449	(3090)
4.0	0.10	516	(3560)	559	(3850)
2.9	0.07	605	(4170)	616	(4250)

Table 14.3.   Commercial High-Strength Steel Wire (Ref. 21)

Description	UTS Range (ksi)	(MN/m²)	Use
Piano wire	260–350	(1790–2410)	Piano strings, springs
Mandolin wire	290–400	(2000–2760)	Small stringed instruments
Bobby pin wire	170–230	(1170–1590)	Bobby pin manu- facture
Brassiere wire	240–280	(1650–1930)	Woman's apparel
Hose reinforcement wire	310–360	(2140–2480)	High pressure hose
Rope wires	to 320	(to 2210)	Wire cables
Galvanized suspension cable wire	200–230	(1380–1590)	Suspension bridges
Mechanical spring wires	to 430	(to 2960)	Springs

nose of the T-T-T diagram prior to any nucleation and then transform to pearlite, as illustrated in Fig. 14.6(*b*). This procedure would cause the pearlite to form at maximum rate and thus have the finest spacing. An estimate of the actual cooling path in a commercial patenting operation is shown on Fig. 14.6(*b*).

The source of the very high strength of patented and drawn pearlitic wire has been investigated with the aid of the transmission electron micros-cope.[20,22,23] The lamellar pearlitic structure is not destroyed by the extreme

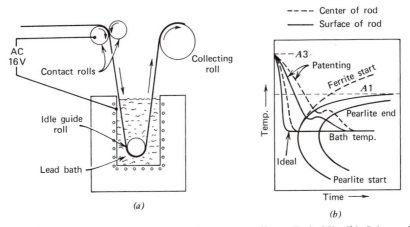

Figure **14.6**   (*a*) A laboratory patenting process (from Ref. 22). (*b*) Schematic indication of cooling curves for producing fine pearlite (from Ref. 21).

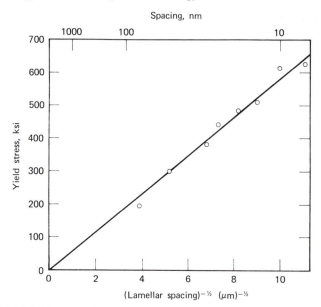

Figure **14.7**   Yield stress of drawn pearlite as a function of lamellar spacing (from Ref. 23).

reduction in area upon drawing. Apparently a large plastic deformation occurs in both the ferrite and cementite lamellae, and the mean lamellar spacing decreases approximately in proportion to the reduction in wire diameter. The flow stress is found to follow the Hall–Petch relation down to the smallest spacings measured; see Fig. 14.7. The initial spacings of around 1000 Å (100 nm) are reduced to spacings just under 100 Å (10 nm) in the most severely drawn and highest strength wire. It is apparent that the very high strength results from microstructural hardening brought about by the extremely fine lamellar spacing and the effectiveness of the lamellar interfaces at blocking dislocation motion. The results suggest that if one could reduce the spacing further, even higher strengths would be achieved; also, if one could somehow produce a technique for preparing bulk samples with such fine spacing, these high strength levels would be available for bulk applications.

Experiments on severely cold drawn wire of both bcc and fcc metals have shown that the cellular substructure produced by dislocation tangles draws out into laths of continuously decreasing thickness as the amount of drawing is increased. The flow stress in this case also follows the Hall–Petch equation with the cell thickness being used as the structure dimension.[23] Strength levels comparable to those of drawn pearlite have been demonstrated in wires drawn from the tempered martensite condition.[22]

However, in commercial practice the combination of strength and ductility achieved in pearlitic wire has not been obtainable by heat treatment or by the addition of alloying elements.[24]

B. THE STRENGTH OF MARTENSITE

With the exception of drawn pearlitic wire, high-strength steels generally owe a significant fraction of their high strength to the presence of martensite. The source of the high strength of martensite has been a subject of considerable interest to metallurgists for many decades. At the present time there is still some debate on this subject, but a limited discussion is presented to show basically why steel martensite has such high strength. The interested student may refer to Refs. 25–30 for further information. Reference 25(a) presents an excellent introduction to the subject with an interesting historical perspective and Ref. 29 summarizes work up to 1970.

In Section 14.1 we summarized four major strengthening mechanisms. Research has shown that the strength of martensite cannot be attributed to just one of these mechanisms, but that in fact all four mechanisms contribute to some extent.

1. Solid-Solution Hardening. In some respects this mechanism is the most interesting because of the way in which it is achieved. It has been shown that solid-solution strengthening from substitutional solutes in martensite is quite small, and that essentially all of this type of strengthening comes from interstitial carbon atoms. (It should be realized, however, that substitutional alloying elements indirectly affect the strength of steels by altering the hardenability, the M_s temperature, and the amount of retained austenite.) If you want to strengthen most alloys by solid-solution hardening you are limited by the solvus line of the phase diagram (see Fig. 11.11) as to the amount of solute that you can force into solid solution. However, in iron we have the very fortuitous occurrence that the high-temperature fcc phase will dissolve up to 2 wt % carbon, whereas the low-temperature bcc phase will only dissolve 0.02 wt % carbon. And furthermore, because we are able to transform from the high-temperature phase to the low-temperature phase by a diffusionless martensitic transformation, we can force carbon concentrations into solid solutions of the low-temperature phase that are much higher than the values given by the equilibrium solvus line. And, in addition, you may see by reference to Fig. 13.14 that the carbon atoms in the octahedral interstitial voids will produce a dipolar strain field in the resultant martensite, which is advantageous for tying up dislocations. As pointed out by Cohen[25a] it is this happy sequence of events provided by nature in the Fe–C system that leads to such a rich variety of possible structural control in steels.

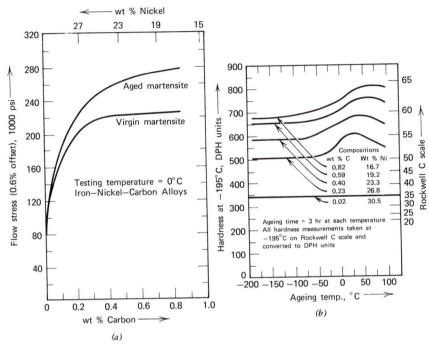

Figure 14.8 (*a*) Flow stress of 100% martensite at 0°C (32°F) versus carbon content. Aging treatment: 3 hr at 0°C (32°F) (from Ref. 25a). (*b*) Hardness of Fe–Ni–C martensites at −196°C (−320°F) after aging for 3 hr at temperatures shown (from Ref. 25a).

In spite of the fact that martensite forms by a rapid quench, there is still time for carbon to precipitate from solid solution after the martensite phase forms, unless the M_s temperature is below room temperature. By addition of Ni to Fe–C alloys, Winchell and Cohen[25a] have been able to lower the M_s sufficiently to eliminate this problem and they have determined the strength of virgin martensite as a function of carbon content, as shown by the lower curve on Fig. 14.8(a) (Ni content was shown to have negligible effect on strength). These and similar data have been analyzed in terms of the solid-solution strengthening models discussed in Section 14.1.[29] It is generally found that the solid-solution strengthening increment increases with the square root of the carbon content as predicted by Eq. 14.4 (see Fig. 14.9), but there is not complete agreement here. There is agreement, however, that the mechanism for strengthening results from elastic interaction between the dislocation and the strains introduced by individual carbon atoms.

It is interesting that the strengthening effect drops off so abruptly above

0.4 wt % carbon, see Fig. 14.8(a). Roberts[29] has suggested that this is due to a change in mode of plastic deformation from slip to twinning at carbon contents above 0.4% C, assuming that the stress needed for twinning is independent of carbon content.

2. **Precipitatation Hardening.** Since the martensite phase is highly supersaturated with carbon, there will be a strong tendency for the carbon atoms to segregate to interfaces or crystal defects and to precipitate from solution as a carbide. Assuming that an average carbon atom would have to migrate 10 Å (1nm) for a measurable amount of carbon to be removed from solution, Cohen[25a] has calculated the required diffusion times from Eq. 6.49 based on data for the diffusion coefficient of carbon in martensite. The results indicate that a measurable precipitation effect will be observed in milliseconds at 200°C (390°F) and in minutes at room temperature. These calculations are supported by data such as presented in Fig. 14.8(b). The martensites here all had subzero M_s temperatures and they were quenched directly to −196°C so that precipitation of carbon on quenching should be negligible. Upon aging at increasing temperatures above −196°C (−320°F) for 3 hours the hardness increased as shown. Notice that the hardness at −196°C (−320°F) is a function of the carbon content and that the increment in hardness upon aging goes to zero as carbon content drops to 0.02. Based on such data Cohen[25a] has concluded that the increase in hardness upon aging is associated with carbon migration. It is amazing that significant carbon migration occurs at temperatures as low as −20 to −40°C (−4 to −40°F). These results support the diffusion calculations based on 10 Å (1 nm) migration distance. Consequently when the M_s temperature is well above room temperature, as in plain carbon steels, precipitation effects will occur upon quenching because, as shown by the results of the above calculations, only milliseconds are needed for significant diffusion. This effect is sometimes called autotempering or quench tempering. Therefore, the contribution of precipitation hardening in a quenched martensite will depend upon the M_s temperature, the quenching rate, and the carbon content. It is generally quite difficult to measure this contribution in high M_s martensites, but it has been measured in low M_s martensites after an aging treatment. The increment in hardening due to precipitation may be seen by the comparison of the two curves in Fig. 14.8(a).

3. **Structure Hardening.** Since the first formed martensite plates grow clear across the individual grains (except in some large-grained cases) it is possible to reduce the martensite plate size by decreasing the prior austenite grain size. As expected, the reduced martensite plate size does increase the strength and, fortunately, it is done so without loss in ductility.

For example, in a low-carbon Fe–32% Ni martensite, reducing grain size from 110 to 33 μm increased the yield strength from 72,000 to 87,000 psi (496 to 600 MN/m^2) with a constant elongation of 22%.[25b] Similar strengthening without loss of ductility is obtained by reducing prior austenite grain size in lath-type martensites. Over the range of grain refinement commercially available, the increase in yield strength in an AISI 8650 steel amounted to only around 4%, whereas using experimental techniques to achieve very small grain sizes (~2 μm), an additional strengthening increment of 16% was achieved at constant elongation.[25b] The small grain sizes were achieved by either controlled recrystallization utilizing the sorts of ideas discussed in Chapter 10, or by a cyclic austenization treatment where the temperature was cycled up and down through the A_s temperature.

The spacing of the twin stacks within plate martensites in steels is frequently on the order of 100 Å (10 nm). By comparison with the structural hardening achieved in drawn pearlite wires having structures of this dimension, see Fig. 14.7, one would expect a very large structural hardening component in twinned martensite plates. However, this does not appear to be the case. Figure 14.9 presents data on the flow stress of both lath and plate martensites in Fe–Ni–C alloys, which show the plate martensite to be only very slightly stronger. This is true in spite of the fact that the dislocation cell structure in the lath martensite has dimensions on

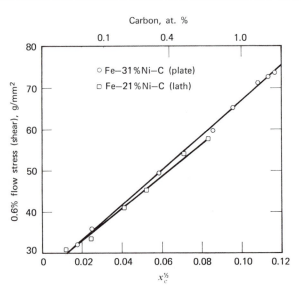

Figure 14.9   Flow stress of the two types of Fe–Ni–C martensites as a function of carbon content (from Ref. 30).

the order of 5000 Å (500 nm) compared to the 100 Å (10 nm) twin stacks in the plate martensites. Apparently, the twin boundaries within the plate martensite are not very effective at blocking dislocation motion.

4. Dislocation Hardening. The dislocation density in plate martensites is relatively low. However, in lath martensite the dislocation density is found to increase to extremely high values as the carbon content increases; see Fig. 14.2. Furthermore, experiments have shown[8] that in these lath martensites most of the carbon is *not* in solution. Apparently, this carbon segregates to dislocations and lath boundaries during quenching and is extremely effective at causing the dislocation density to be increased. Consequently, in these lath martensites a large portion of the strength is due to the high dislocation density, and only a small portion can be attributed to solid-solution hardening.

5. Summary. It is generally agreed that the major portion of the strength of martensite is due to some combination of solid-solution hardening and precipitation hardening. However, this conclusion can only apply to plate martensites because in lath martensites the solid-solution hardening is replaced with dislocation density hardening. The amount of strengthening due to precipitation depends on the degree of autotempering in any particular case and, in general, is quite difficult to experimentally determine. The interstitial solute, carbon, plays a key role in these mechanisms and little additional strength is obtained from substitutional solutes. A small increment in strength without loss to ductility may be achieved by reduction of the prior austenite grain size.

C. TEMPERING

A steel contains relatively high levels of residual stress immediately following the quench. These stresses arise from the fact that the surface regions cool faster than the interior regions and also because the specific volume of the martensite phase is on the order of a few percent larger than that of the austenite phase. Also, microstresses arise due to the supersaturation of carbon produced on quenching. These residual stresses can be sufficiently high that quench cracks will appear either during quenching or following the quench at room-temperature aging. In addition, it is sometimes observed that as-quenched plate martensites contain numerous microcracks traversing the plates. This microcracking occurs in Fe–C alloys at compositions above around 0.8 wt % C and is believed to result from impingement of intersecting martensite plates.[31,32] Because of these effects, quenched steels are extremely brittle and are, therefore, rarely used in their as-quenched state. By performing the simple experiment described on p. 521 you may quickly show yourself why it is necessary to temper

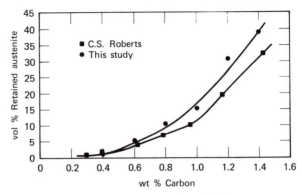

Figure 14.10   Retained austenite in quenched Fe–C alloys at room temperature (from Ref. 31).

quenched martensite structures. By heating to 100°C (212°F) the residual stresses are substantially relieved by localized plastic flow and carbon redistribution and the tendency for spontaneous cracking is removed. This initial tempering increases the ductility with little or no loss in strength.

As indicated in the previous chapter the quenched structure is not necessarily 100% martensite, depending on the location of M_s and M_f relative to the quench temperature. As the carbon content is increased in Fe–C alloys the amount of retained austenite following a quench to room temperature increases as shown in Fig. 14.10. Therefore, in our discussion of tempering we must be concerned about what happens to both the martensite and the retained austenite. The following presentation is based largely upon Refs. 19(c), and 33–35; in addition, a good basic discussion of tempering may also be found in Ref. 13 and a detailed practical discussion in Ref. 36.

1.   Lath Martensite.   Speich[33,35] has recently presented a comprehensive study on the tempering of Fe–C lath martensites. His results are summarized in Fig. 14.11, which serves as an excellent guide to the discussion of the structural changes occurring during tempering. In addition to relief of residual stresses, initial tempering action consists mainly of migration of carbon atoms from their interstitial sites to lattice defects, primarily dislocations and lath interfaces. This action actually starts during the quench as discussed above and, with carbon contents below 0.2 wt % C, it has been estimated that 90% of the carbon is removed from solid solution during rapid quenching (10,000°C/sec).[35] If the carbon content is below 0.2 wt % tempering at temperatures up to 150–200°C (300–390°F) only produces additional carbon migration. However, at higher carbon contents an extremely fine precipitate forms within the laths. The dispersion is so fine that direct confirmation of the precipitate structure has not been possible by electron diffraction. However, it has been deduced from other information that this precipitate is a metastable iron carbide called ε carbide, which has a hcp structure and a composition of roughly $Fe_{2.5}C$. At temperatures of 200–250°C (390–480°F) a rod or plate-shaped precipitate forms within the laths in both the high and low carbon content martensites of Fig. 14.11. In the higher carbon content steels the precipitates form on dislocations while in the lower carbon content steels they form as Widmanstatten arrays. These precipitates are also thought to be ε carbide and they form as plates or rods upon the {100} martensite planes. At temperatures beginning around 250–300°C (480–570°F) the equilibrium Fe_3C^* precipitate begins to form. It forms as a plate-shaped precipitate, often at lath boundaries. Upon reaching the temperature range of 400–600°C (750–1110°F) as shown on Fig. 14.11, the plate-shaped Fe_3C begins to grow and spheroidize while the ε carbide formed at the lower temperature dissolves and the martensite is converted to ferrite. The ϵ carbide is clearly

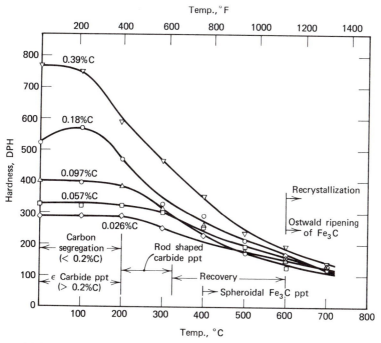

Figure 14.11 The hardness and the sequence of reactions occurring upon tempering Fe–C lath martensites for 1 hr at various temperatures (from Ref. 35).

*Graphite is actually the equilibrium phase but, from the practical point of view, Fe_3C is considered the equilibrium phase here since graphite will not form.

an intermediate metastable precipitate and the precipitation sequence that occurs on tempering is quite similar to those discussed in Chapter 11, and the same principles apply. Recovery processes become important at around 400°C (750°F) and recrystallization occurs around 600°C (1110°F). Both processes occur in the manner and follow the principles described in Chapter 10. Tempering just below the eutectoid temperature causes the cementite to coarsen by the Ostwald ripening process described in Chapter 11 into roughly spherical shapes within a ferrite matrix. This is generally termed a *spheroidized* structure.

2. Plate Martensite. These martensites occur at higher carbon contents in Fe–C alloys and have the internally twinned structure with a relatively low dislocation density. Initially, ε carbide precipitates form as very fine Widmanstatten arrays at temperatures from room temperature to 200°C (390°F) and, apparently, the carbon remains in solid solution until the precipitates form. These ε-carbide precipitates are in the size range of GP zones and in an Fe–20 Ni–0. 8C alloy have been measured to be 10–25 Å wide by 50–200 Å long (1–2.5 nm wide by 5–20 nm long).[37] As the tempering temperature increases the precipitates grow in size. Above 200°C (390°F), however, the ε carbide begins to dissolve and cementite forms plate-like precipitates upon the internal twins.[34] At this point the structure has an appearance almost indistinguishable from lower bainite. At higher temperatures the twin structure disappears and the evolution of events becomes essentially identical to that in lath martensites. There is some evidence that the carbide formed between 200 and 300°C (390–480°F) in these higher carbon steels is a second intermediate precipitate of the form Fe_5C_2.[33]

3. Retained Austenite. It may be seen from Fig. 14.10 that retained austenite is present in Fe–C alloys only for carbon compositions above around 0.4% C. In these alloys, then, the metastable retained austenite will decompose upon tempering. It is observed that between 200 and 300°C (390–480°F) this retained austenite transforms into bainite containing ε carbide accompanied by a significant increase in volume. It has been customary to subdivide tempering into three stages. The first stage is the precipitation of ε carbide at temperatures up to 200°C (390°F). This stage only occurs in Fe–C martensites with carbon contents above 0.2%. It is accompanied by a marked decrease in specimen volume and sometimes a slight increase in hardness in high-carbon steels. The second stage is the formation of bainite from retained austenite and it is necessarily restricted to Fe–C alloys with carbon contents above 0.4%. The transformation occurring in this stage produces an increase in specimen volume and an increase in hardness. The third stage is the dissolution of the intermediate ε carbide and formation of cementite. This stage exhibits a decrease in volume and a decrease in hardness and it occurs at all carbon compositions. Hence, the traditional three stages of tempering are restricted to only the high-carbon martensites of binary Fe–C alloys; consequently, there is a recent trend to drop the three-stage classification.

4. Effect of Alloying Elements. The decrease in hardness during tempering is due to a loss of carbon from solid solution and/or a decrease in dislocation density. It is possible, however, to regain nearly the original hardness of the quenched alloy by addition of carbide forming alloying elements such as Ti, Mo, Cr, and V. This is illustrated in Fig. 14.12(a) for the addition of V to an 0.3% C steel. In Fig.14.12(b) the individual effects of adding Co, Cr, and Mo to an Fe–10Ni–0.13C alloy may be seen. The strengthening peak is generally found to occur in the 500–600°C (930–1110°F) range and it results from the precipitation of fine carbide precipitates. The added strength here results from formation of very fine coherent precipitates typical of the age-hardening alloys discussed in Chapter 11. The vanadium carbide forms as plates of about 10 Å (1nm) thickness by 100 Å (10 nm) diameter, and the molybdenum carbide as needles of about 100 Å (10 nm) length.[33] Chromium is an example of an element that under certain conditions forms a carbide by directly replacing Fe from the coarse Fe_3C precipitate. This effect does not promote strengthening because of the coarse precipitate distribution that results. The carbide responsible for the peak hardness in Fig. 14.12(b) is a finely dispersed $(Mo, Cr)_2C$ carbide. This alloy displays good ductility, having elongations on the order of 16–21%. The initial increase in strentth upon carbide formation is actually accompanied by an increase in ductility. This result is apparently a consequence of the coarse Fe_3C precipitate being replaced by the much finer $(Mo,Cr)_2C$ precipitate. The large Fe_3C precipitates are thought to nucleate voids more easily, and these voids lead to stress concentrations and subsequent fracture.[38]

Apparently the presence of carbide forming solutes does not alter the early stages of tempering from that of plain carbon steels. But the carbide formers do tend to stabilize ε carbide and retard recrystallization.[19c] Sometimes the formation of the alloy carbides and the consequent strengthening is called a fourth stage of tempering.

5. Embrittlement upon Tempering. Figure 14.13 presents data on the variation of the mechanical properties of AISI 4340 steel upon tempering. Notice that the impact toughness as measured by the impact test experiences a significant drop between around 200–400°C (390–750°F) while the

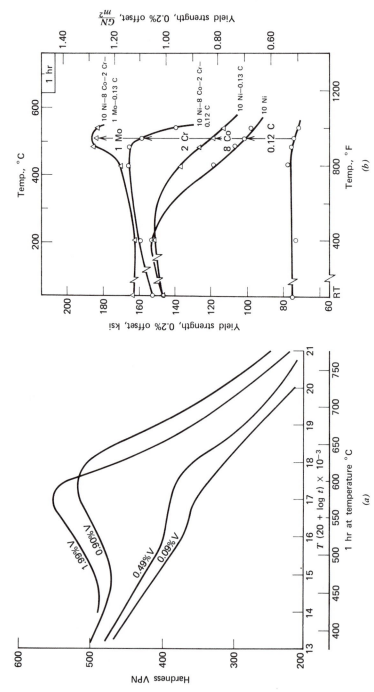

Figure 14.12 (*a*) The effect of vanadium on tempered hardness of a 0.3% C steel (from Ref. 34). (*b*) The effect of C, Co, Cr, and Mo on tempered strength of 10% Ni steels (from Ref. 38).

strength also continues to fall.* This detrimental embrittlement occurs in many AISI steels and it is often called 500°F *embrittlement,* see, for examples Refs. 33 and 36. It is believed to be associated with the initial nucleation of Fe_3C and it is generally thought to result from the formation of film-like carbides upon grain boundaries and subboundaries.[33] Because of the embrittlement many low alloy steels such as AISI 4340 of Fig. 14.13 are not tempered above 200°C (390°F). The combination of strength and ductility that can be obtained from alloy steels is significantly superior to that of plain carbon steels at the 200°C (390°F) temper temperature. A comparison of the strengths as a function of carbon content is shown in Fig. 14.14(*a*), and a comparison of impact values at given strengths in Fig. 14.14(*b*). The increased strength is due to a slightly higher initial strength for the alloys and a smaller loss of strength upon tempering in the alloys. The alloying elements themselves tend to retard the tempering reactions. It is found that Si is particularly effective in this regard. For example, addition

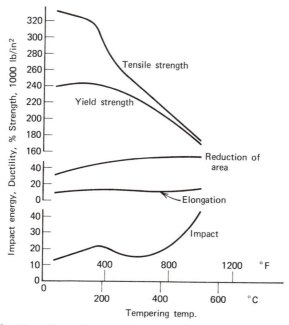

Figure 14.13   The effect of tempering on mechanical properties of AISI 4340 steel (from Ref. 39).

* Note that the ductility as measured by both elongation and reduction in area do not show a corresponding decrease. This fact illustrates that elongation tests on smooth tensile specimens will often not measure the toughness of a material to impact loading and that one must be careful in using elongation to judge the resistance of a material to embrittlement. See Ref. 2, p. 1 for a specific example.

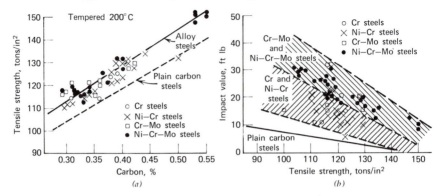

Figure 14.14 (*a*) The effect of C content on strength of alloy and plain carbon steel (from Ref. 45). (*b*) Impact strength at given tensile strengths for alloy and plain carbon steels (from Ref. 45).

of Si is found to shift the embrittlement temperature up from around 500 to 800°F (260–430°C) and, therefore, allows the use of these alloys to higher temperatures.[36] Recent experiments indicate that the Si slows down growth of carbide particles due to rejection of Si by the growing carbide particles and the consequent diffusion-controlled growth.[33]

High-speed tool steels depend upon the above-mentioned secondary hardening peak at 500–600°C (930–1110°F) due to carbide formation to achieve high-temperature strength. Here adequate ductility is generally achieved by slightly overaging at the sacrifice of some strength. However, this depends on the particular steel; see Fig. 97 of Ref. 36.

There is another type of embrittlement encountered in some steels, called *temper embrittlement*. It occurs in the range of 370–565°C (700–1050°F) and failure occurs by loss of cohesion at prior austenite grain boundaries. It is believed to be due to grain-boundary segregation of Sb, P or perhaps As or Sn. See Ref. 33 or 36 for further discussions of temper embrittlement.

As discussed in Section 14.3B and shown again in Fig. 14.15, the strength of martensite is a function of carbon content. However, the ductility drops rapidly as carbon content is increased and the strengths at the higher carbon contents shown on Fig. 14.15 are not usable for engineering applications because the steels do not possess adequate ductility at the higher strength levels. The general effect of alloying elements on strength is shown in Fig. 14.15. It is apparent that these alloy additions allow one to achieve higher strength levels at lower carbon contents, and, as shown explicitly in Fig. 14.14(*b*), this allows higher strength at the minimum acceptable level of ductility.

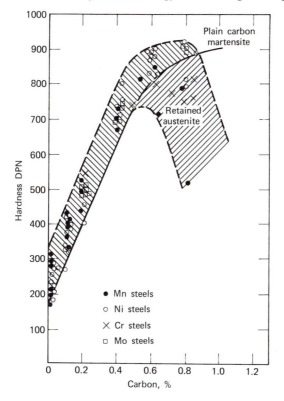

Figure **14.15**   The effect of carbon content and alloying elements on the strength of martensite (from Ref. 45).

D. AUSFORMED STEELS

The highest strength bulk steels with adequate toughness for engineering applications are the ausformed steels; see Fig. 14.1. These steels were developed in the late 1950s, but for a variety of reasons they are not widely used at the present time. Nevertheless, they present a very interesting technique for increasing strength levels above that of conventional steel alloys without a substantial loss in ductility. For additional reading on the ausforming process see Refs. 2 and 40–44.

Figure 14.16(*a*) shows a T-T-T diagram for a steel that has a significant gap between the pearlite-start and the bainite-start curves. This region is often called the *bay* on the T-T-T diagram. If the steel of Fig, 14.16(*a*) is hot rolled at 900°F following quenching from about 1600°F, the austenite will be deformed without introduction of pearlite or bainite, and it may then be quenched to form a fully martensitic steel. This treatment is shown

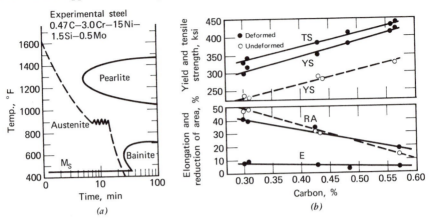

Figure 14.16 (*a*) The T-T-T diagram for an ausformed steel (from Ref. 2). (*b*) The effect of ausforming upon mechanical properties as a function of C content. A 3% Cr alloy deformed 91% and tempered at 625°F (330°C) (from Ref. 2).

schematically by the dashed line on Fig. 14.16(*a*). It illustrates the basic principle of the ausforming process: deformation of austenite above the M_s at times and temperatures insufficient to form pearlite or bainite or to cause recrystallization, followed by quenching to achieve martensite and then tempering. It is apparent that ausforming requires a steel with a significant bay on its T-T-T diagram. A number of alloy steels meet this requirement and in many cases[41] a significant improvement in mechanical properties is obtained by ausforming. Figure 14.16(*b*) illustrates the effect of the ausforming deformation on strength and ductility as a function of carbon content in a 3% Cr steel. These steels were deformed 91% at around 482°C (900°F), oil quenched to form martensite, liquid-nitrogen quenched to minimize retained austenite, and then tempered for 1 hour at 330°C (625°F). As illustrated, the yield strength shows an increase of around 75,000 psi (517 MN/m²) with little change in ductility. Experiments on conventional high-hardenability martensitic steels have shown strength increases after ausforming on the order of 125,000 psi (862 MN/m²) with little or no loss in ductility.[2] Results on a type H-11 hot-work die steel are presented in Fig. 14.17(*a*), and it may be seen that the large strength increases require considerable deformation, a fact that detracts from practical applications of ausforming this steel. Some ausformable steels display a maximum strength increase at lower percent deformations. The influence of the ausforming process on the tempering of two commercial steels is shown in Fig. 14.17(*b*). It is seen that hardness is significantly increased and the secondary hardening present in conventional treatment of these steels is absent.

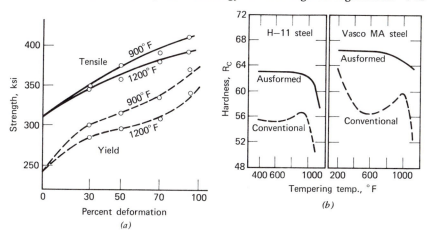

Figure 14.17 (a) Variation of strength with amount of deformation for two deformation temperatures. H-11 steel (0.4C, 5Cr, 1.2 Mo, 0.5 V) tempered at 510° (950°F) (from data in Ref. 2). (b) The effect of tempering temperature on hardness for two steels (from Ref. 2).

There seems to be general agreement on the main features of the strengthening effect produced by ausforming.[42,44] During deformation of the austenite a fine carbide dispersion is produced in the austenite matrix, which may be alloy carbides or iron carbides depending on the alloy content. The carbide particles lead to a high rate of dislocation multiplication during austenite deformation and a high dislocation density. This high dislocation density is inherited by the martensite subsequently formed. These inherited dislocations, plus any dislocations formed by the martensite transformation itself, result in an extremely high dislocation density within the martensite. The grain refinement produced by the deformation (long skinny or flattened grains) gives rise to a refinement in the martensite plate size. The normally twinned substructure of plate martensite is changed by ausforming to a substructure consisting of fine precipitates in a dense dislocation array. Hence, there are three primary ways in which the microstructure of quenched ausformed steels differ from the conventional case. The dislocation density is higher, precipitates formed during deformation are present, and martensite plate sizes are smaller. All three of these changes will lead to additional strengthening, but there is debate on the relative magnitude that should be attributed to the three effects. Nevertheless, one scheme[44] is presented in Fig. 14.18 because it illustrates the sorts of ideas involved in understanding how the microstructural changes affect strength upon quenching and during tempering. In the quenched condition shown at the extreme left, added strength is present in

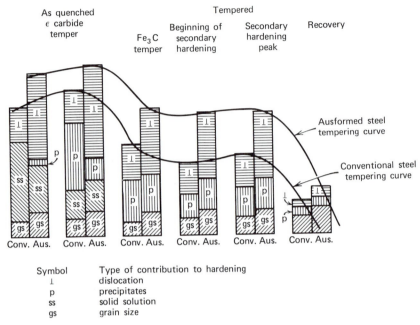

Figure 14.18 A schematic representation of the major contributions to the strength of conventional and ausformed steel during tempering (from Ref. 44).

the ausformed case from dislocations, precipitates, and grain size, while less strength results from solid solution. Tempering to produce ε carbide (stage 1) increases the overall strength by adding more strengthening from precipitate hardening than is lost to solid-solution hardening. The net increase in strength is slightly greater in the conventional case. Upon forming Fe_3C (stage 3) the solid-solution hardening is lost. At still higher tempering temperatures the secondary hardening peak due to alloy carbide formation (stage 4) is much less in the ausformed case. Finally, as recovery processes take over, the dislocation and precipitate hardening fall off dramatically in both cases. There are two other interesting points: The grain size contribution remains essentially constant during tempering since boundary migration does not occur, and the dislocation density in the ausformed case is consistently higher.

The answer to the question of why the ductility of ausformed steels is higher than conventional steels at equivalent strength levels is not well understood. One suggestion (A. Tetelman, Ref. 42, p. 325), is that the martensite plates produced by the ausforming process contain a high density of *unpinned* dislocations. Crack nucleation is favored by deformation twinning and a high dislocation density reduces the twinning mode of deformation in bcc metals, thereby reducing crack nucleation. Also, crack propagation may be reduced because the high dislocation density will allow local plastic relaxation at the crack tip.

E. TRIP STEELS

Figure 14.19(a) shows a typical engineering stress–strain diagram for a uniaxial tensile test on a steel, and Fig. 14.19(b) gives the corresponding true stress–true strain diagram. (See Ref. 46 or 47 for discussions of differences in these two types of diagrams.) When the ultimate tensile stress is reached the specimen begins to "neck down" at a local region as indicated on Fig. 14.19(a). Since the cross-sectional area at the neck drops, the true stress increases locally in the neck, rapidly accelerating the necking process and leading to failure. After necking begins, virtually all of the elongation occurs in the neck region. In high-strength steels the reduction in area at fracture is on the order of 50%, which indicates that these metals have been relatively ductile in spite of failure. This is indicated in better perspective in Fig. 14.19(b), which shows considerable strain after necking begins. Prior to necking the gauge length of the tensile specimen has been undergoing a uniform strain as shown on Fig. 14.19(a). The necking is really a kind of plastic instability; all of a sudden, at one point along its surface the metal cylinder begins *locally* to flow and thereby locally reduce its area. Prior to this local flow there was a steady uniform plastic flow along the entire gauge length. It is apparent that if one could prevent local plastic flow the plastic instability could be avoided and much greater elongations would be available for practical uses. At the region of local plastic flow an increase in strain is experienced. If this local strain increase

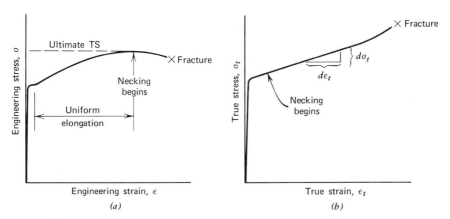

Figure 14.19   (a) Engineering stress–strain diagrams for a steel. (b) Corresponding true stress–true strain diagram.

were to produce a corresponding local strength increase, the plastic flow would be inhibited locally, and stability thereby restored. Hence, the key to plastic stability here is the work hardening rate, which is given by the slope shown on Fig. 14.19(b), $d\sigma_t/d\varepsilon_t$.

In our previous discussions we have seen that work hardening is caused by dislocation interactions that promoted the restriction of dislocation motion during plastic flow. One method of increasing this work-hardening rate was to introduce particles that were effective at tying up dislocations (see p. 411). A more effective and an extremely clever technique has recently been promoted by Zackay and Parker,[48,49] based on a principle that has been operative for nearly 100 years in Hadfield-manganese steels. The basic idea is to cause the local strain to induce the local formation of martensite, which in turn will give local strengthening to prevent plastic instability. The values of $d\sigma_t/d\varepsilon_t$ will be increased by the plastically induced martensite formation. Because the *t*ransformation is *i*nduced by *p*lastic flow the authors have chosen to call these steels TRIP steels.

Zackay et al.[48] have developed a series of TRIP steels, one of which will be described here, having composition 8.9 wt % Cr, 8.3 Ni, 3.8 Mo, 2.0 Mn, 1.9 Si, and 0.3 C. The treatment of this steel was as follows:

1. The alloy was solution treated at 1120°C (2050°F) for 1 hour and then quenched to room temperature. The alloy was now all austenite with both M_s and M_D below room temperature.

2. The alloy was deformed 80% at 450°C (840°F) and then cooled to −196°C (−320°F) after which it was still nearly all austenite. This treatment served two main purposes.

a. It produced a fine dispersion of carbides and a high dislocation density in a matrix that was essentially all austenite. This gave rise to a significant strengthening by a mechanism similar to that occurring in ausformed steel.

b. The loss of solid solution carbon upon carbide formation raised the M_D temperature above room temperature but left the M_s temperature still quite low.

Deformation was now carried out at room temperature. Since this temperature is now below the M_D temperature any local plastic flow was found to be effective at producing local transformation to martensite. The values of $d\sigma_t/d\varepsilon_t$ were considerably above conventional tempered martensitic steels (for example, AISI 4340) as hoped for, and very high ductilities were observed in room-temperature tests as shown in Table 14.4. Treatment I was as described in steps 1 and 2 above and treatment II was identical, except that prior to testing, the samples were given a 15%

Table 14.4. Mechanical Properties of Two TRIP Steels (Ref. 48)

Treatment	YS (ksi)	YS (MN/m²)	UTS (ksi)	UTS (MN/m²)	Elong (%)
I	222	(1530)	254	(1750)	41
II	293	(2020)	293	(2020)	26

deformation at room temperature followed by a 30-minute temper at 400°C (750°F). Zackay et al.[48] have compared these and other results to available data on high-strength steels as shown in Fig. 14.20. At the 250,000–300,000 psi (1720–2070 MN/m²) stress level these TRIP steels appear to offer a definite ductility advantage over other classes of high-strength steels. Unfortunately, TRIP steels have not found widespread use, probably because they require large prior hot deformations and close composition control and are limited to low-temperature applications. However, these steels do illustrate quite clearly the importance of a high work-hardening rate with regard to producing useful ductility in metals that exhibit significant reduction in area. From the measured reduction in

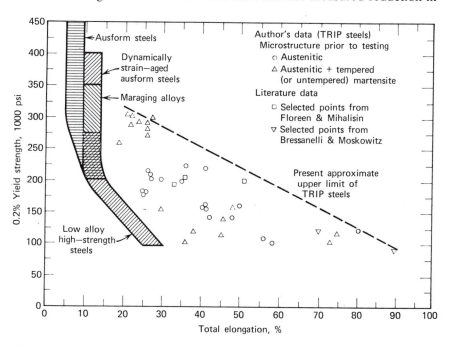

Figure 14.20 Range of 0.2% yield stress versus elongation for some special classes of high-strength steels (from Ref. 48).

area in several high-strength steels, Parker and Zackay[49] have calculated the potential elongation at fracture if plastic instability could be eliminated; they find that up to yield stress levels of 450,000 psi (3000 MN/m²) elongations of 95% should be possible (2 inch gauge length).

F. MARAGING STEELS

This class of high-strength steels was developed at International Nickel Company in the years around 1960; they are discussed here in some detail because they present an interesting example of applied physical metallurgy. These alloys are basically Fe–Ni alloys utilizing the Fe–Ni martensite transformation that was mentioned at a number of points in the previous chapter. Precipitates are formed by aging of the martensite phase below its decomposition temperature and, hence, the name, maraging. The alloys are Fe–18% Ni with additions of Co, Mo, Ti, and Al in the ranges:

$$\text{Co, 8–12\%} \qquad \text{Ti, 0.2–1.7\%}$$
$$\text{Mo, 3.2–5\%} \qquad \text{Al, }\sim0.1\%$$

Notice that there is no carbon in these alloys, although around 0.03% C is present as an impurity. For additional information on these alloys see Refs. 50–53, but particularly the excellent review paper by Floreen.[52]

1. Phases. The iron-rich portion of the iron–nickel phase diagram is shown in Fig. 14.21(a). It may be seen that binary alloys containing 18% Ni would be two phase mixtures of bcc α ferrite and fcc γ austenite at room-temperature equilibrium conditions. At compositions of 0–10% nickel the austenite will decompose into the equilibrium α ferrite upon slow cooling. However, at higher Ni compositions a martensitic reaction occurs even upon slow cooling of the austenite. This reaction transforms the γ phase into a bcc structure that we will call α martensite. Hence, of the two possible transformations upon cooling γ austenite,

    1. γ→α ferrite (bcc)   [equilibrium]

    2. γ→α martensite (bcc)   [non equilibrium]

the second one will occur even at very slow cooling rates in an Fe–18% Ni binary alloy.

2. Martensitic Reaction. The martensitic reaction has a large shear strain component as shown in Table 13.3. It therefore has a large hysteresis as was shown on Fig. 13.24. Notice that contrary to the Fe–C martensites, the reverse martensite to austenite reaction does occur in this system even though the α martensite is metastable. However, the decomposition of α martensite into the equilibrium composition of α ferrite does occur at

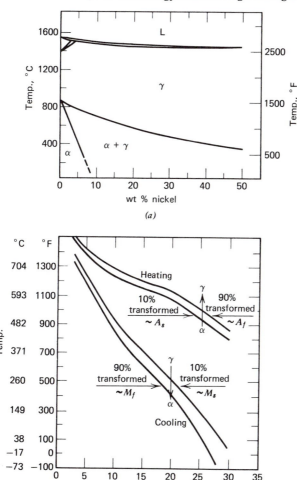

Figure 14.21 (a) The iron-rich portion of the Fe–Ni phase diagram (from Ref. 52). (b) The M_s, M_f, A_s, and A_f temperatures for the Fe–Ni martensite transformation (from Ref. 50).

temperatures just below A_s. Fortunately this decomposition is slow at the aging temperature of the maraging steels. Figure 14.21(b) presents data that approximate the composition dependence of the M_s, M_f and A_s, A_f temperatures for Fe–Ni binary alloys. It can be seen that both the M_s and A_s temperatures are decreased as Ni content rises. The addition of the alloying elements Co, Mo, and Ti also acts to lower M_s and A_s,[53] and these effects are quite important to the development of maraging alloys for two reasons. First, it is desirable to have the austenite to martensite reaction complete at room temperature. Second, it is desirable for the ageing process to occur in the martensite. Hence, the ageing temperature must be at or below the A_s temperature, but this temperature must be high enough to allow sufficient diffusion for the ageing reactions to occur in a reasonable time.

As may be seen from Table 13.4 the substructure of the binary Fe–Ni martensite switches from lath to plate at Ni compositions of 25–29 wt %. As expected, therefore, the martensite in maraging alloys is the lath type. There is evidence that this lath martensite with its high dislocation density produces better toughness than does the plate martensite with its twinned substructure.[52] The transition from plate to lath martensite occurs in both Fe–C and Fe–Ni alloys as the solute content in the iron increases; and this effect is thought to result from the decrease in the M_s temperature produced by the solutes. Since the alloying elements, Co, Ti, and Mo, also drop the M_s temperature, a further restriction is placed on their total additions if the lath morphology is to be maintained.

3.   Heat-Treat Cycle.   A typical heat treat cycle for a maraging alloy is shown in Fig. 14.22, which also indicates the dilational changes in the specimen through this heat-treat cycle. The first step is a 1-hour heat treatment at 815°C, (1500°F) which is sometimes referred to as *ausaging*. This treatment produces the γ phase and perhaps recrystallizes the structure depending on prior deformation. It also causes some precipitation reactions to occur that alter the matrix composition and therefore affect the M_s temperature. After air cooling to form the martensite phase the second step of the cycle is a 3-hour ageing treatment at 480°C (900°F),

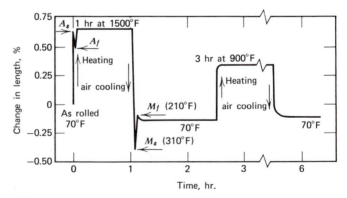

Figure 14.22   Heat-treat cycle for maraging process illustrating dilational changes in specimen (from Ref. 51).

which is followed by air cooling to room temperature. This ageing temperature is below the A_s temperature so that the martensite phase remains stable with respect to austenite. In addition, the 3-hour hold is apparently short enough so that the α martensite is also stable with respect to the equilibrium α ferrite at this ageing temperature. Consequently, precipitation occurs from the martensite matrix during ageing.

4. Ageing Reactions. There have been numerous studies to clarify the nature of the precipitation reactions in these alloys; they are summarized in Ref. 52. The main action appears to involve two precipitates, Ni_3Mo and Ni_3Ti. The standard 480°C (900°F) heat treatment produces Ni_3Mo, but if overaged or aged at higher temperatures this precipitate is replaced by Fe_2Mo. The Ni_3Mo appears to nucleate heterogeneously at either dislocations or lath boundaries. Some studies have found GP zones of Ni_3Ti that were later replaced by an η-Ni_3Ti (different structure). In general, the precipitates formed during maraging are found to be uniformly distributed over the martensite laths and to have a size of several hundred angstroms. They often display a preferred orientation and are believed to be coherent with the martensite matrix as is typical of age-hardening precipitates. Co has not been detected in the precipitates and yet, unless Co is added, the strengthening increment on ageing is not nearly as large. The exact role of Co remains a mystery, but there is good evidence that it influences the Mo precipitate in some way. Since it cannot be found in the precipitate, it is thought to cause the precipitate to have a finer dispersion, perhaps by increasing the dislocation density in the lath martensite and providing more uniform nucleation, or by decreasing the Mo solubility in the martensite matrix. In addition to forming Ni_3Ti, the Ti is thought to tie up any C impurities, since it it is a very strong carbide former, and thus avoid loss of Mo in the form of a carbide. Aluminum is beneficial to toughness perhaps because of deoxidation and formation of a secondary precipitation reaction.[53]

5. Mechanical Properties. Three grades of maraging alloys, (200), (250), and (300) were originally developed and later[53] a fourth alloy, (350), was developed by careful adjustment of composition variables. Increased strength is achieved in the higher grades by increasing the amounts of the Co and Ti additions. The mechanical properties are shown in Table 14.5. Two points become obvious from examination of the data in this table. First, the maraging alloys can be strengthened to quite high levels and, second, the strength added by precipitation is quite appreciable. The data do not illustrate, however, the most outstanding mechanical property of these steels, their high fracture toughness. It is generally found that maraging steels display a higher fracture toughness at comparable strength

Table 14.5. Data on Some Mechanical Properties of the Four Grades of Maraging Alloys (Refs. 51 and 53)

Alloy	0.2 Yield Strength		Tensile Strength		Elongation in 1 inch
	(ksi)	(MN/m²)	(ksi)	(MN/m²)	
200 annealed	117	(807)	145	(1000)	17
200 maraged	190–225	(1310–1550)	195–230	(1340–1590)	11–15
250 annealed	116	(800)	146	(1010)	19
250 maraged	240–265	(1650–1830)	245–270	(1690–1860)	10–12
300 annealed	115	(793)	147	(1013)	17
300 maraged	260–300	(1790–2070)	265–305	(1830–2100)	7–11
350 annealed	—		—		—
350 maraged	315–350	(2170–2410)	330–360	(2280–2480)	7–8

levels than do the conventional tempered martensite steels. Figure 14.23 compares Charpy V-notch impact data of maraging steels with quenched and tempered AISI 4340 steels. The relative fracture toughness of the 350 maraging grade is, however, not generally superior to conventional alloys. Carter[54] has recently shown that the critical crack depth in a commercial 350 grade alloy is five times smaller than for a silicon modified 4340 steel commonly used in aircraft landing gear.

It appears that about 50% of the strength of martensite results from the martensite structure and 50% from the particle strengthening produced in

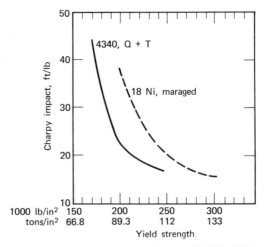

Figure 14.23 Charpy impact data of maraging and AISI 4340 steels at various strength levels (from Ref. 53).

the ageing process. The bulk of the particle strengthening is thought to come from the Ni_3Mo precipitate while η-Ni_3Ti is the major secondary hardening precipitate. The reason for the high fracture toughness of these steels is not well understood. It is thought to be associated with the Ni_3Mo precipitate and it has been postulated that somehow the precipitate does not prevent dislocation motion but merely slows it down.[51] For a discussion of this question see p. 127 of Ref. 52.

6. Characteristics.   The maraging steels possess a number of characteristics that offer advantages over conventional tempered martensite steels.

1.   As mentioned above, at some of the strength levels these steels have superior fracture toughness.

2.   Since the age-hardening component is so large, these alloys are relatively soft and, therefore, easily machined prior to ageing.

3.   As illustrated in Fig. 14.22 there is little shape change following the ageing treatment so that parts can be machined to finished sizes prior to ageing.

4.   Parts are easily heat treated. No quenching is required. There is no decarburization problem. The parts harden uniformly throughout the section size, that is, they have high hardenability.

5.   These steels display good weldability characteristics. They may be welded in the fully heat-treated condition without preheating, and the properties of the heat affected zone may be restored by a post-weld ageing treatment.

Their fatigue properties are not substantially different from many high-strength steels. Their corrosion resistance is comparable to low-alloy high-strength steels (such as 4340) under sea water conditions but is slightly better under other conditions.[51] However, the corrosion resistance is poor compared to stainless steels and, also, these maraging steels are subject to stress corrosion cracking under many environments. A suitable maraging alloy is available for casting and it provides a high-strength casting alloy with yield strengths in the 200,000 psi (1380 MN/m^2) range.[50]

In closing this chapter it seems appropriate to mention that the high-strength alloys discussed here represent only a small specialized fraction of available high-strength steel alloys. A general summary of high-alloy high-strength steels may be found in Ref. 55.

There are three subjects that we have not discussed that are very important in applying the principles of physical metallurgy to the understanding and control of the mechanical properties of metals. It should be clear from the above discussions that *fracture* is one of these subjects. A second subject is *fatigue*. Many metal parts fail in service at applied stress levels below their yield stress due to the cyclic reversal of stress, a process which is known as fatigue. The third subject, *creep*, involves the slow deformation of a metal at high temperatures at stresses below the yield stress. A good discussion of these three subjects, fracture, fatigue, and creep, may be found in Chapters 19 and 20 of Ref. 13., and in most books on mechanical metallurgy.

Similarly, in applying the principles of physical metallurgy to control of the electrical and magnetic properties of alloys it is very helpful to have an understanding of the physics of metals. This is a relatively complex subject and its mastery requires considerable study. For treatments of this subject slanted toward metallurgy see Refs. 56 and 57.

Throughout this book and particularly in this last chapter the use of structure control as a means of manipulating the mechanical properties of alloys has been emphasized. Although the elecrical and magnetic properties of alloys are generally not quite as structure sensitive as the mechanical properties, extensive use of physical metallurgy principles is made in the electronics industry in the preparation of devices. One current and particularly exciting area of the application of physical metallurgy principles is in the area of superconducting alloys. The critical current and magnetic field above which a given alloy is no longer a superconductor is strongly dependent on the structure of the alloy. Many clever and interesting techniques have been used to manipulate structure in order to maximize the critical current and magnetic field.[58]

REFERENCES

1. G. Thomas, V. F. Zackay, and E. R. Parker, in *Strengthening Mechanisms, Metals and Ceramics*, J. J. Burke, N. L. Reed, and V. Weiss, Eds., Syracuse Univ. Pr, Syracuse, 1966, pp. 3–39.

2. V. F. Zackay and W. M. Justusson, in *High-Strength Steels*, Report 76, Iron and Steel Institute, London, 1962, pp. 14–21.

3. E. Hornbogen, in *Physical Metallurgy*, R. W. Cahn, Ed., 2nd revised ed. North-Holland, Amsterdam, 1970, pp. 589–653.

4. P. M. Kelly, J. Aust. Inst. Metals **16**, 104 (1971).

5. A. Kelly, *Strong Solids*, Chapter 4, Clarenden, London, 1966.

6. *The Strengthening of Metals*, D. Peckner, Ed., Reinhold, New York, 1964.

7. *Strengthening Mechanisms in Solids*, American Society Metals, Metals Park, Ohio, 1962.

8. M. Kehoe and P. M. Kelly, Scripta Met. **4**, 473 (1970).

9. R. Armstrong, I. Codd, R. M. Douthwaite, and N. J. Petch, Phil. Mag. **7**, 45 (1962).

10.   W. B. Morrison, Trans. ASM **59,** 824 (1969).

11.   C. J. Phillips, Am. Sci. **53,** 20 (1965).

12.   J. G. Morley, P. A. Andrews, and I. Whitney, Phys. Chem. Glasses **5,** 1 (1964).

13.   R. E. Reed-Hill, *Physical Metallurgy Principles,* 2nd ed., Van Nostrand, New York, 1973.

14.   H. C. Rogers, in *Ductility,* Chapter 2, American Society of Metals, Metals Park, Ohio, 1968.

15.   *Fracture Toughness,* I.S.I. Publication 121 1968.

16.   E. A. Steigerwald, Met. Prog. **92,** 97 (1967).

17.   D. S. Clark and W. R. Varney, *Physical Metallurgy for Engineers,* 2nd edition, Appendix G, Van Nostrand, New York, 1962.

18.   J. S. Hirschhorn, *Introduction to Powder Metallurgy,* American Powder Metallurgical Institute, New York, 1969, p. 263.

19.   J. Iron Steel Institute, June 1969:
   (a) K. J. Irvine, The Physical Metallurgy of Steel, pp. 837–853.
   (b) W. E. Duckworth and J. D. Baird, Mild Steels, pp. 854–871.
   (c) J. Nutting, Physical Metallurgy of Alloy Steel, pp. 872–893.
   (d) V. F. Zackay, Anticipated Developments in Physical Metallurgy Research, pp. 894–901.

20.   J. D. Embury and R. M. Fisher, Acta Met. **14,** 147 (1966).

21.   F. J. Harbone, J. Aust. Inst. Met. **16,** 73 (1971).

22.   V. K. Chandhok, A. Kasak, and J. P. Hirth, Trans ASM **59,** 288 (1966).

23.   J. D. Embury, A. S. Keh, and R. M. Fisher, Trans AIME **236,** 1252 (1966).

24.   E. A. Shipley, in Ref. 2, pp. 93–99.

25.   M. Cohen, (a) Trans AIME **224,** 638 (1962); (b) J. Iron Steel Inst. **201,** 883 (1963).

26.   *Physical Properties of Martensite and Bainite,* Iron Steel Institute Report **93,** (1965):
   (a) P. M. Kelly and J. Nutting, pp. 166–170.
   (b) M. J. Roberts and W. S. Owen, pp. 171–178.

27.   J. M. Chilton and R. M. Kelley, Acta Met. **16,** 637 (1968).

28.   W. C. Leslie, Ref. 1, pp. 43–61; Trans. ASM **60,** 459 (1967).

29.   M. J. Roberts, in *Martensite Fundamentals and Technology,* E. R. Petty, Ed., Chapter 6 Longman, London, 1970.

30.   M. J. Roberts and W. S. Owen, J. Iron Steel Inst. **206,** 377 (1968).

31. A. R. Marder and G. Krauss, Trans. ASM **60,** 651 (1967).

32. A. R. Marder, A. O. Benscoter, and G. Kraus, Met. Trans. **1,** 1545 (1970).

33. G. R. Speich and W. C. Leslie, Met. Trans. **3,** 1043 (1972) [see also *Metals Handbook*, Vol. 8, p. 202 (1973)].

34. F. G. Wilson, in Ref. 29, Chapter 7.

35. G. R. Speich, Trans. AIME **245,** 2553 (1969).

36. P. Payson, *The Metallurgy of Tool Steels*, Chapter 6, Wiley, New York, 1962.

37. E. Tekin and P. M. Kelly, in *Precipitation from iron-base alloys*, G. R. Speich and J. B. Clark, Ed., Gordon and Breach, New York, 1965, pp. 173–221.

38. G. R. Speich, D. S. Dobkowski, and L. F. Porter, Met. Trans. **4,** 303 (1973).

39. J. D. Murray, Ref. 2, pp. 41–50.

40. V. F. Zackay, M. W. Justusson, and D. J. Schmatz, in Ref. 7, Chapter 7.

41. W. E. Duckworth and P. R. Taylor, *High-Alloy Steels*, ISI Report No. 86, pp. 61–70 (1964).

42. G. Thomas, D. Schmatz, and W. Gerberich, in *High-Strength Materials*, V. F. Zackay, Ed., Wiley, New York, 1965, pp. 251–297.

43. E. B. Kula, in Ref. 1, Chapter 4.

44. R. Philips and W. E. Duckworth, in Ref. 42, pp. 307–324.

45. K. J. Irvine et al., J. Iron Steel Inst. **196,** 66 (1960).

46. J. Marin, *Mechanical Behavior of Engineering Materials*, Chapter 1, Prentice-Hall, Englewood Cliffs, N.J., 1962.

47. W. J. McGregor Tegart, *Elements of Mechanical Metallurgy*, Macmillan, New York, 1966, p. 14.

48. V. F. Zackay, E. R. Parker, D. Fahr, and R. Busch, Trans. ASM **60,** 252 (1967).

49. E. R. Parker and V. J. Zackay, Sci. Am. **219,** 36 (1968).

50. W. Steven, in Ref. 41, pp. 115–124.

51. G. P. Contractor, J. Met. **18,** 938 (1966).

52. S. Floreen, Metals and Materials **2** (Met. Rev. **13,** 115 (Sept. 1968).

53. G. W. Tuffnell and R. L. Cairns, Trans. ASM **61,** 798 (1968).

54. C. S. Carter, Met. Trans. **1,** 1551 (1970).

55. R. Brook and J. E. Russell, in Ref. 41, pp. 19–33.

56. C. Wert and R. Thomson, *Physics of Solids*, McGraw-Hill, New York, 1964.

57. P. Wilkes, *Solid State Theory in Metallurgy*, Cambridge Univ. Pr., Cambridge, 1973.

58. W. D. Gregory et al., Editors, *The Science and Technology of Superconductivity*, Vol. 1, Plenum, New York, 1973, p. 289.

APPENDIX
SI UNITS

It now appears evident that the scientific and technological communities will achieve the adoption of a standardized system of units. These are the International System of Units (Systeme International d'Unites), which are commonly referred to as S.I. units. A brief discussion of the S.I. units commonly encountered in physical metallurgy is presented here. For a more complete presentation one is referred to the ASTM publication E380–72, Metric Practice Guide or *Metallurgical Transactions*, Vol. 3, pp. 355–58, January 1972.

The major basic units we encounter are given in the following table.

Quantity	Unit	SI Symbol
length	meter	m
mass	kilogram	kg
time	second	s
electric current	ampere	A
temperature	degree kelvin	°K
amount of substance	mole	mol

A system of prefixes for fractions and multiples has been established.

Fraction	Prefix	Symbol	Multiple	Prefix	Symbol
10^{-1}	deci	d	10	deka	da
10^{-2}	centi	c	10^2	hecto	h
10^{-3}	milli	m	10^3	kilo	k
10^{-6}	micro	μ	10^6	mega	M
10^{-9}	nano	n	10^9	giga	G
10^{-12}	pico	p	10^{12}	tera	T

As examples of the use of these prefixes consider two fairly common measures of length, the mil (0.001 inches) and the angstrom (10^{-8} cm). In both cases we convert to meters with the appropriate prefix to make the conversion factor a relatively small number:

$$1 \text{ mil} = 25.4 \ \mu\text{m}$$
$$1 \text{ Å} = 10 \text{ nm}$$

Derived S.I. units commonly encountered in physical metallurgy are listed below.

Quantity	Unit	S.I. Symbol	Formula
Area	square meter	—	m^2
Density	kilogram/cubic meter	—	kg/m^3
Energy	joule	J	$N \cdot m$
Force	newton	N	$kg \cdot m/s^2$
Power	watt	W	J/s
Stress	pascal	Pa	N/m^2

The conversion between some commonly encountered traditional units and S.I. units are

$$1 \text{ dyne} = 10^{-5} \text{ N}$$
$$1 \text{ erg} = 10^{-7} \text{ J}$$
$$1 \text{ calorie} = 4.184 \text{ J}$$
$$1 \text{ psi} = 6.8948 \text{ kN/m}^2 \text{ (kPa)}$$
$$1 \text{ ksi} = 6.8948 \text{ MN/m}^2 \text{ (MPa)}$$
$$1 \text{ kg/(mm)}^2 = 9.8066 \text{ MN/m}^2 \text{ (MPa)}$$
$$1 \text{ hbar} = 10 \text{ MN/m}^2 \text{ (MPa)}$$

Perhaps the most difficult conversion to become accustomed to is the stress conversion. It appears that most stress levels will now be given as MN/m^2 (mega pascals). The following table is presented to assist those of us who think in terms of psi or ksi in converting to the new units of meganewtons per square meter, MN/m^2.

ksi	MN/m^2	ksi	MN/m^2
1	6.9	5	34.5
10	68.9	15	103.4
20	137.9	25	172.4
30	206.8	35	241.3
40	275.8	45	310.3
50	344.7	55	379.2
60	413.7	65	448.2
70	482.6	75	517.1
80	551.6	85	586.1
90	620.5	95	655.0
100	689.5	110	758.4
120	827.4	130	896.3
140	965.3	150	1034.2
160	1103.1	170	1172.1
180	1241.1	190	1310.0
200	1379.0	210	1447.9
220	1516.9	230	1585.8
240	1654.7	250	1723.7
260	1792.6	270	1861.6
280	1930.5	290	1999.5
300	2068.4	350	2413.2
400	2757.9	450	3102.7
500	3447.4	600	4136.9
700	4826.4	800	5515.8
900	6205.3	1000	6894.8

INDEX

Major Basic Units

Quantity	Unit	SI Symbol
Length	meter	m
Mass	kilogram	kg
Time	second	s
Electric current	ampere	A
Temperature	degree kelvin	°K
Amount of substance	mole	mol

Prefixes for Fractions and Multiples

Fraction	Prefix	Sumbol	Multiple	Prefix	Symbol
10^{-1}	deci	d	10	deka	da
10^{-2}	centi	c	10^2	hecto	h
10^{-3}	milli	m	10^3	kilo	k
10^{-6}	micro	μ	10^6	mega	M
10^{-9}	nano	n	10^9	giga	G
10^{-12}	pico	p	10^{12}	tera	T

Derived SI Units

Quantity	Unit	SI Symbol	Formula
Area	square meter	—	m^2
Density	kilogram/cubic meter	—	kg/m^3
Energy	joule	J	$N \cdot m$
Force	newton	N	$kg \cdot m/s^2$
Power	watt	W	J/s
Stress	pascal	Pa	N/m^2

Conversion of Customary Units to SI Units

$$1 \text{ dyne} = 10^{-5} \text{ N}$$
$$1 \text{ erg} = 10^{-7} \text{ J}$$
$$1 \text{ calorie} = 4.184 \text{ J}$$
$$1 \text{ psi} = 6.8948 \text{ kN/m}^2 \text{ (kPa)}$$
$$1 \text{ ksi} = 6.8948 \text{ MN/m}^2 \text{ (MPa)}$$
$$1 \text{ kg/(mm)}^2 = 9.8066 \text{ MN/m}^2 \text{ (MPa)}$$
$$1 \text{ hbar} = 10 \text{ MN/m}^2 \text{ (MPa)}$$